EINSTEIN'S THEORIES OF RELATIVITY

EVERYONE'S GUIDE TO SPECIAL & GENERAL RELATIVITY

BY

ROBERT L. PICCIONI, PH.D.

Real Science Publishing

3949 Freshwind Circle, Westlake Village, CA 91361 USA

Edited by Joan Piccioni

Second Edition

Visit our web site:

www.guidetothecosmos.com

Front cover image: artist's conception of two collapsed stars, rapidly orbit one another in a compact binary system, producing gravitational waves that alternately stretch and compress the fabric of spacetime. The energy lost in radiating these waves causes the stars to spiral inward, eventually resulting in a cataclysmic collision. Image by LIGO.

Back cover image: computer simulation of two black holes, in a compact binary system, radiating gravitational waves that propagate through space like ripples on a pond. Image by NASA / Caltech.

See Chapter 34 for more about these images.

Einstein's Theories of Relativity

Everyone's Guide to Special & General Relativity

Albert Einstein's theories of special and general relativity were the most profound revolution in science in over 200 years. Einstein's theories forever changed our concepts of space, time, atoms, light, gravity, mass, motion, and the universe itself. For physicists, that's about all there is.

Sir Isaac Newton published *Philosophiae Naturalis Principia Mathematica* in 1687, establishing physics as a rigorous, rational, achievement of human intellect. Newton's laws of motion and gravity are comprehensive and mathematically perfect. They explained every physical phenomenon that anyone had ever observed, and provided numerical predictions that were more precise than anyone could measure. Newton's theories were the greatest advance in science in at least 2000 years, if not the greatest ever.

Unfortunately, nature is more complex than Newton imagined. By the beginning of the 20th Century, several puzzles eluded solution.

In 1905, Einstein published his theory of special relativity, supplanting Newton's law of motion, and refuting his concepts of space, time, and simultaneity. This forever changed our concepts of reality.

In 1915, Einstein published his theory of general relativity, supplanting Newton's law of gravity. Einstein opened our eyes to our wondrous universe as it truly is. General relativity again shattered our concepts of reality, and revealed the secrets of black holes, gravitational waves, and the expansion of the cosmos.

My goal is to reach as many eager readers as possible, and bring Einstein's genius to a wider audience. For those who were reluctant to face that challenge, this book is for you. I make Einstein's theories easier to understand, without watering down his brilliant insights.

As an aid to students, I provide every step in every derivation of every prediction of Einstein's theories of both special and general relativity.

Physics is one of the greatest adventures of the human mind, but with adventure comes challenge. While I strive to make this adventure as easy as possible, special and general relativity will be two of the greatest intellectual challenges of your entire life. Give it your very best, and you will be rewarded.

Enjoy Exploring.

The topics we explore in this book include:

Gravitational Waves, Theory and Observation
What $E = m c^2$ Really Means
The Nature of Light
Why Time and Distance are Relative
Why Mass is a Form of Energy
What is Relative and What is Not
Twin Paradox and Other Fun Puzzles
Gravity Acting on Starlight
Einstein's Equivalence Principle

Elementary Particle Physics
Expansion of the Universe and Dark Energy
Black Holes, Simple and Complex
Relativistic Astrophysics and Space Travel
Einstein's Field Equations of General Relativity
Galileo and Relativity
Kepler's Laws of Planetary Motion
Newton's Laws of Motion and Universal Gravity
Four-Dimensional Spacetime
Motion in Curved Spacetime
Gravitational Time Dilation
Tensor Calculus, everything you need to know
Precession of Mercury
Hulse-Taylor Binary Pulsar
Linearized General Relativity in Weak Gravity
… and much more

Let's talk math. Some love it and some are afraid of it; I'm not sure why, but I know this is true. I therefore strive to make my books understandable, interesting, and enjoyable for ***Everyone***.

Mathophobes will find simple English explanations for the most profound discoveries of modern physics. I make it easy to skip challenging math by placing it in separate Sections and Appendices.

Hard-core mathophiles, eager to master the tensor calculus of differential topology, will love this book— I provide and derive ***every*** equation. No other book is better; trust me. While this book does not contain problem sets per se, it does provide numerous examples, particularly in the Appendices. Use them for practice. See how I calculate part of a problem; then try to calculate the rest yourself, and compare your results with my complete solutions.

Chapter 6 reviews special relativity; if you already know the subject, start here for a quick reminder.

A list of physical constants, with their numerical values, is at the end of this book.

Appendix 1 provides a complete list of mathematical symbols.

Appendix 2 provides a brief introduction to calculus.

To learn more about my books, to receive updates, and send us your comments, visit:

www.guidetothecosmos.com.

To further ***Simplify*** your adventure, check out ***Feynman Simplified 4A Math for Physicists***, ebook or paperback, in which I explore and explain all the math physicists need to master.

If you enjoy this book please do me the great favor of rating it on Amazon.com.

Table of Contents

Chapter 1
Development of Special Relativity

Newton's laws of nature were the gold standard of science for over 200 years.

In 1905, Einstein proclaimed that Newton's laws of motion were wrong, particularly for extremely high velocities, and he provided a new, correct set of laws of motion — his theory of *special relativity* that forever changed our perception of reality.

§1.1 Principles of Special Relativity

Einstein's theory of special relativity is one of the most beautiful and profound theories in physics. It is also the theory that is most comprehensively and precisely confirmed by countless experiments.

The two fundamental principles from which all of special relativity derives are:

- Only relative velocities are physically meaningful.
- The speed of light is the same for all observers.

The first statement is the principle of relativity, first conceived by Galileo, and later incorporated into Newton's laws. The second statement is a shocking departure from Newtonian physics, and is due entirely to Albert Einstein.

§1.2 Principle of Relativity

The principle of relativity was known long before Einstein.

Galileo Galilei, an Italian scientist of the late 16th and early 17th Century, was the first to realize the necessity of testing scientific ideas with experiments. He was also the first to emphasize the importance of relativity in physical laws.

Before Galileo, science was dominated by the approach of the ancient Greeks: philosophical discourse. Scientific questions were decided by debate and by weight of authority. Aristotle was considered the ultimate authority on science.

The ancient Greeks contributed enormously to the advance of human knowledge and culture, but they also clung to some mistaken notions of natural science. They preferred to divine through discourse the ideals to which nature *should conform*, instead of observing what *actually happens* in nature. They were skeptical of physical observation that spiteful demigods might manipulate, and believed logic was the more certain path to understanding ultimate, and perhaps hidden, Platonic truths. They believed everything had its own inherent condition—heavy objects "should" fall, thus it made sense that heavier objects fall faster. Had they tested that idea, they might have discovered that all objects fall at the same rate, scooping Galileo by 2000 years. Their approach dominated science for two millennia.

Galileo was the first to systematically apply mathematics and experimentation to science. He may have drawn inspiration from his father, a famous lutenist, who discovered that the pitch of a lute string was proportional to the square root of its tension.

Galileo said science "is written in this grand book, the universe … in the language of mathematics." He wanted to study this "book" and discover what really happens in nature, not what Aristotle thought should happen. He wondered: how fast do objects fall, and do different objects fall at different rates?

Galileo developed instruments and created special circumstances to probe more deeply and carefully. Before modern clocks, he devised clever techniques to quantify time, using his pulse to measure the period of pendulums. He rolled balls down inclined planes to slow their fall, thereby enabling more precise measurements of their motion.

Galileo discovered that when any object falls, the distance it travels increases with the square of time. He also discovered the principle of inertia: objects move with constant velocities unless acted upon by an external force.

His rejection of the blind acceptance of authority created a revolution that separated science from dogma. This made Galileo the first modern scientist and the originator of experimental science.

One of his discoveries was:

The Principle of Relativity

- The laws of nature appear the same to all observers moving at constant velocity.
- The concept of "absolute velocity" has no significance in nature.
- Only relative velocities are physical meaningful.

Special relativity is "special" in that it applies only to special circumstances: systems with constant velocities, those without accelerations. General relativity applies generally: to any state of motion. Einstein's theory of general relativity extends the principle of relativity to observers and objects whose velocities are changing, including those accelerated by gravity.

Let's explore what the principle of relativity means with several examples.

Galileo observed that all objects fall in the same manner — straight down — on a moving ship as they do on land. The modern version of this might be: if you spill your coffee while "dining" in an airplane, the coffee falls straight down onto your lap even though the plane is flying west at 1000 km/hr (600 mph). At first, it is not surprising that coffee falls straight down; everything does. But consider how this looks from outside the airplane: the coffee is not just falling down, it is also moving west at 1000 km/hr along with everything else in the plane, even though nothing is pushing it forward as it falls. Aristotle would be surprised.

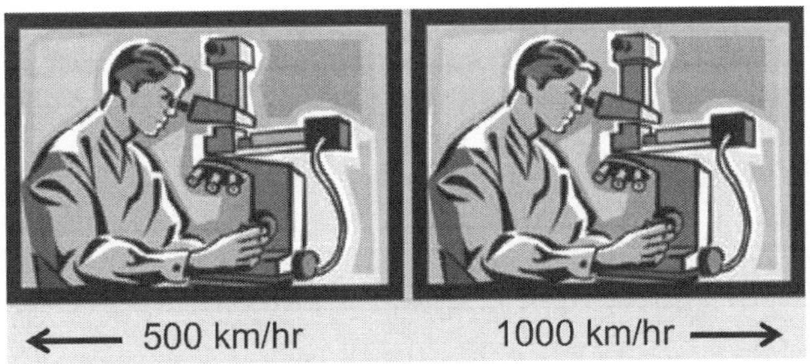

Figure 1-1 Scientists Cannot Measure Their Velocities

Consider another example: imagine two laboratories, each with a scientist and all the equipment they might desire. *Stop!* That is impossible; make that "a lot of equipment." Seal the labs so the scientists cannot see anything or measure anything outside their own lab. Then put each lab in its own airplane, with one flying west at 500 km/hr and the other flying east at 1000 km/hr, as shown in Figure 1-1.

The principle of relativity states there is no test or measurement the scientists can perform to determine their planes' velocities. Every test they make gives exactly the same results in both planes, as long as they cannot detect anything outside their labs. They could find their planes' speeds with GPS, but detecting an external signal is cheating.

Special relativity says all observers see the same physical phenomena, the same laws of nature, regardless of their own velocities, as long as their velocities are constant.

Absolute velocity is not detectable. But we can detect when one object's velocity changes or when two objects have different velocities. If one of the above airplanes suddenly stops (let's not ask why), we could measure the velocity change (deceleration) with a pendulum: the pendulum would tilt toward the plane's nose, just as airline passengers have the sensation of being pulled forward when the pilot hits the brakes after landing. Velocity changes are physically meaningful and measurable. So are velocity differences; from the ground, we can measure an airplane's speed relative to us.

It is a very good thing that the laws of nature do not depend on our "true" or "absolute" velocity, because we do not really know what that is. You might think that while you are sitting in a chair reading this book your velocity is zero, but is it? As Earth spins on its axis once daily, it carries us eastward at up to 1600 km/hr (1000 mph). As Earth orbits the Sun, we are being carried along at 100,000 km/hr. As the Sun orbits the Milky Way galaxy, we are moving at 800,000 km/hr. And we are moving at 1,400,000 km/hr through the *cosmic microwave background* ("CMB") radiation, the first light of our universe. So, how fast are you really moving? (For the complete CMB story, see my ebook ***Our Universe 3: CMB, Inflation & Dark Matter*** or my print book ***Our Universe: The Known & The Unknown***.)

An observer moving with constant velocity — constant speed and direction — is said to be in an *inertial frame*. Any change in speed or direction is a change of velocity that requires acceleration. Both Newton's laws of motion and special relativity are valid only in inertial frames. In inertial frames, physics is simpler, and who doesn't want simpler physics?

Earth's surface is generally considered to be an inertial frame, except in high precision applications. This is because we can generally ignore or compensate for Earth's gravity and rotation. An example of a non-inertial frame is a 78-rpm vinyl record (what music was recorded on before cassettes, CDs, and iPods). If we stand next to a record player (in Earth's inertial frame), a ping-pong ball rolling across the record would appear to us to move in a straight line (absent friction). For us, the laws of physics are simple: no force means no acceleration, hence no change in the ball's speed or direction. But in the non-inertial frame of the rotating record, the ball's path is curved. Here the laws of physics are more complex: to apply Newton's laws we must first invent a fictitious force, the *Coriolis force*, to explain the ball's apparently curved trajectory. Whenever possible, it is better to avoid non-inertial frames.

If all this was known before Einstein, why is relativity such a big deal? Since Newton's laws conform to Galileo's principle of relativity, there was no problem. At least not until James Clerk Maxwell published the theory of electromagnetism in 1861, and said light was an electromagnetic wave that moved at a fixed speed.

That was a problem.

§1.3 What Inspired Special Relativity?

Before 1905, everyone believed that the mass of a body was constant, the same whether it was moving or not. This was consistent with all experiments of the day, and was assumed in Newton's laws.

Einstein's theory made the astonishing claim that a body's mass increases at high speeds. But, for "normal" speeds, the mass increase predicted by Einstein is extremely small, far below the weighing precision of the day. At 48,000 km/h (30,000 mph), a body's mass increases by only one part per billion.

There was no observational evidence to motivate Einstein to develop a theory of mass increase. Einstein's motivation for special relativity was driven by something other than the laws of moving bodies — he wanted to make relativity and electromagnetism compatible. Einstein's concerns were more philosophical than practical. One issue that particularly bothered Einstein was that Maxwell's theory had two equations for one problem.

Let's consider a wire next to a magnet, as shown in Figure 1-2. If the wire moves up, through a stationary magnetic field, an electric current is induced in the wire, according to one of Maxwell's equations. But, if the magnet moves down while the wire is stationary, the moving magnetic field creates an electric field that induces a current in the wire, according to a different one of Maxwell's equations.

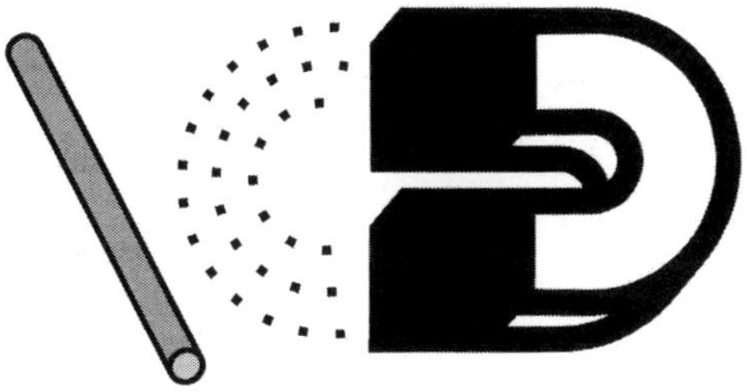

Figure 1-2 Magnet and Wire

Both of Maxwell's equations yield the same current, so no one cared that there were two equations. No one that is, except Einstein.

Einstein's 1905 paper on special relativity is one of the greatest triumphs of science. He began by explaining his concern about Maxwell's electromagnetic equations. Einstein accepted the principle of relativity: only relative velocities have physical significance. Hence, it must make no difference whether the wire moves up or the magnet moves down. Einstein said these are not *two different* problems — there is only *one* problem: a wire and a magnet move *relative to one another*. Einstein thought having two different equations for one problem was ugly, and he was sure nature is not ugly.

Since nature does not distinguish between the two, Einstein said, neither should physicists.

Others realized that the equations of electromagnetism were not consistent with relativity. Let's understand why by using the Galilean transformation, the standard of Newtonian physics, that translates coordinates measured in a stationary reference frame into coordinates measured in another reference frame moving at a constant velocity. Assume the velocity of the moving frame is u in the +x-direction, and the origins of the two frames are perfectly aligned at time t = 0. If an object's position has value x in the stationary frame and value X in a moving reference frame, these quantities are related by:

$$X = x - ut$$

Additionally, the measured speed of any object moving in the x-direction is different in these two frames. Since speed is distance divided by time, velocities measured in the two frames are related by:

$$X/t = x/t - u$$

Newton's laws are equally valid in both the stationary and moving reference frames. Indeed they are valid in any frame compatible with the Galilean transformation. This is because Newton's laws do not specify *absolute* positions or velocities; they depend only on *relative* positions and velocities. For example, in $F = G M m / r^2$, r is the distance between M and m, not the absolute position of either. In additional, Newton's equations tell us how much the velocities of M and m *change* over time, not what their absolute values are at any specific time. Nor do Newton's equations depend on prior absolute velocities.

We describe this by saying: Newton's laws are *invariant* under any Galilean transformation, and are therefore consistent with the principle of relativity. This makes it impossible to determine which frame is "actually" moving by observing any phenomenon of mechanics — forces, work, kinetic energy, gravitational potential energy, and such.

However, Maxwell's equations are *not invariant* under Galilean transformations. Maxwell's equations specify the speed of light in one particular reference frame — the frame in which light's purported medium, the *luminiferous ether,* is stationary. Light's speed would be different in any frame moving relative to the luminiferous ether; hence Galilean transformations change not just the values of position and velocity, but also the form of Maxwell's equations.

It appeared physicists would be forced to choose between two seemingly wonderful pillars of science: the principle of relativity, and Maxwell's equations. Most physicists believed someone would one day detect the luminiferous ether, forcing everyone to agree that relativity is not a universal principle.

§1.4 Search For Ether

Before Einstein, most physicists believed light required a medium to travel through.

For more than a century, it had been well established that light was a wave phenomenon. Light exhibits interference and diffraction that are definitive signatures of wave behavior.

All other waves are clearly the organized motions of some medium — ocean waves are the motion of water molecules, and sound waves are the motion of air molecules.

What was the medium of light? What moved as light passed by? No one knew, but they named this unidentified medium the *luminiferous ether*.

Physicists launched major efforts to find this mysterious material, often simply called "ether".

It was quite clear that ether must fill the entire universe, because we can see the light of extremely distant galaxies. This is different from sound — we cannot hear the sounds of stars and galaxies, so no one claims the universe is filled with air (it is not). Hence, light and ether posed a new and unique challenge.

It was also clear that if the universe is filled with ether, we must be moving through it. People once thought Earth was the center of all existence, with everything else revolving around us. Now we know that even our own galaxy is but a small part of a much greater universe. We also know Earth is moving relative to everything else in the cosmos. It would be absurd to claim that the Earth is not moving through the ether — that somehow all the ether in the universe moves in lock-step with Earth.

If Earth is moving through light's medium, we should see light having different speeds in different directions. Why? Consider the following analogy. When a rock is thrown into a pond, ripples spread across the water moving out at the same speed in all directions. When a rock is thrown into a river, a person on the bank sees the ripples moving faster going downriver and slower going upriver, because the current carries everything downriver. If Earth moves through the ether, a "river" of ether must flow past us, and we must see light moving faster downriver than upriver.

Maxwell's equations of electromagnetism (see my book ***Feynman Simplified Part 2***) say the speed of light is a fixed number that is the same in all directions. Hence, Maxwell's equations would be correct only in the frame in which the ether is stationary, which is not Earth's frame. We could then measure our absolute velocity, our velocity relative to the ether, contradicting the principle of relativity.

Ether would make Maxwell's equations and relativity incompatible.

Physicists set out to discover if light's speed really does vary. High precision, meticulous experiments measured the speed of light in different directions. They all found the same light speed in all directions. And Earth's motion seemed to make no difference — the same results were obtained whether these measurements were made in daytime or at night, or during different months as Earth turned around the Sun.

Experiments generally cannot prove two quantities are exactly equal; there are always limitations to the precision of our instruments. The most that experiments can say is that two quantities are equal within a specified level of precision. If we measure two meter-sticks, we might be able to say they are the same length to 1 mm. With better instruments, we might be able to say they are equal to 0.001 mm, but we can never say they are *exactly* the same length. While exactitude is impossible, experimental physicists can sometimes achieve amazing levels of precision.

American physicists Albert Michelson (1852–1931) and Edward Morley (1838–1923) performed the most famous of these experiments in 1887. For this experiment, Michelson developed the first *interferometer*. More than a century later, its descendants are still the gold standard of precise optical measurements, and are still making ground breaking discoveries, including *gravitational waves* that we will explore in Chapter 34.

Michelson's interferometer, illustrated in Figure 1-3, had two orthogonal arms, joined at one end, each with a mirror at its far end. He sent light back and forth along each arm and very precisely measured the *difference* in travel times along the two paths.

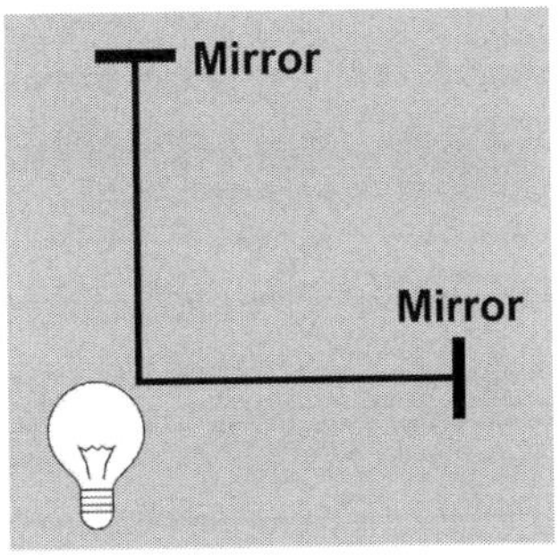

Figure 1-3 Michelson-Morley Interferometer

Next, Michelson rotated the entire interferometer, while keeping the arms perpendicular. To minimize vibrations, the interferometer lay on a large marble table that floated on a pool of mercury. (Fortunately scientists have long since learned to avoid such toxic setups.) He also watched his interferometer as Earth orbited the Sun, changing our motion relative to the purported ether river flowing past us (indicated by light gray arrows), as sketched in Figure 1-4.

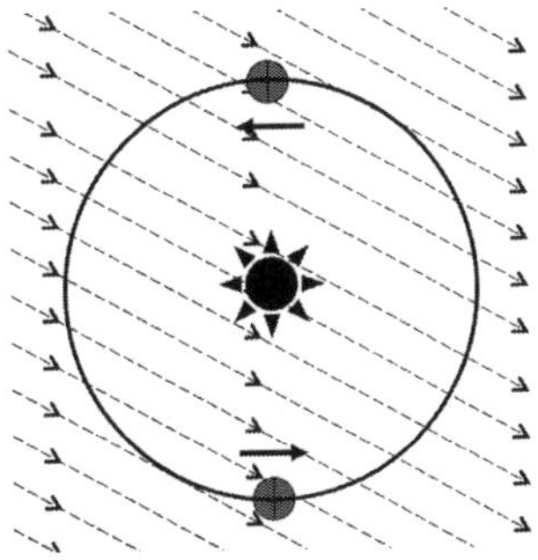

Figure 1-4 Earth Orbiting Sun in a Flowing Ether "River"

If ether did exist, the light travel times would have changed as the direction of the light beams rotated through the ether. However, Michelson and Morley found no changes in travel time differences to a precision of 2.5% of what was expected from Earth's orbital speed — pretty impressive.

Does this prove ether does not exist?

After Michelson and Morley published their dramatic result in 1887, theorists George FitzGerald (in 1889) and Hendrik Lorentz (in 1892) suggested that material objects are compressed as they pass through the ether. This became known as the *Lorentz-FitzGerald contraction*, or simply the *Lorentz contraction*. Specifically, they said objects are foreshortened in the direction of motion as follows:

$$L = \mathrm{L}\ \sqrt{(1 - \mathrm{v}^2 / \mathrm{c}^2)}$$

Here, L is the shortened length at velocity v, and L is the length at zero velocity. While this equation gave the needed adjustment that "explained" Michelson-Morley's null result (no light speed change moving through ether), it was widely seen as contrived, having no meaningful rationale, and invented solely to "plug a hole".

In 1905, special relativity provided a cogent and comprehensive understanding of why lengths appear contracted.

In 1907, Michelson became the first American to receive a Nobel Prize in Physics. Surprisingly, despite being acclaimed for proving Earth was not in a "river" of ether, Michelson continued to believe that ether really did exist. Long after Einstein's rejection of ether became widely accepted, Michelson struggled to understand how his marvelous experiment had gone wrong.

The truth is, Michelson's experiment was not wrong. It has been repeated and confirmed many times, with ever-greater precision. The latest experiments find light's speed is the same in all directions to a precision of 1 part in a billion, billion.

If ether does exist, why can't we detect it?

§1.5 Einstein Understands Light

Before Einstein, physicists universally believed that there were two completely separate and entirely different entities in nature: particles and waves. Particles were simpler, just very small versions of golf balls. Waves were much more complex. The essential differences between particles and waves, as believed in 1900, are listed here.

Property	Particles	Waves
Quantized	YES	NO
Localized	YES	NO
Diffraction	NO	YES
Interference	NO	YES
Needs a Medium	NO	YES
Speed	unrestricted	set by medium

Particles always come in integer numbers and are precisely localized. They require no medium to travel through, and can move at any speed. Conversely, waves can have any amplitude; with waves it is how much, not how many. Waves spread throughout all available space, and exhibit complex diffraction and interference effects that particles never do. Waves are the organized motion of a medium, such as air molecules for sound waves. Waves cannot exist without a medium, and their speed is determined by the properties of their medium, not by their source.

If all this does not convince you that waves and particles are completely different, just compare their equations of motion (for now, there is no need to worry about what the math means):

$$\text{Particle: } 0 = \boldsymbol{a}$$

$$\text{Wave: } 0 = d^2A/dx^2 + d^2A/dy^2 + d^2A/dz^2 - d^2A/dt^2 / v^2$$

It was obvious to everyone that particles and waves had nothing in common. Obvious to everyone, that is, except Einstein. One might say Einstein thought "outside the box." But, it may be more accurate to say that he never even noticed there were boxes that he should stay within.

Einstein believed the principle of relativity and Maxwell's theory of electromagnetism were both too *elegant* to be wrong. He realized that if both are valid, it must be our understanding of light that is wrong. He became convinced that there is no preferred reference frame in which ether is stationary.

Ether simply had to go.

In 1905, Einstein radically changed our understanding of light by proclaiming:

- Light is *both* a particle and a wave.
- Light has no medium, unlike all other waves.
- Light always travels at the same speed (through empty space).

Einstein's assertion that light is both a particle and a wave led to the concept of *particle-wave duality*, the fundamental principle of quantum mechanics (see my book ***Feynman Simplified Part 3***). The particles of light are called *photons*. Einstein also said that, unlike all other waves, light needs no medium, such as the luminiferous ether, to travel through. Light is an electromagnetic wave that travels through empty space.

Light is comprised of oscillating electric ***E*** and magnetic ***B*** fields that give light its wave properties. Light is not a material object; it contains no matter, just fields. No physical object moves as light travels. The ***E*** and ***B*** fields oscillate together, cresting and troughing simultaneously.

The directions of ***E***, ***B***, and light's velocity are all mutually perpendicular. Figure 1-5 attempts to portray light's 3-dimensionality: the ***E*** field (solid curve) goes up and down in the plane of the page, the ***B*** field (dotted curve) goes in and out of the page, as light moves to the right across the page.

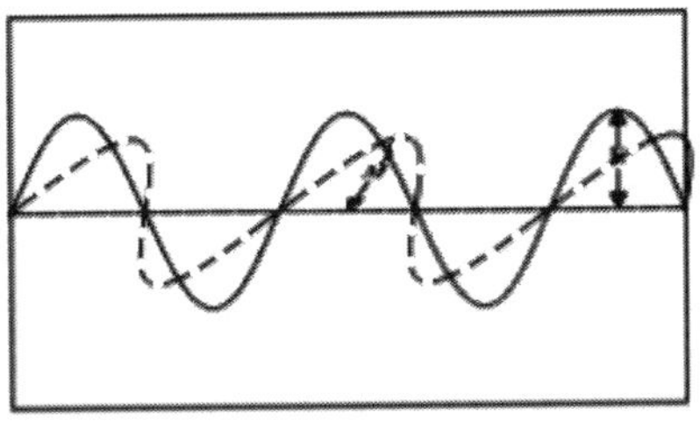

Figure 1-5 Oscillating E and B Fields of Light

The ***E*** and ***B*** fields can oscillate at any frequency whatsoever. Physicists work with light frequencies from a few Hz (cycles per second) to even more than a trillion, trillion Hz. Light's energy increases with frequency. To my knowledge, 16 TeV is the highest photon energy ever observed. That exceeds the energy of our most powerful particle accelerator, the LHC at CERN near Geneva. (TeV means trillion electron-volts; 1 eV is the energy gained by an electron traversing a 1-volt potential.)

Photons are characterized by wavelength λ and frequency f. Wavelength is the distance traveled in one full cycle, and frequency is the number of full cycles the electromagnetic fields oscillate through per second. Frequency determines a photon's energy E, according to Einstein's equation $E = h\,f$, with h being *Planck's constant*. For any wave, the product of its wavelength and its frequency equals the wave's velocity $\lambda\,f = v$. For light moving through empty space $\lambda\,f = c$.

Various portions of the light frequency / wavelength spectrum have names:

f (Hz)	λ (m)	Name
$> 10^{19}$	$< 3\times10^{-11}$	gamma rays
3×10^{18}	1×10^{-10}	x-rays
3×10^{15}	1×10^{-7}	ultraviolet
7.5×10^{14}	4×10^{-7}	blue } visible
4.3×10^{14}	7×10^{-7}	red } light
1×10^{13}	3×10^{-5}	infrared
$< 3\times10^{11}$	$> 10^{-3}$	radio waves

While these names are commonly used, they have no natural significance. Light is a single phenomenon that comes in a continuous spectrum of all wavelengths and frequencies. No natural dividing line separates gamma rays from x-rays, or any of the other named types of light.

Human eyes are sensitive to a certain range of frequencies that we call "visible light." Of course that light is vitally important to us. But other species can see light of different frequencies.

Assigning different names to light of different frequencies is like having different names for a car moving at different speeds. Ferraris and Fiats truly are different, but two Ferraris are not different just because their speeds are different. In fact, a specific frequency is not an intrinsic property of any particular photon. Light's frequency depends on the observer as well as the source. Betelgeuse, the star at the left shoulder of the Orion constellation, looks red to us on Earth, but it might look blue to the crew of a rocket speeding toward it. For these reasons, I use the word "light" to describe photons in every part of the electromagnetic spectrum, regardless of frequency or wavelength.

While specific frequencies are not intrinsic properties, light's wave properties are essential characteristics of its nature, leading to the phenomena of interference and diffraction.

For other types of waves, such as sound waves, the wave's speed is determined by the properties of its medium. All sounds, whether from a human voice or an aircraft, move through air at the same speed — "the speed of sound".

This is because sound waves are the motion of air molecules, and the speed of a sound wave is determined by how rapidly air molecules move, which varies with temperature and pressure. The speed of the original source of a sound wave — a person or aircraft — is irrelevant.

For waves other than light, wave speed is also fixed relative to the motion of its medium. In standard conditions, sound waves move at 1234 km/hr (768 mph) in the reference frame in which air is stationary. Ripples in a smoothly flowing stream move at a certain speed relative to the water's rest frame.

In Figure 1-6, a stationary Einstein and the pilots of two jets observe a lightning bolt. The higher jet is flying toward the lightning at 1000 km/hr, and the lower jet is flying away from the lightning at 1000 km/hr.

Figure 1-6 Lightning & Three Observers

The three observers sense thunder passing them at different speeds:

Einstein:	1234 km/hr	
Upper jet:	2234 km/hr	(1234 + 1000)
Lower jet:	234 km/hr	(1234 − 1000)

Before 1905, physicists believed that these three observers would also perceive the light flash from the lightning pass by them at different speeds, just as the sound waves do. But, Einstein declared that light waves are uniquely different; they propagate without any material medium. Equivalently, we can consider the medium of light to be the vacuum of empty space. Maxell's equations of electromagnetism relate the speed of light, denoted by c, to properties of vacuum: $1/c^2$ = (vacuum dielectric permittivity) × (vacuum magnetic permeability). Light's speed c is exactly 299,792,458 meters (about 186,282 miles) per second.

Since space is at rest in every reference frame, Einstein reasoned the speed of light has the same value in all frames, unlike every other type of wave. Einstein and both jet pilots observe the lightning's flash pass them at the same speed, despite their differing velocities. This is a remarkable and unique property of light.

With his theory of special relativity, Einstein proclaimed that the laws of nature are *universal*, the same for all observers in uniform motion, while measurements of certain quantities are *relative*, different for different observers. By "uniform motion" we mean non-accelerating. In his later theory of general relativity, Einstein was able to remove the "uniform motion" requirement, extending the universality of nature's laws to all observers regardless of their motion.

So far we have discussed the speed of light c through empty space, a perfect vacuum. Light's speed through glass, water, or even air, is *apparently* less than c. This is because photons are absorbed and reemitted by atoms in these materials. Photons travel between atoms at speed c, but absorption and reemission introduce delays that reduce their average effective speed. This is a bit like racecars, whose average speed for an entire race is reduced by pit stops.

A material that reduces light's effective speed to c / N is said to have an *index of refraction* of N. For air N = 1.0003, thus air slows light only slightly to 0.9997 c. For glass N = 1.52, and for water N = 1.33. For a girl's best friend, N = 2.42 (one reason diamonds are dazzling). That said, I will not continue repeating "in empty space" when discussing light's speed.

For the sake of brevity and to reduce clutter, many physicists, myself included, often use *natural units* in which the speed of light in empty space c equals 1. Astronomers, for example, typically measure time in years and distances in light-years, where one light-year is the distance light travels in one year (about 9.46 trillion km or 5.88 trillion miles). This makes the speed of light: c = 1 light-year / 1 year = 1.

For your convenience, I will keep all the c's in every equation in the chapters on special relativity.

In those chapters, we will explore the remarkable consequences of Einstein's postulates about light's constant velocity and lack of medium.

§1.6 Special vs. General Relativity

Relativity was developed in two stages. Einstein published special relativity in 1905 and general relativity in 1915. Special relativity is the simpler theory, covering simpler, more restrictive, special situations — those in which everything moves with a constant velocity. That restriction precludes the consideration of forces, which cause accelerations that change velocities. Particularly precluded is gravity. General relativity removes all these restrictions and hence applies generally, but at the cost of much more complex mathematics.

Perhaps surprisingly, special relativity is adequate for a vast range of applications despite its restrictions. In particle physics and quantum mechanics, for example, special relativity is used throughout, with particle interactions (the forces) being treated as instantaneous events separated by long stretches of force-free motion. Even situations with constant accelerations can be handled with special relativity, with sufficiently careful mathematics.

Special relativity supersedes Newton's laws of motion and his concepts of space and time. It is not so much that Einstein's ideas contradict Newton's, rather Einstein's ideas address a broader range of circumstances. Newton's laws are still adequate for almost all day-to-day applications—bridge design, planetary motion, and billiards. But in more advanced applications, particularly when velocities approach the speed of light, nature deviates from Newtonian mechanics. It is in those cases that we need the greater accuracy and more profound understanding provided by special relativity.

If the speed of light were infinite, nothing else could approach its speed and there would be no need for special relativity. In that sense, light's speed c sets the scale at which special relativity becomes essential.

Chapter 2
Time & Distance in Special Relativity

Before Einstein, physicists were confident they understood distance, time, and mass. They believed any skilled person could measure distance with a ruler, tell time with a clock, weigh mass with a scale, and get the TRUE values. With his theory of special relativity, Einstein proved this was wrong — there are no TRUE values.

Einstein shook the very foundations of science.

In this chapter, we explore how Einstein's postulate that the speed of light is constant affects measurements of time, and distance. But first, let's derive the *Lorentz transformation* — the coordinate transformation of relativity.

§2.1 Lorentz Transformation

If we accept that the speed of light c is the same in all reference frames, we must replace the Galilean transformation — we need a new rule.

Let's define two reference frames, with time as the fourth dimension: a stationary frame (t, x, y, z) and another frame (T, X, Y, Z) moving with relative velocity u in the +x-direction. Assume both frames align perfectly at (0, 0, 0, 0). Figure 2-1 shows how these frames appear at some later time.

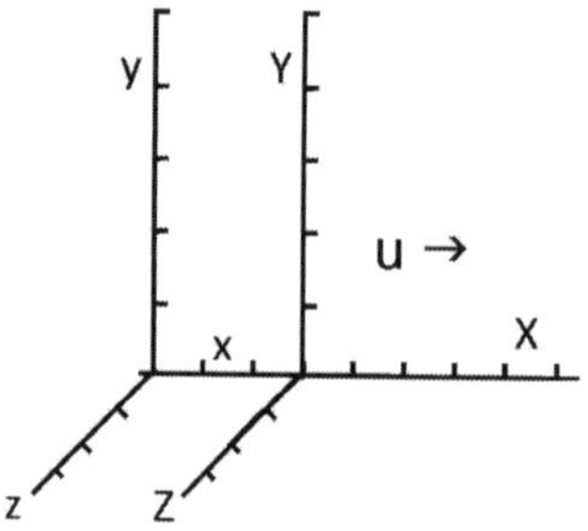

Figure 2-1 Two Reference Frames with Relative Motion

Since space is homogenous, the transformation must be linear in each coordinate — any terms like x^2 would result in different transformations in different locations. We also know that if the x-transform equation is modified at relativistic velocities, so must be the t-transform equation, since x / t, the speed of a photon moving in the +x-direction, must equal c in all frames. The most general linear transformation rule for motion in the x-direction is:

$$X = A\,(x + B\,c\,t)$$

$$c\,T = D\,(c\,t + C\,x)$$

$$Y = y$$

$$Z = z$$

Here, A, B, C, and D are dimensionless functions of u to be determined. I explain at the end of this section why Y and Z must be unchanged when the relative velocity between the frames is entirely in the x-direction. In the limit of u << c (u much less than c), our new rule must approach the Galilean rule:

$$A = D = 1$$

$$B = -\,u\,/\,c$$

$$C = 0$$

Now consider an object moving at the same velocity as frame (T, X, Y, Z). It is stationary in that frame, meaning its X coordinate has the same value for every time T.

In the stationary frame, that object's motion equation is $x = u\,t$. Hence in the moving frame:

$$X = A\,(u\,t + B\,c\,t)$$

For the right side of this equation to be constant for all values of t, B must equal $-u/c$. This means the general rule for X must be:

$$X = A\,(x - u\,t)$$

We must also require that in any frame, the speed of light must equal c. Hence in both frames, for photons moving in either the +x or –x directions, the rate of change of a photon's position must be the same in both frames. This means:

$$c = x/t = X/T$$

$$x\,T = X\,t$$

$$x\,D\,(t + C\,x/c) = A\,(x - u\,t)\,t$$

$$D\,x\,t + D\,C\,x^2/c = A\,x\,t - A\,u\,t^2$$

$$(D - A)\,x\,t + D\,C\,x^2/c + A\,u\,t^2 = 0$$

For motion toward +x, $x = +c\,t$: $\quad + (D - A)\,c\,t^2 + (A\,u + c^2\,D\,C/c)\,t^2 = 0$

For motion toward –x, $x = -c\,t$: $\quad - (D - A)\,c\,t^2 + (A\,u + c^2\,D\,C/c)\,t^2 = 0$

Subtracting these two equations yields: $\quad 2\,(D - A) = 0$

$$D = A$$

Adding the prior two equations yields: $\quad 2\,(A\,u + D\,C\,c) = 0$

With D = A, this is: $\quad C = -u/c$

Our transformation equations are: $\quad X = A\,(x - u\,t)$

$$c\,T = A\,(c\,t - u\,x/c)$$

Finally, we require that the transformation be symmetric: to transform from frame (T, X, Y, Z,) to frame (t, x, y, z), which is moving with relative velocity – u, the rule must be:

$$x = A\,(X - [-u]\,T)$$

$$x = A\,(\,[A\,(x - u\,t)\,] + u\,[A\,(t - u\,x/c^2)\,]\,)$$

$$x = A^2\,(x - u\,t + u\,t - u^2\,x/c^2)$$

$$1 = A^2\,(1 - u^2/c^2)$$

$$A = 1/\sqrt{(1 - u^2/c^2)}$$

This relativistic factor comes up in nearly every equation in both special and general relativity. It is conventionally denoted by γ, the Greek lower case gamma. The factor γ equals 1 for stationary objects with $u = 0$. It remains very close to 1 for velocities much less than the speed of light ($u \ll c$), and increases rapidly for velocities approaching light speed. At light speed ($u = c$), γ is infinite; this can occur only for massless particles, like photons.

We have our rule. The *Lorentz transformation* for relative velocity u in the +x-direction is:

With $\gamma = 1/\sqrt{(1 - u^2/c^2)}$: $\quad X = \gamma\,(x - u\,t)$

$$T = \gamma\,(t - u\,x/c^2)$$

$$Y = y$$

$$Z = z$$

By assuming the constancy of the speed of light, we derived a transformation rule that makes Maxwell's equations consistent with the principle of relativity.

Now why are Y and Z unchanged by a Lorentz transformation in the x-direction?

Let's imagine two identical space shuttles that are flying toward you, as sketched in Figure 2-2. Assume the left shuttle is moving much faster than the right one, and that the tops of their tail fins barely miss one another's wing tips as they pass.

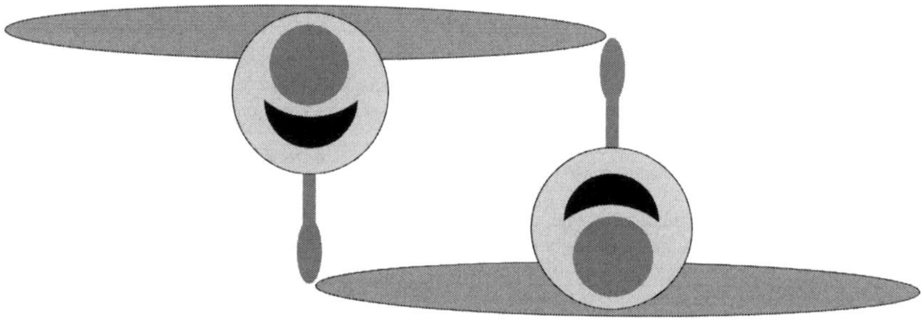

Figure 2-2 Space Shuttles Pass

The principle of relativity says only relative velocities have physical meaning. Each crew sees the other shuttle passing at the same relative velocity, but in opposite directions. Therefore each crew must have the same perception of the other shuttle's lateral dimensions. Both crews can unambiguously compare the two shuttles' wing lengths and tail heights. The symmetry required by relativity would be broken if one shuttle's wing appeared longer or its tail shorter. Thus wing lengths and tail heights must appear equal to both crews for any combination of shuttle velocities.

The Lorentz transformation, therefore, cannot modify dimensions perpendicular to the relative velocity between frames.

§2.2 Adding Velocities

Let's use the Lorentz transformation to determine how velocities transform.

First, choose a frame (t, x, y, z) in which an object moves in the +x direction with velocity u. Define our coordinates so that $x = 0$ at $t = 0$. The object's equation of motion in this frame is: $x = u\,t$. Next, choose a frame (T, X, Y, Z) that is moving in the –x direction with velocity v relative to the first frame. Let the origins of the two frames be perfectly aligned at $t = T = 0$. What is the object's velocity, call it w, as measured in the second frame?

Since the object is moving toward +x, and the second frame is moving toward –x, the object appears to be moving faster in the second frame; for non-relativistic velocities, $w = u + v$.

Let's do the relativistic calculation of w by transforming to the moving frame using the Lorentz transformation equations with velocity –v. In the moving frame, $w = X / T$, the distance traveled divided by the travel time.

$$X = \gamma(x + v\,t) = \gamma(u\,t + v\,t)$$

$$T = \gamma(t + v\,x / c^2) = \gamma(t + v\,u\,t / c^2)$$

$$w = X / T$$

$$w = \gamma(u\,t + v\,t) \;/\; \gamma(t + v\,u\,t / c^2)$$

$$w = (u + v) \;/\; (1 + u\,v / c^2)$$

This is the relativistic equation for the addition of *collinear* velocities u and v.

As with all equations of special relativity, this reduces to the Newtonian equivalent for velocities much less than c, the speed of light.

Here are some examples of adding collinear relativistic velocities u and v:

u / c	v / c	w / c
0.5	0.5	0.8
0.8	0.8	0.976
0.9	0.9	0.994
0.99	0.99	0.99995

You might wish to confirm that if either u or v equals c, the collinear sum becomes w = c.

Now, let's try adding velocities that are orthogonal. Let the object's velocity in the first frame be u in the +y direction. Thus, the equation of motion is y = u t, x = z = 0. In the second frame, again moving with velocity v in the –x direction and $\gamma^2 = 1/(1-v^2/c^2)$:

$$X = \gamma\,(0 + v\,t) = \gamma\, v\, t$$

$$Y = u\,t$$

$$Z = 0$$

$$T = \gamma\,(t + 0) = \gamma\, t$$

$$X / T = v$$

$$Y / T = u / \gamma$$

Thus, in the second frame, the object's velocity is u / γ in the Y direction and v in the X direction.

Some useful relationships are:

$$\gamma = 1 / \sqrt{(1 - u^2 / c^2)}$$

$$1 / \gamma^2 = 1 - u^2 / c^2$$

$$1 = \gamma^2 - \gamma^2 u^2 / c^2$$

$$1 + \gamma^2 u^2 / c^2 = \gamma^2$$

For u << c, we can approximate γ as: $\gamma = 1 + u^2 / 2\,c^2 + 3\,u^4 / 8\,c^4 + \ldots$

For u close to c, we can approximate γ as:

$$1 / \gamma^2 = 1 - u^2 / c^2$$

$$1 / \gamma^2 = (c - u)\,(c + u) / c^2$$

$$1 / \gamma^2 = 2\,(1 - u / c)$$

§2.3 Time Dilation

Let's next examine how the constancy of the speed of light impacts time.

Time is a very elusive quantity. We have many different perceptions of what time is — all rather are nebulous. Einstein cut through all that fog. He was a practical person, as a good physicist must be. He did not dwell over the metaphysics of what is TIME?

Einstein simply said: "Time is what a clock measures."

Let's put that quote to work by doing a *thought experiment* suggested by Einstein. Einstein loved thought experiments; they were the only experiments he ever did. By shunning real experiments, he avoided high voltage shocks, liquid nitrogen burns, and ionizing radiation — in other words, all the things that made me the man I am today.

In thought experiments, we focus on key principles, ignore messy real-world issues such as friction, and imagine what nature does in a decisive situation.

To discover the impact of relativity on time, let's imagine a perfect clock, one with no moving mechanical parts that wear, age, or vary with temperature. The ideal clock, Einstein said, has two parallel mirrors and one photon bouncing endlessly between them. This being a thought experiment, the mirrors are perfectly parallel and 100% reflective. We count "tick" when the photon hits one mirror and "tock" when it hits the other mirror.

Now imagine two identical photon clocks, as illustrated in Figure 2-3, with the mirrors shown as solid horizontal bars and the photon paths shown as dotted lines. The stationary clock on Earth is at the left, while the other clock is in a jet flying to the right at 90% of the speed of light ($v = 0.9\ c$). In the stationary clock on the left, the photon bounces up and down, as time flows normally.

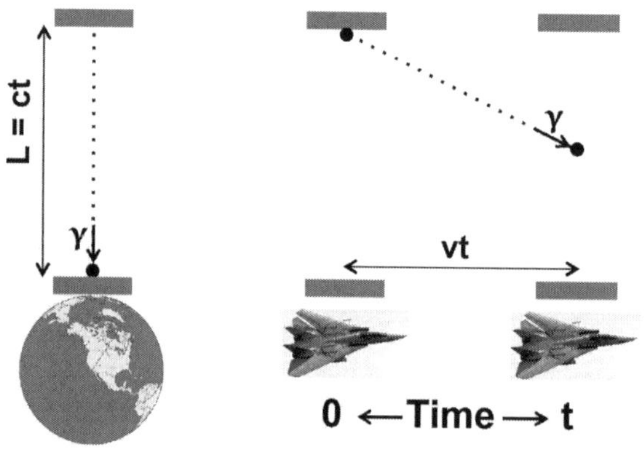

Figure 2-3 Stationary & Moving Photons Clocks

The moving clock on the right is shown twice: when the photon begins to descend at time 0, and at a later time t. While the photon in the jet's clock bounces straight up and down in the moving jet's frame, we on Earth observe it moving along a diagonal path with the jet. To us, that diagonal path is clearly longer than the photon path in the stationary clock.

By convention we denote photons with γ, the lower case Greek letter gamma. (Yes, we use γ with different meanings. Sorry, but one must learn to infer from context whether γ means photon, specific heat ratio, the relativistic factor, or even something else.)

Einstein said all photons are seen by any observer to have the same speed c in all circumstances. Thus, to stationary observers on Earth, the photon in the moving clock appears to take more *time* to bounce between its mirrors than the photon in the stationary clock, due to the former's longer path. This means the moving clock takes longer to count tick-tock, and therefore seems to run slower. Let's calculate how much slower.

Let L be the vertical distance between the mirrors, and let T be the moving clock's photon transit time from its upper mirror to its lower mirror, as we observe it in our stationary frame. During time T, the lower mirror of the jet's clock moves a distance v T. (This distance is not shown above, since the figure does not show the photon reaching the lower mirror.) That photon's complete diagonal path length is:

$$\text{right photon path length} = \sqrt{(L^2 + v^2 T^2)}$$

That length equals elapsed time T multiplied by photon speed c, hence:

$$c^2 T^2 = L^2 + v^2 T^2$$

$$T^2 (c^2 - v^2) = L^2$$

$$T = L / \sqrt{(c^2 - v^2)}$$

$$T = L / \{ c \sqrt{(1 - v^2 / c^2)} \} = \gamma\ L / c$$

Now let's compare T, the apparent transit time of the photon in the moving clock, to t, the photon transit time in Earth's clock. Since Earth's clock is deemed stationary, the distance between its mirrors is L, and t is simply L / c, L divided by the speed of light. The two times are related by:

$$T = \gamma \, L / c \; = t \, \gamma$$

Since $\gamma > 1$, $T > t$. We in the stationary frame observe the photon transit time in the moving frame to be longer than in ours. We observe the clock in the moving frame counting at a slower rate by the relativistic factor γ.

This is called *time dilation*. For v = 0.9 c, time dilates (slows) by a factor of about $\gamma = 2.3$.

Should the jet's crew get a better clock, one that keeps the *right* time? Einstein said that is impossible. If another clock kept the *right* time, we could use photon clocks to tell us who is *really* moving and who is *really* stationary, in violation of the principle of relativity.

Einstein said everything in the moving frame runs at exactly the same slower rate: the crew's hearts beat slower, their brain waves wave slower, their fingernails grow slower, and they age slower — all at the same slower rate.

What this really means is that *time* itself runs slower in the moving frame, as we observe it from the stationary frame.

Relativistic effects are symmetric. We are not special just because we call ourselves "stationary." The principle of relativity says there are no special reference frames.

The jet's crew, looking back at our photon clock, sees our mirrors moving at v = 0.9 c to the left, our photon having the diagonal path, and our clock running 2.3 times slower than theirs.

We say their clock runs slower than ours. They say ours runs slower than theirs. Who is right?

We both are! Each set of observers is correct, but *only in their own frame of reference*.

This may seem impossible, but it is really true, and it all boils down to how measurements are actually made. Because the speed of light c is the same in all inertial frames, and because c is not infinite, observers moving at different speeds necessarily measure different values for such basic quantities as time and distance.

There really is no single right answer to: "what time is it?"

§2.4 Time Dilation & Cosmic Rays

One real world consequence of time dilation involves cosmic rays reaching Earth's surface, and us. Cosmic rays from the Sun and from deep space continuously rain down on Earth; these are called *primaries*. About 89% of primaries are protons, the nuclei of hydrogen atoms, with neutrons comprising 9% and the remainder being electrons and heavier nuclei.

The higher energy cosmic rays create *showers* of *secondaries* when they collide with atoms in our atmosphere. This rapidly reduces the average particle's energy. Only a few in 10,000 primaries reach Earth's surface. However, many secondaries decay into muons, which have greater penetrating power because they interact only weakly with other particles. Muons do not produce secondary showers and thus lose energy more slowly. Of all the particles of cosmic ray origin that penetrate our atmosphere and reach Earth's surface, about 98% are muons.

Cosmic ray muons account for between 10% and 25% of the average person's total radiation exposure. Each individual's actual exposure varies substantially, and depends on their occupation, the altitude where they live, their water source, and other factors.

The average person's total annual radiation exposure is 3.0 mSv (millisievert). Of that total, cosmic rays deliver 0.4 to 1.0 mSv. Other natural sources amount to 2.0 mSv (radon gas, food, heavy element decays in the ground and in building materials), and 0.6 mSv from medical testing and therapies. Nuclear power plants, prior atomic bomb tests, and occupational exposures (mostly mining) average 0.01 mSv. All non-cosmic ray factors are highly variable; some exposures exceed 10 times the averages.

Radiation can cause genetic mutations, which are almost always harmful. But in rare instances, mutations are beneficial. Mutations drive evolution; without them, humans would not exist.

What does all this have to do with Einstein and relativity? Without time dilation almost no cosmic ray muons would reach life on Earth's surface, thus reducing the pace of evolution and raising a question of whether humans would have evolved by now without time dilation.

Muons have a mean lifetime of 2.2 microseconds, corresponding to 660 m at light speed, only about 5% of the thickness of the dense portion of our atmosphere. But, there are enough muons with $\gamma > 20$ and lifetimes > 44 microseconds that do reach the surface to account for 10% and 25% of our total radiation exposure, for better or worse.

Time dilation also plays a critical role in particle physics, where particle velocities are often near the speed of light. My own thesis experiment, for example, employed a calibration from decays of *short-lived neutral kaons*, whose mean lifetime is 0.09 nanoseconds. Even light travels only 2.7 cm (1.1 inches) during that lifetime. Fortunately, at velocity v = 0.992 c, γ equals 8, making that kaon lifetime long enough for them to travel 21.6 cm, a much more practical distance.

§2.5 Length Contraction

Let's re-examine the Michelson-Morley interferometer experiment, now assuming the speed of light is constant.

Recall the setup: two orthogonal arms are joined at one end, and have mirrors at their far ends, as sketched in Figure 2-4. Light travels along both arms, and the interferometer measures the difference in light travel time between the two arms.

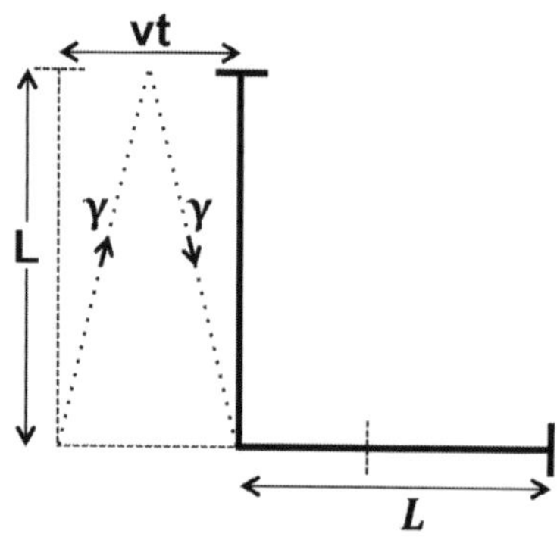

Figure 2-4 Vertical Light Path

When the device is stationary, assume both arms have length L. Let's define the horizontal arm length to be *L* when the device moves horizontally at velocity v, thereby allowing for the possibility that this length changes.

Now, what is the travel time along the vertical path? In Figure 2-4, the dashed lines indicate the interferometer's position when light starts its round trip along the vertical path, and the solid lines show the interferometer's position at the instant light returns to its starting corner. Due to the device's motion, we in a stationary reference frame observe light traveling along a diagonal path to reach the upper mirror and return to the starting corner. If the total up-and-down travel time is t, the interferometer moves horizontally a distance v t, and the total length of both diagonal paths is:

$$2 \sqrt{(L^2 + v^2 t^2 / 4)}$$

The travel time t along both diagonal paths is that length divided by c:

$$t = [\,2\,\sqrt{(L^2 + v^2 t^2 / 4)}\,] \,/\, c$$

$$c^2 t^2 = 4 L^2 + v^2 t^2$$

$$t^2 (c^2 - v^2) = 4 L^2$$

$$t = 2 L / \sqrt{(c^2 - v^2)}$$

Now let's examine the horizontal path. In Figure 2-5, the dashed lines indicate the interferometer's position when light starts its round trip along the horizontal path, and the solid lines show the interferometer's position at the instant light returns to the starting corner.

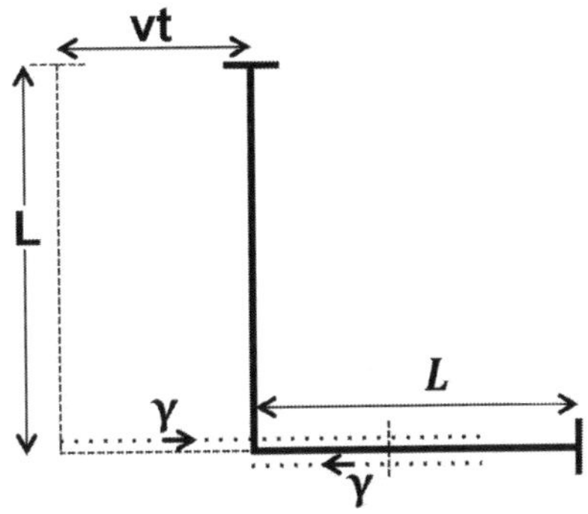

Figure 2-5 Horizontal Light Path

In the stationary frame, while light is moving to the right, the mirror is moving away at speed v. The *closing speed*, the rate at which the separation between the light beam and the mirror decreases, is c – v. Conversely, while light is moving back to the left, the starting corner is moving toward it at speed v. The closing speed then is c + v. Hence the total horizontal travel time along the arm of length *L* is:

$$t = L / (c - v) \; + \; L / (c + v)$$

$$t = L \{ (c + v) \; + \; (c - v) \} \; / \; (c - v)\; (c + v)$$

$$t = L \{ 2\, c \} / (c^2 - v^2)$$

Since Michelson found no change in travel time difference, regardless of the interferometer's motion, we must conclude that the horizontal and vertical travel times are equal for any velocity v. This means:

$$2\, c\, L / (c^2 - v^2) = 2 L / \sqrt{(c^2 - v^2)}$$

$$L = L \sqrt{(1 - v^2 / c^2)} = L / \gamma$$

This is called *length contraction*: to stationary observers, the lengths of moving objects appear shorter along the direction of motion.

§2.6 Everyday Relativity

How much does relativity matter in our everyday lives? After all, few things travel 600 million mph.

Surprisingly, relativity does matter, even for slowpokes. Many people have GPS (Global Positioning System) navigation systems in their cars to guide them to their destinations — maps are so 20th Century. GPS receivers determine where we are, where we are headed, and how fast we are going from timing signals emitted by a fleet of 24 GPS satellites. These satellites orbit Earth at a speeds over 13,000 km/hr (8000 mph). The clocks in GPS satellites run at a different rate than do stationary clocks on Earth due to both special relativistic and general relativistic effects. If GPS clocks were not corrected for these relativistic effects, the GPS system would be worthless.

So if you get where you want to go, you should thank the man who could never find his own keys.

Chapter 3
Dynamics in Special Relativity

In this chapter, we explore how Einstein's postulate that the speed of light is constant affects measurements of mass, and the dynamics of momentum and energy.

§3.1 Relativistic Mass

Let's begin with the impact of a constant speed of light on measurements of mass. We will employ the conservation of momentum and the equation for adding relativistic velocities derived in Section §2.2. Newtonian momentum p equals m v, the product of mass and velocity. Velocity is not modified by relativity, so relativity can modify momentum only by changing mass. The relativistic equation for momentum must therefore have the form:

$$p = v\ m(v)$$

Here, m(v) is a function of velocity to be determined. To match Newtonian momentum at v = 0, m(0) must equal the rest mass that we denote m_0.

Let's do another thought experiment: consider two identical particles of rest mass m_0 that collide and bounce off one another, as shown in Figure 3-1. In the left image, the lower particle enters with velocity w directly up along the y-axis, and after the collision moves directly down the y-axis. In the drawing, its up and down paths are slightly separated for clarity. The upper particle enters from the right with speed v at glancing angle ø, and exits to the left with the same speed and angle. Its vertical component of velocity before the collision is v sinø in the –y-direction, and v sinø in the +y-direction afterwards. Its horizontal component of velocity is v cosø in the –x-direction, both before and after the collision.

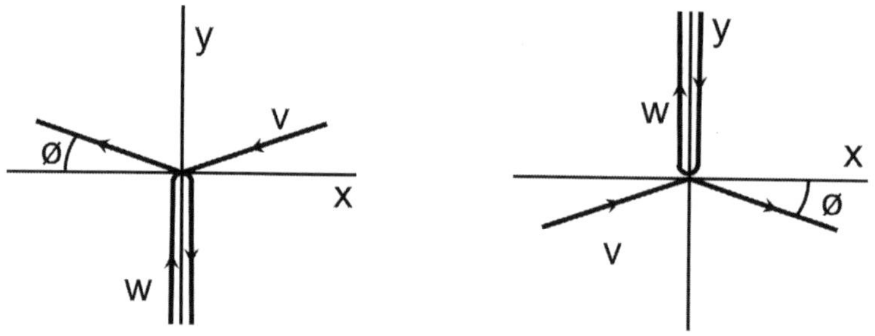

Figure 3-1 Two Views of a Collision

In the right image of Figure 3-1, we observe the same collision in a frame moving to the left (toward –x) with velocity v cosø, the horizontal velocity of the upper particle in the left image. In this frame, the horizontal velocity of the upper particle is zero. Note that the right and left images are symmetric — they are identical to one another with a 180° rotation, ensuring all corresponding speeds and angles are equal.

In Section §2.2, we derived the equation for a velocity entirely in the y-direction (here that velocity is w) that is transformed to a frame moving with relative velocity entirely in the x-direction (here that velocity is –v cosø). The equation is:

Eqn. #1: $$v \sin\text{ø} = w \sqrt{(1 - v^2 \cos^2\text{ø} / c^2)}$$

Note that for w = 0 and ø = 0, v can have any value. Squaring both sides of this equation yields:

$$v^2 (1 - \cos^2\text{ø}) = w^2 (1 - v^2 \cos^2\text{ø} / c^2)$$

$$v^2 \cos^2\text{ø} (- 1 + w^2/c^2) = w^2 - v^2$$

Eqn. #2: $$v^2 \cos^2\text{ø} = (v^2 - w^2) / (1 - w^2/c^2)$$

The next step is deriving an equation relating the relativistic masses of the two identical particles that collide in Figure 3-1. From conservation of momentum, we know that the change in the lower particle's y-momentum plus the change in the upper particle's y-momentum must equal zero. From the left image, this means:

$$2\ w\ m(w) = 2\ v \sin\phi\ m(v)$$

Recall Eqn. #1: $$v \sin\phi = w \sqrt{(1 - v^2 \cos^2\phi / c^2)}$$

Substituting v sinø: $$w\ m(w) = m(v)\ w \sqrt{(1 - v^2 \cos^2\phi / c^2)}$$

$$m(w) = m(v) \sqrt{(1 - v^2 \cos^2\phi / c^2)}$$

From Eqn. #2: $$m(w) = m(v) \sqrt{\{ 1 - (v^2/c^2 - w^2/c^2) / (1 - w^2/c^2) \}}$$

$$m(w) = m(v) \sqrt{\{[1 - w^2/c^2 - (v^2/c^2 - w^2/c^2)] \;/\; (1 - w^2/c^2) \}}$$

$$m(w) = m(v) \sqrt{\{ [1 - v^2/c^2] / (1 - w^2/c^2) \}}$$

$$m(w) \sqrt{(1 - w^2/ c^2)} = m(v) \sqrt{(1 - v^2/ c^2)}$$

We have successfully put all the w-dependent quantities on the left, and all the v-dependent quantities on the right.

Note that the velocities w and v are substantially independent. For any selected value of w, v can have a broad range of values, each associated with its own angle ø. And for all those v, ø combinations, the right side of the last equation must have the same value, because the left side has the same w. This independence means the above equation can be true only if each side has a constant value — call it K — for any velocities w and v. For the special case of v = 0, we know that K must equal m_0. Thus, each side of this equation must equal m_0 for any velocities v and w.

$$m(v) \sqrt{(1 - v^2 / c^2)} = m_0$$

$$m(v) = m_0 \;/\; \sqrt{(1 - v^2 / c^2)} = m_0\, \gamma$$

We have derived the equation relating relativistic mass m(v) at velocity v to rest mass m_0.

Special relativity thus nullifies a previously long-cherished principle: the conservation of mass. Before scientists studied nuclear reactions, they were only familiar with chemical reactions that have much lower energy. Those low-energy reactions change masses by only parts per billion, far less than the weighing precision of the day. As far as 19th Century chemists could tell, mass was conserved. But as science pushed into higher energy domains, they discovered many reactions that change mass dramatically.

§3.2 E = m c²

By re-examining the concept of force from a relativistic prospective, we can now derive the most famous of all equations: $E = m\ c^2$. To do this properly, we need a concept from calculus: we define dQ/dt to be the *rate of change* — the *time derivative* — of any quantity Q. See Section §A2.1 for a more complete explanation.

Newton's force law is usually written F = m a, where mass m is assumed constant, and acceleration a is the rate of change of velocity. Since we now know that mass is not constant, it is more effective to define force as the rate of change of momentum, velocity v multiplied by relativistic mass $m_0\, \gamma$:

$$F = d(m_0\, \gamma\ v)/dt$$

This redefinition has no effect on Newtonian mechanics, which assumes m is constant. Hence, we can continue employing equations like F v = power, the product of force and velocity equals power, the rate of change of energy. If we apply a constant force F to an object of rest mass m_0 that is moving with velocity v, its rate of change of energy is given by:

$$dE/dt = F\ v$$

Combining the prior two equations yields:

$$dE/dt = v\ d(m_0\,\gamma\,v)/dt$$

$$dE/dt = m_0\,\gamma\,v\ dv/dt + m_0\,v^2\ d\gamma/dt$$

From Appendix §A2.5:

$$c^2\ d\gamma/dt = \gamma^3\ v\ dv/dt$$

$$dE/dt = m_0\,\gamma\,(\gamma^{-3}\,c^2\,d\gamma/dt) + m_0\,v^2\ d\gamma/dt$$

$$dE/dt = (\gamma^{-2}\,c^2 + v^2)\ m_0\ d\gamma/dt$$

$$dE/dt = (c^2 - v^2 + v^2)\ d(m_0\,\gamma)/dt$$

$$dE/dt = c^2\ dm/dt$$

$$E = m\,c^2$$

Here is Einstein's famous equation. Without the c^2, Einstein's message is more vivid: *mass is energy*.

The highest velocity charged particles ever produced by a particle accelerator were electrons from SLAC, the Stanford Linear Accelerator Center. SLAC electrons had $\gamma = 40{,}000$, and a velocity of 0.999,999,999,7 c — nine 9's and a 7. If one of those electrons raced a photon (in a straight evacuated tube) from San Francisco to New York, the electron would trail the photon by only the thickness of one sheet of paper. Protons in the LHC in Geneva have $\gamma = 4300$; their velocities are seven 9's and a 7. Since protons are much more massive than electrons, LHC's beam energy is 200 times higher than SLAC's.

For modest speeds, we can rewrite $E = m\,c^2$ using a series expansion presented in Section §2.2.

$$E = m_0\,c^2 + m_0\,v^2/2 + 3\,m_0\,v^4/8\,c^2 + \ldots$$

The first term on the right hand side is the rest mass energy, while the second term is the kinetic energy of Newtonian mechanics. The higher order terms are relativistic corrections.

Let's now derive an important relationship between E, p, and rest mass m_0.

$$m_0^2\,c^4 + p^2\,c^2 = m_0^2\,c^4 + c^2\,m_0^2\,\gamma^2\,v^2$$

Per Section §2.2, $c^2 + \gamma^2\,v^2 = c^2\,\gamma^2$:

$$m_0^2\,c^4 + p^2\,c^2 = m_0^2\,c^4\,\gamma^2$$

For any particle at any velocity v:

$$m_0^2\,c^4 + p^2\,c^2 = E^2$$

§3.3 Relativity & Photons

How do the equations of special relativity relate to particles moving at the speed of light? At light speed, γ = infinity. In physics, infinity is generally a problem. Let's find the equation for velocity in terms of energy and rest mass m_0.

$$E = m_0\,c^2\,\gamma$$

$$m_0\,c^2/E = 1/\gamma = \sqrt{(1 - v^2/c^2)}$$

$$m_0^2\,c^6/E^2 = c^2 - v^2$$

$$v^2 = c^2 - m_0^2\,c^6/E^2$$

$$v = c\sqrt{\{1 - m_0^2\,c^4/E^2\}}$$

In the limit that m_0 goes to zero and E remains nonzero, velocity v goes smoothly to c, the speed of light.

Now let's examine $\gamma\,m_0$:

$$\gamma\,m_0 = E$$

$$\gamma\,m_0 = E\,(m_0/m_0)$$

No surprises here, but this shows that the right hand side of the last equation has a well-behaved limit as m_0 goes to zero. Therefore, although m_0 goes to zero and γ goes to infinity, $m_0\,\gamma$ is well behaved at $v = c$.

Note that if $m_0 = 0$, and $v < c$, γ is finite and $E = m_0 \gamma = 0$. Since nothing exists that has zero energy, any real massless particle *must* move at the speed of light, as photons do.

From $E^2 = c^2 p^2 + c^4 m_0^2$, we have for photons: $E = p c$. This means light has momentum as well as energy; hence the dream of propelling spaceships with solar sails.

Since photons only exist at velocity c, we cannot transform to their "rest" frame, as we can for any massive object. When Einstein was 16, he tried to imagine how a light beam would appear if one could race along side it at light speed. A decade later, his theory of special relativity made clear why this is impossible.

We can consider a limiting case: the rest frame of a particle of mass m_0 as m_0 approaches zero. In that limit, γ approaches infinity, and all time intervals dilate to zero: $T / \gamma \rightarrow 0$. In that limit, a particle created in the Big Bang that traveled for nearly 14 billion years to reach our eye "sees" its entire existence as a single instant in time — simultaneously created and absorbed in our retina, with all the cosmos along its path compressed to a point of zero size. For that particle, time and distance do not exist.

Most commonly, physicists use m to denote an object's rest mass, and write its momentum and energy as:

$$E = \gamma m c^2$$

$$p = \gamma m v$$

$$\gamma = 1 / \sqrt{(1 - v^2 / c^2)}$$

§3.4 What is Not Relative?

As we have shown, measurements of certain quantities are relative — they are different for observers in different states of motion. Relative quantities include: mass, time, distance along the direction of relative motion, and quantities that depend on these, such as energy and momentum.

Other phenomena are absolute — all observers in any state of motion will agree on absolute phenomena, including these:

- Existence of specific objects
- Quantities and identities of objects in a system
- Distances orthogonal to direction of motion
- Amount and polarity of electric charge
- Causal relationships: causes always precede effects
- Interactions among objects
- Contents of data and documents

Some specific examples of absolute phenomena are:

- Earth orbits the Sun.
- A carbon-14 atom radioactively decays.
- An electron and a positron annihilate, producing two photons.
- A supernova occurs before a nearby planet vaporizes.
- In your wallet is a picture of someone you love.

Chapter 4
Spacetime in Special Relativity

Hermann Minkowski, Einstein's college math professor was unimpressed with young Albert, calling him a "lazy dog." He may have reconsidered after the publication of special relativity. Minkowski realized Einstein's theory changed our understanding of the geometry of our universe.

Minkowski said the three dimensions of space can no longer be considered distinct from time; space and time must be viewed as a combined four-dimensional entity that he named *spacetime*.

In four-dimensional spacetime, points are called *events*. Four quantities — such as time, longitude, latitude, and altitude — are required to completely specify an event.

Measurements of all dimensions of spacetime are relative (may be different for different observers).

§4.1 Four-Vectors

In relativistic 4-D spacetime, *4-vectors* are effective mathematical tools. A 4-vector is comprised of a normal 3-vector defined in 3-D space plus a fourth quantity that is most commonly called component number 0. The 3-vector parts are called components 1, 2, and 3. A properly constructed 4-vector is *Lorentz covariant*, meaning that it transforms from one reference frame to another in accordance with the Lorentz transformation. Let's meet some important covariant 4-vectors:

position of event A:	$(c\,t_A,\ x_A,\ y_A,\ z_A)$
separation B – A:	$(c\,t_B - c\,t_A,\ x_B - x_A,\ y_B - y_A,\ z_B - z_A)$
4-velocity:	$(c\,\gamma,\ \gamma\, v_x,\ \gamma\, v_y,\ \gamma\, v_z)$
4-momentum:	$(E/c,\ p_x,\ p_y,\ p_z)$

The B – A separation 4-vector is called the *displacement* 4-vector. Adding displacement (B – A) to the A position 4-vector produces the B position 4-vector: (B – A) + A = B.

Given any covariant 4-vector, we can obtain another covariant 4-vector by multiplying each component by any constant q, as in:

$$q\,(a,\ b,\ c,\ d) = (q\,a,\ q\,b,\ q\,c,\ q\,d)$$

This is because ordinary numbers have the same values in all reference frames. But because γ is *not* the same in all frames:

$(c\,\gamma,\ \gamma\, v_x,\ \gamma\, v_y,\ \gamma\, v_z)$	is covariant
$(c,\ v_x,\ v_y,\ v_z)$	is *not* covariant

The momentum 4-vector $(E/c,\ p_x,\ p_y,\ p_z)$, is the proper covariant union of energy and the 3-vector momentum. However, $(E/c,\ x,\ y,\ z)$ and $(c\,t,\ p_x,\ p_y,\ p_z)$ are *not* Lorentz covariant 4-vectors, because they do not transform properly.

The dot product of two 4-vectors is similar to the dot product of two 3-vectors, except for a *critical minus sign* that is the signature of relativity. For example, the dot product of the momentum and position 4-vectors is:

$$(E/c, p_x, p_y, p_z) \bullet (c\,t, x, y, z) = -t\,E + x\,p_x + y\,p_y + z\,p_z$$

The minus sign in the product of the 0-components distinguishes relativity from Newtonian physics and ensures the constancy of the speed of light.

The dot product of any two covariant 4-vectors is *invariant*, the same number in every reference frame. In particular, the dot product of any covariant 4-vector with itself is invariant. For example, the dot product of the momentum 4-vector with itself is:

$$(E / c, p_x, p_y, p_z) \bullet (E / c, p_x, p_y, p_z) = -E^2 / c^2 + p^2 = -m_0{}^2 c^2$$

This dot product equals minus the square of (rest mass multiplied by c) in every reference frame.

One way to tell if a 4-vector is covariant is calculating its dot product with itself, and determining whether the result has the same value in all reference frames, as does rest mass. For example, the following is not invariant:

$$(c\,t,\ p_x,\ p_y,\ p_z) \bullet (c\,t,\ p_x,\ p_y,\ p_z) = -c^2 t^2 + p^2$$

§4.2 Einstein Conventions

To manage the complexity of relativistic equations, Einstein established several conventions for writing 4-vectors and their dot products. For brevity, we write the components of the position 4-vector as x_μ, with index μ having any of the values 0, 1, 2, 3, and thereby specifying components according to:

$$x_0 = c\,t$$

$$x_1 = x$$

$$x_2 = y$$

$$x_3 = z$$

The 4-D dot product of the position and momentum 4-vectors is concisely written:

$$x_\mu\, p_\mu = -t\,E + x\,p_x + y\,p_y + z\,p_z$$

This illustrates the Einstein summation convention: the repeated index μ directs us to sum over all four values of index μ — the summation sign Σ_μ is implicit. (Remember the minus sign in component 0.)

The dot product of the separation 4-vector with itself, ds^2, is called the *invariant interval*:

$$ds^2 = dx_\mu\, dx_\mu = -c^2\, dt^2 + dx^2 + dy^2 + dz^2$$

Here, we define the *differential*: dq means a small, typically infinitesimal, change in the variable q. This equation and all that follow in this chapter are valid for infinitesimal separations in any spacetime; they are also valid for macroscopic separations in spacetime geometries without curvature. The use of the differentials for both is conventional. Although each of the four components of separation varies from frame to frame, ds^2 remains constant. Because of the leading minus sign, ds^2 may be positive, zero, or negative. Perhaps this is why it is called the *interval* rather than some type of *distance* between events.

In any reference frame, the interval between two events along the path of a photon is always zero. Let's see why.

3-D distance between events A and B: $dr = \sqrt{(dx^2 + dy^2 + dz^2)}$

Time dt for light to travel dr: $dt = dr / c$

Interval between A and B for light: $ds^2 = -c^2\, dt^2 + dr^2 = 0$

With so much changing from one frame to another, it is important to focus on what is the same in all reference frames — those things that are invariant. In both special and general relativity, the interval between any two events in 4-D spacetime is the most important invariant. Intimately related to the invariant interval is *proper time*, $d\tau$, given by:

$$d\tau^2 = -ds^2$$

For events A and B with separation 4-vector (c dt, dx, dy, dz):

$$c^2\,d\tau^2 = +c^2\,dt^2 - dx^2 - dy^2 - dz^2$$

$$c^2\,d\tau^2 = dt^2\,(c^2 - dx^2/dt^2 - dy^2/dt^2 - dz^2/dt^2)$$

Substituting v^2 for the sum of its three components yields:

$$c^2\,d\tau^2 = dt^2\,(c^2 - v^2)$$

$$d\tau = dt\,/\,\gamma$$

Proper time $d\tau$ is what an ideal clock measures in its own rest frame.

§4.3 Rotations & Boosts

In both three and four dimensions, the components of vectors change when transformed from one reference frame or coordinate system to another. Yet, essential aspects of these vectors are unchanged — their lengths do not change, and, neither do the orientations of vectors relative to one another.

Let's examine this more closely. Shown in Figure 4-1 are two 2-D coordinate systems, rotated relative to one another, but each with the same vector ***AB***. Clearly the vector has different component values in the xy system compared with the vu system.

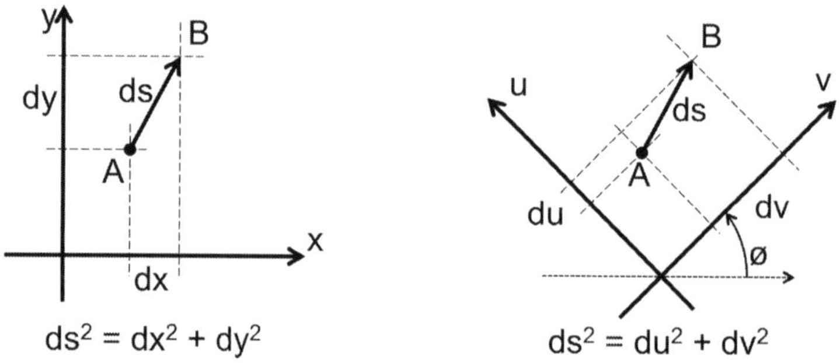

Figure 4-1 Rotating Coordinate Systems

In Figure 4-1, the vu axes are rotated in the positive sense (counterclockwise) by angle ø. The equations for transforming xy coordinates to vu coordinates are:

$$v = +x\cos ø + y\sin ø$$

$$u = -x\sin ø + y\cos ø$$

Note there is one minus sign in the above equations. The differentials are:

$$dv = +dx\cos ø + dy\sin ø$$

$$du = -dx\sin ø + dy\cos ø$$

Let the components of vector ***AB*** be (dx, dy) in the xy system and (dv, du) in the vu system. The length of ***AB*** in v-u is:

$$ds^2 = dv^2 + du^2$$

$$ds^2 = dx^2\cos^2 ø + 2\,dx\,dy\cos ø\sin ø + dy^2\sin^2 ø$$
$$+ dx^2\sin^2 ø - 2\,dx\,dy\sin ø\cos ø + dy^2\cos^2 ø$$

$$ds^2 = dx^2 + dy^2$$

The length of vector ***AB*** is the same in both coordinate systems for any rotation angle ø. We can clearly extend this analysis to three dimensions, since any 3-D rotation can be obtained with at most three 2-D rotations employing Euclid angles (see my book ***Feynman Simplified Part 4***, Section 4A§10.2).

Let's now compare rotational transformations to Lorentz transformations. Consider a stationary xt frame and a XT frame moving in the +x-direction with velocity w and $\gamma = 1 / \sqrt{(1 - w^2 / c^2)}$.

The equations for transforming xt coordinates into XT coordinates, often called a *boost*, are:

$$X = \gamma x - \gamma w t$$

$$T = \gamma t - \gamma w x / c^2$$

Note the two minus signs above. If we transformed with velocity –w, there would be zero minus signs. The number of minus signs is the key distinction between boosts that have zero or two, and rotations that have one.

In 4-D spacetime, the square of the length of 4-vector ***AB*** is its invariant interval:

$$ds^2 = (B - A)_\mu \; (B - A)_\mu$$

Unlike 3-D, the square of a 4-vector's length in 4-D spacetime can be negative. Let's calculate ds^2 in the XT frame:

$$ds^2 = - c^2 \, dT^2 + dX^2$$

$$ds^2 = - \gamma^2 (c^2 \, dt^2 - 2 \, w \, t \, x + w^2 \, dx^2 / c^2) + \gamma^2 (dx^2 - 2 \, w \, x \, t + w^2 \, dt^2)$$

$$ds^2 = - \gamma^2 (c^2 \, dt^2 - w^2 \, dt^2) + \gamma^2 (dx^2 - w^2 \, dx^2 / c^2)$$

$$ds^2 = - \gamma^2 c^2 \, dt^2 (1 - w^2 / c^2) + \gamma^2 \, dx^2 (1 - w^2 / c^2)$$

$$ds^2 = - c^2 \, dt^2 + dx^2$$

The ***AB*** vector's invariant length is the same in both frames, as promised.

The ***AB*** vector is shown in Figure 4-2 in both the xt frame and the boosted XT frame. The boost is, pardon the expression, graphically different from the rotation of Figure 4-1.

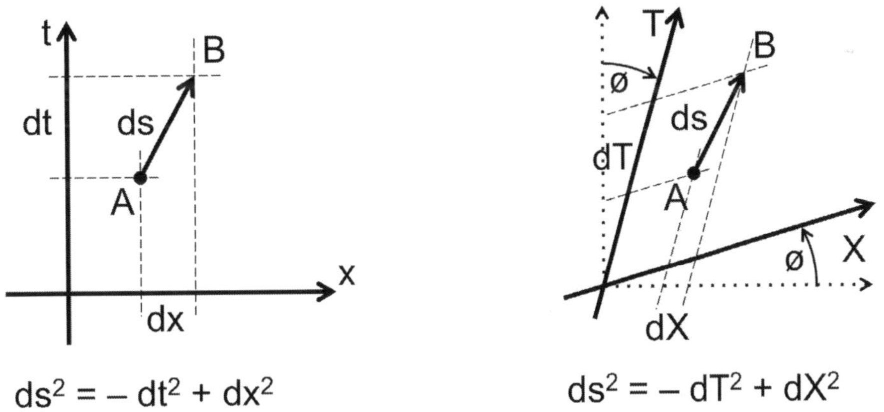

Figure 4-2 Boosting Coordinate Systems

In a rotation by angle ø, both axes turn in the same direction by the same angle ø. In a boost by velocity w, both time and space axes turn inward toward the X = T line at 45° to the x and t axes; the X and T axes rotate by the same angle ø, but in opposite directions, with:

$$\tanh(ø) = w / c$$

Here, tanh is the hyperbolic tangent.

For velocity w in the +x-direction, the Lorentz transformation of 4-vector $(E / c, p_x, p_y, p_z)$ produces the 4-vector $(E^* / c, p_x^*, p_y, p_z)$:

$$E^* = \gamma E - \gamma w p_x$$

$$p_x^* = \gamma p_x - \gamma w E / c^2$$

§4.4 Simultaneity

Simultaneity is an important aspect of our everyday experience. We expect concert musicians to play their parts at the same time; football players must move together when the ball is snapped; and we expect clocks to read the same time.

Yet special relativity tells us universal simultaneity is impossible — observers in different states of motion will never agree on which events are simultaneous. Let's see why.

Let's imagine two clocks in our stationary reference frame, with each clock being a distance D from a light source that synchronizes them, as shown in Figure 4-3. Each second the light flashes, and the clocks increment their displays when they receive the flash. In our stationary frame, each clock detects the flash at the same time, a time D / c after the flash occurs at the source. The clocks advance absolutely simultaneously.

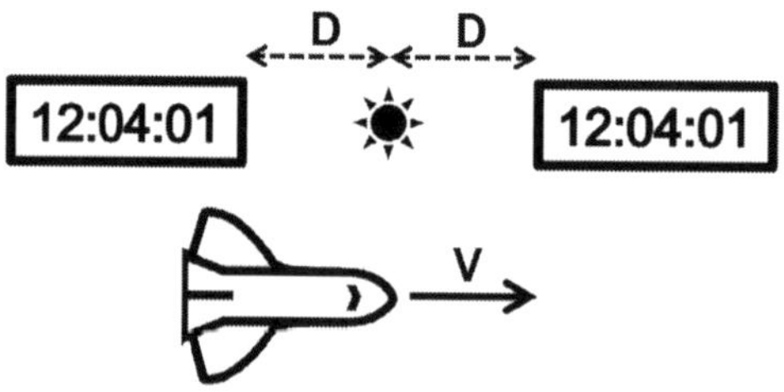

Figure 4-3 Clocks Viewed in Stationary Frame and by Passing Spaceship

How does all this appear to the crew of a spaceship on a parallel course traveling at speed v? The first difference is that the spaceship crew observes the distance between the light source and each clock contracted to D/γ. Now let's determine *when* each clock advances in the spaceship frame.

Consider the flash arrival times at each clock after the source flashes at time T = 0. The spaceship crew sees the left clock moving left at speed v, *away* from the oncoming flash. The closing velocity, the rate at which distance D/γ is being reduced, equals c – v, and light arrives at the left clock at time:

Flash arrival time at left clock: $$T_L = (D/\gamma)\ /\ (c-v)$$

The spaceship crew also sees the right clock moving left at speed v, but this is *toward* the oncoming light flash. The closing velocity equals c + v, and light arrives at the right clock at time:

Flash arrival time at right clock: $$T_R = (D/\gamma)\ /\ (c+v)$$

The two clocks do not advance simultaneously, as seen in the spaceship's frame: the right clock advances earlier by:

$$T_L - T_R = (D/\gamma)\ \{\ 1/(c-v)\ -\ 1/(c+v)\ \}$$
$$T_L - T_R = (D/\gamma)\ \{\ (c+v)\ -\ (c-v)\ \}\ /\ (c^2-v^2)$$
$$T_L - T_R = 2\,v\,D\,\gamma^2 / c^2\,\gamma = 2\,v\,D\,\gamma / c^2$$

This time difference is positive because the flash arrives at the right clock before the left one. No observer, moving relative to the clocks, sees these clocks advance simultaneously. Note that if D = 0, the clocks would increment simultaneously in all reference frames, but zero separation really means there is just one clock. With full generality, special relativity states that observers moving at different speeds will not agree on which *separated* events are simultaneous.

Let's step back from the equations and explore the physical reason that the above clocks do not advance simultaneously. The stationary observer and the spaceship observer see the clocks and the light source moving at different velocities, just as Newton would expect — the former sees these objects at rest, while the latter sees them moving at velocity v to the right. But contrary to Newtonian mechanics, these observers in different frames both see the light flashes moving at the same velocity c. The constancy of light's speed leads to different closing velocities on the left compared with the right, and therefore different light flash arrival times.

We can explore simultaneity graphically by imagining 4-D spacetime as being a salami, as shown in Figure 4-4. In our reference frame in the upper image, each thin slice perpendicular to the long axis is an xy-plane of constant time. Time progresses, from slice to slice, increasing toward the right. For us, all events on each slice are simultaneous, and the two black dots on different slices occur at different times.

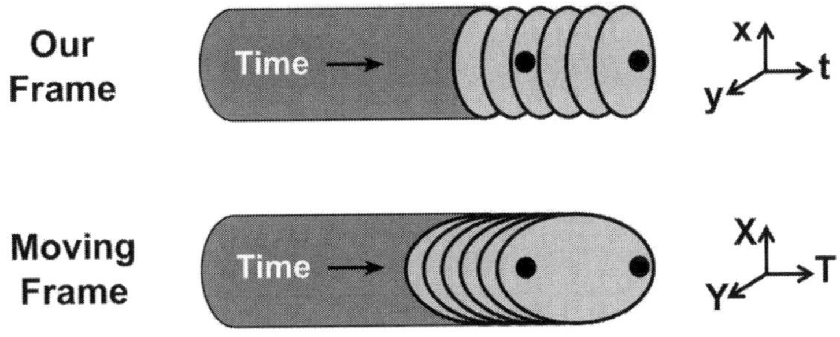

Figure 4-4 Slicing Spacetime like Salami

In the moving frame in the lower image, the slices of constant time seem askew to us. The two dots occur at the same time, and other events that we consider simultaneous seem to moving observers to occur at different times. Only when two events occur at the same time and place will all observers agree they are simultaneous. But that is a trivial case, in which only one physically distinct event really exists. (The z-axis is not shown in these images; my drawing ran out of dimensions.)

Observers in relative motion see one another's time and space mix together.

We must increase the list of observables that are measured differently by observers in different states of motion, adding *simultaneity* to the *length* of a meter-stick, the amount of *time* in one Earth orbit of the Sun, and the *mass* of 1 kg.

§4.5 Space-Like & Time-Like Intervals

We discovered in Section §4.2 that the interval between two events along the path of a photon is always zero in any reference frame. This is because photons move a distance dr = c dt during time dt, so $dr^2 - c^2 dt^2$ is always 0. For any object with velocity less than light speed, $v < c$, the intervals along its path are all less than zero, because dr = v dt < c dt. If anything could move faster than light (nothing can), the intervals along its path would all be greater than zero, because dr = v dt would be > c dt.

Two events with interval greater than zero are said to be *space-like separated* (their spatial separation dr exceeds c dt, their temporal separation). If A and B are space-like separated, there is a reference frame in which A and B occur at the same time but at different locations.

Two events with interval less than zero are *time-like separated* (c dt > dr). If A and B are time-like separated, a rocket can travel between them at a constant velocity $v < c$. In the rocket's rest frame, A and B occur at the same spatial coordinates but at different times. (Anything whose velocity is constant is stationary in its own rest frame.)

Since ds^2 is invariant, events that are time-like separated in one frame are time-like separated in all frames. Similarly, events that are space-like separated in one frame are space-like separated in all frames. Finally, events with interval zero in one frame have interval zero in all frames.

§4.6 Causality & Light Cones

The *principle of causality* states that causes must precede their effects. Relativity ensures causality. Two space-like separated events cannot have a causal relationship. Even light cannot travel fast enough to span their spatial separation within their time difference, because in some frame they occur at the same time at different locations.

However, two time-like separated events, which must always have $ds^2 < 0$, can have a proper causal relationship. This is because, whichever event occurs first in one frame, occurs first in all frames.

Let's see why events occur in the same time sequence in all frames. In any selected reference frame, call an earlier event A, and a later event B. Define dt to be their time difference.

$$dt = \text{time of event A} - \text{time of event B}$$

By definition, $dt < 0$ in the selected reference frame. Since the Lorentz transformation is continuous (its values vary smoothly without discontinuities), the *intermediate value theorem* of calculus states that dt cannot go from less than zero to greater than zero without passing through zero. Therefore, if we could transform to any frame in which $dt > 0$, there must be an intermediate transformation to a frame in which $dt = 0$. But, if $dt = 0$, ds^2 cannot be negative in that frame, contrary to the invariance of ds^2 and the requirements of time-like separated events. Hence, there can be no frame in which dt is not negative (in which A does not precede B).

All this means that a cause at A can affect B, but no cause at B can affect A.

Next, let's imagine the set of all events, the *locus* of events, that are time-like separated from event O at coordinates $x = y = z = t = 0$. In Figure 4-5, these events lie within the two gray cones. (Again, the *z* axis is not shown.)

Events that occurred before O are said to be in the *past light cone* of O. Any event in O's past light cone can influence O. For example, if a supernova occurred anywhere within O's past light cone, its light will reach O. Conversely, O cannot influence any event in its past light cone.

Events in the *future light cone* of O can be influenced by O — if a supernova occurs at O, its light will reach every event in O's future light cone. Conversely, no event in O's future light cone can influence O.

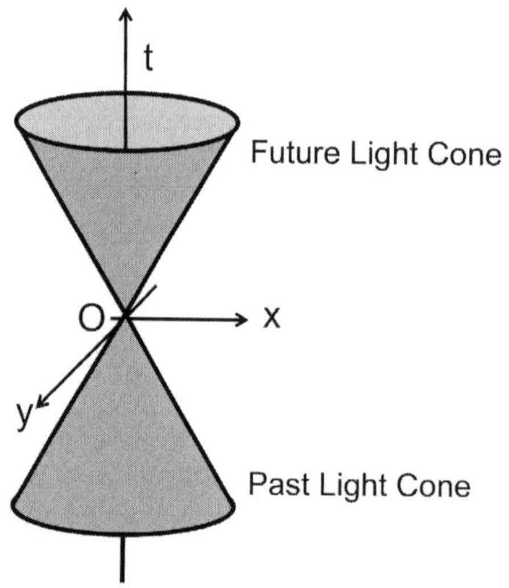

Figure 4-5 Past & Future Light Cones

The future light cone extends from O to $t = +\infty$, while the past light cone extends from O to $t = -\infty$.

All events outside the past and future light cones of O are causally disconnected from O — they can neither affect O, nor can O affect them.

Chapter 5
Relativity: Philosophy & Paradoxes

At cocktail parties and such, someone occasionally announces:

"Einstein said *everything* is *relative*."

Often, the speaker hopes this will be accepted as an excuse for their aberrant social behavior.

Actually, Einstein said no such thing.

His message was entirely different. The laws of nature, he proclaimed, are *universal*, the same for all observers — nothing relative about that — it is the measurements of certain quantities that are definable only *relative* to an observer's state of motion.

§5.1 Philosophy & Special Relativity

Einstein's theory of special relativity addresses nearly all natural phenomena and makes many, absolutely precise statements that are rigorously testable.

Quite remarkably, there are **zero** adjustable parameters in either special or general relativity. If any of Einstein's predictions were definitively proven errant by even one part in a trillion, his theories would be *completely and utterly wrong*.

In fact, the opposite has occurred. Special relativity is the most extensively and precisely tested theory of science, and none of its predictions has ever been convincingly disproved. Occasionally, some have claimed disproof, but invariably, upon closer examination, it was they who were proven wrong. A recent example was the announcement of faster than light neutrinos, which the authors later retracted blaming faulty equipment.

In his ***Lectures on Physics***, Richard Feynman says there are three philosophical lessons that we physicists should learn from the development and validation of special relativity.

Firstly, even our most cherished and extensively verified theories may later be proven inadequate in the face of more and better observations.

Secondly, when presented with strange new ideas — such as time running slower on foreshortened airplanes — whether or not we like these ideas is irrelevant. All that matters is whether or not the strange ideas agree with experiment.

Thirdly, we can learn much from the symmetry of natural laws by examining which transformations leave the form of each law unchanged. I extensively explore the symmetry of natural laws in ***Feynman Simplified Part 1***, Chapter 49, and in ***Feynman Simplified Part 3***, Chapter 23.

§5.2 Physics & Philosophers

There are, unfortunately, many "pop" philosophers who, without any real understanding of Einstein's theories, say: "That all is relative is a consequence of Einstein, and it has a profound impact on our ideas." It is trivial to say that different observers see the world differently — people looks different from the front and back. Some think it obvious that velocities cannot be measured except relative to something else. They ask: why did it take physicists so long to figure out something so simple? Why didn't they just ask us wiser men?

Of course, nature is more subtle. Uniform motion, without any acceleration, can only be measured relative to something else. However, accelerations are easily detected without any need for external reference.

An example of an acceleration that is intrinsically detectable is uniform rotation. We can measure Earth's rotation with a Foucault pendulum, even inside a sealed building. A clever philosopher might respond that perhaps uniform rotation is detectable because Earth is rotating relative to the stars; perhaps some influence from stars creates centrifugal forces. If so, one might wonder why stars care if we rotate, but not if we move straight ahead at constant speed.

In any case, scientific truth cannot be found by parsing English words, or those of any other human language. Words are too imprecise. For example, what does the word "mean" mean? My dictionary lists 24 possible "mean"-ings.

Thankfully, scientists are spared that futility by focusing on what is measurable and computable. Whether or not a motion *should be* detectable without external reference is irrelevant. All that matters is what is observable and testable. Only experiments can determine which quantities are absolute and which are relative; a task for physicists not philosophers.

§5.3 Fun & Value of Paradoxes

There are no true paradoxes in special relativity, general relativity, or quantum mechanics. After a century of exhaustive examination and testing, we are very confident that all these theories are mathematically self-consistent and are consistent with all confirmed measurements. This does not mean that physicists accept these theories without question — quite the contrary. These theories are rigorously retested every day, in every conceivable way.

Nothing we know about the world of our existence has been more completely and precisely confirmed than special relativity and quantum mechanics. The constancy of the speed of light has been validated to 18 decimal digits, one part in a billion, billion. The validity of the Lorentz transformation has been validated to better than 22 decimal digits, 50 parts in a trillion, trillion. Quantum mechanics has been validated to 12 decimal digits, one part in a million, million. General relativity is much more difficult to test, and is validated to about one part per thousand. The limits just quoted are not where the theories fail — there are no confirmed failures — these are simply the limitations of our best instruments. Whenever new technology permits, physicists will perform more stringent tests. Theories are accepted only to the extent that observations confirm them.

Special relativity, general relativity, and quantum mechanics are our best descriptions of nature. Mankind has had 100 years to digest them, yet these theories are sometimes shockingly counterintuitive. Some of the most startling results have become enshrined as purported paradoxes. The best paradoxes feature seemingly simple situations in which the predictions of these theories are bizarre but correct. While *seemingly* simple, these puzzles can be quite challenging. Many cleverly conceal one or more of these common errors:

Mistake #1: Forgetting that some physical quantities are relative

Mistake #2: Applying special relativity to non-inertial frames

Mistake #3: Assuming simultaneity where none exists

Mistake #4: Comparing quantities in different inertial frames

Paradoxes are engaging and intriguing puzzles that serve valuable purposes. They help us better grasp where our human intuitions and incomplete understandings fail. They can also serve as conversation-starters at social events.

Some have even attempted to use paradoxes to impress the opposite gender, although I can't recommend that — success is at best *relative* and *uncertain*.

§5.4 Twin Paradox

The twin paradox is the most famous puzzle of special relativity. Two twins, Al and Bert, claim to understand Einstein's theories. While Al stays home, Bert rockets off to the Delta Quadrant and returns to Earth. At Bert's Welcome-Home party, Al and Bert both claim they are now the older twin.

Al says Bert aged more slowly during his high-speed voyage due to time dilation. Bert responds that velocity is relative; as he perceived it, he was not moving. It was Al and Earth that sped away and back, which Bert says makes Al the younger twin.

Each twin claims the other was moving, was subject to time dilation, and is therefore younger. They both cannot be younger. So which twin is right, Al or Bert?

Continue reading when you want to check your answer.

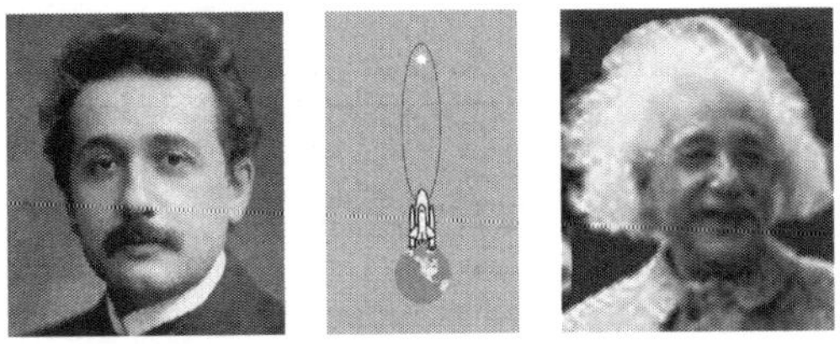

Figure 5-1 Twins after Bert's Odyssey

Recall that inertial frames move at constant velocities, without any acceleration. We assume Earth's motions are negligible, and we generally treat Earth as an inertial frame.

If Bert had rocketed on forever, never turning back, both twins would be right, each in their own reference frame. Time is relative; it passes at different rates in frames moving at different velocities. There is no conflict if Al and Bert remain in different inertial reference frames, because they can only pass one another once, making it impossible to compare their rates of aging.

Only if Al and Bert share the same frame can they legitimately compare how much each has aged. But to share a common reference frame, at least one of the twins must change their velocity. In this case, Bert's rocket decelerated at the Delta Quadrant, turned, accelerated toward Earth, and finally decelerated again to land here. He must have changed velocity at least three times. Thus Bert was not in the same inertial frame throughout.

Bert's mistake is trying to apply special relativity in a non-inertial frame. The equations of special relativity are valid only if one does not change reference frames and the frame's velocity does not change. By the way, the same restrictions apply to Newtonian mechanics.

Therefore Al, the Earth-bound twin, is older and wiser.

§5.5 World Lines & Proper Time

Let's analyze the twin paradox by introducing another important part of the relativity toolbox: *world lines*. In Figure 5-2, the vertical axis is time T, and the horizontal axis is position X, as measured in Earth's inertial frame.

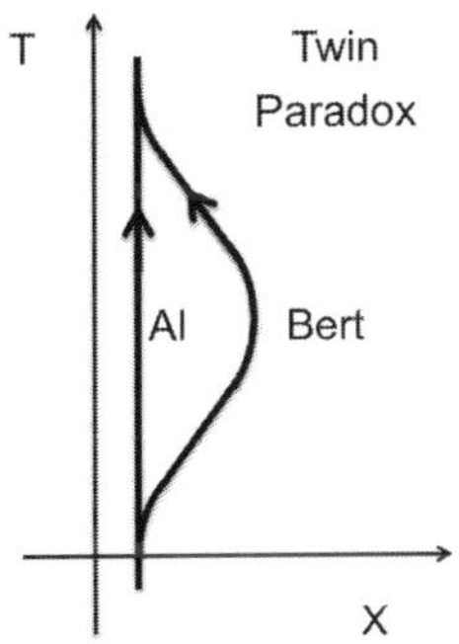

Figure 5-2 Twins' World Lines

Al's path through spacetime, his *world line*, is simple: the straight vertical line in Figure 5-2. Al's position is constant and he ages at the normal rate in Earth's frame. Bert's world line (the bowed one) shows him accelerating away from Earth, coasting at a very high speed, decelerating to a momentary stop, accelerating back toward Earth, coasting again at a very high speed, and finally decelerating to land home. The accelerating portions of Bert's world line are curved, and the coasting portions are straight. The slopes of the straight portions of Bert's world line indicate his coasting speed is very high. The speed of light corresponds here to a slope of $v = \pm c = \pm 1$, a line at ± 45°. Zero velocity corresponds to a vertical line, like Al's. Bert's coasting line segments have a slope of ± 53°, corresponding to about 75% of the speed of light.

Now let's compute how each twin ages. For a velocity entirely in the x-direction, the equation for proper time is:

$$d\tau^2 = + dt^2 - dx^2 / c^2$$

$$d\tau^2 = dt^2 (1 - dx^2/dt^2 / c^2)$$

$$d\tau^2 = dt^2 (1 - v^2 / c^2)$$

$$d\tau = dt / \gamma$$

Since proper time τ is invariant, calculating τ in any inertial frame gives us its value in all inertial frames. We next compute the amount of proper time τ that elapses for each twin between Earth time $t = A$, when Bert lifts off, and Earth time $t = B$, when Bert lands:

For Al: $$\tau = \int_A^B dt / 1 = B - A$$

For Bert: $$\tau = \int_A^B dt / \gamma < B - A$$

Since γ is always > 1 for any velocity $v > 0$, less proper time has elapsed for Bert than for Al.

Now we are all older but wiser.

§5.6 Starships & String Paradox

Two Federation starships, at rest alongside a hexagonal space dock, are tied together with a string of length L, represented by the solid line at the bottom of Figure 5-3. Both starships face the same direction, and are separated by the same distance L.

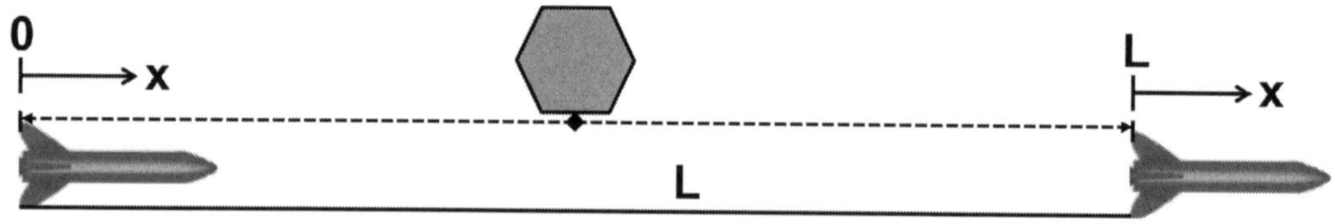

Figure 5-3 Starships at Space Dock

At time $t = 0$, GO signals (dotted lines) from the space dock arrive at the antennas on the tail fins of each starship. Both ships immediately blast off with constant acceleration, and both stop accelerating when their velocities relative to the space dock reach u, a velocity close to the speed of light.

Define the left starship's position to be $x(t)$ at time t, and let $x = 0$ at $t = 0$. The right starship is therefore at $x = L$ at $t = 0$. All this is defined in the space dock frame.

Our question is: what happens to the string?

As we discovered in Section §2.4, lengths contract along the direction of motion by the relativistic factor γ, hence, in the space dock frame, the moving string's length appears to be L / γ. But does the separation of the starships also contract, and if so, does it contract by the same factor? Some say Yes, others say No. Even some Ph.D. physicists have gotten this wrong.

Let's first examine this situation from the point of view of the Starfleet Admiral in the space dock.

In the space dock frame, both starships start at rest and blast off at the same time with the same acceleration. Hence, the starships are always moving at the same velocity, which means the starships' separation does not change, and the right starship's position is always given by: x(t) + L. As the starships accelerate, the Starfleet Admiral sees the length-contracting string break.

A side note for space travel enthusiasts: Appendix 3 provides an extensive discussion of the possibilities of exploring the cosmos, including flight durations and fuel requirements.

The starships' crews might protest the Starfleet Admiral's conclusion that the string breaks. When viewed in the starships' reference frame, neither the starship separation nor the string contract, so the string does not break. And, if the string does not break in one frame, it cannot break in any other frame. (String breaking is an absolute event not a relative one.)

Hence the paradox. What do you think? Ready for the answer?

The starships' crews made two critical mistakes: they falsely assume universal simultaneity, and their analysis was performed in a non-inertial frame.

Let's address simultaneity first. As discussed in Section §4.4, two spacetime events that occur simultaneously in one inertial frame but at different positions are not simultaneous in other frames in relative motion. There can be no universal agreement on which events are simultaneous.

It is all about the timing of the GO signals. Assume the space dock is equally distant from each starship's tail fin antenna, and assume the GO signals travel at the speed of light. In the space dock frame, the GO signals arrive at each starship at the same time, as we assumed above. However, let's view this in a reference frame moving in the +x-direction (to the right) with constant velocity v. In this frame, and before the starships accelerate, both starships and space dock are moving in the –x-direction at velocity v, as shown in Figure 5-4.

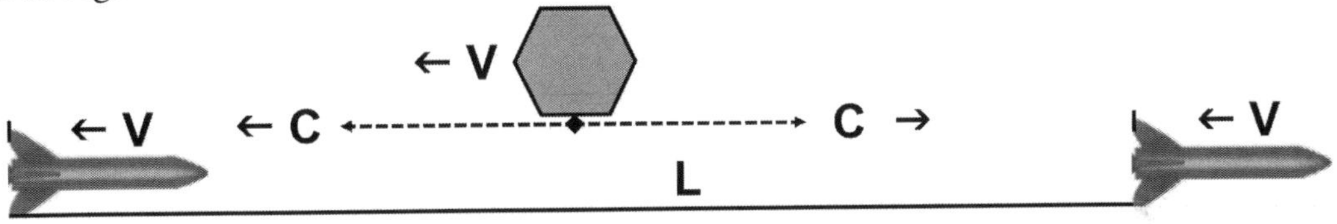

Figure 5-4 Starships awaiting GO Signal in Moving Frame

In this moving frame, a GO signal races toward the left starship in the –x direction at velocity c. The left starship moves in the same direction with velocity v, away from the oncoming GO signal, hence the closing velocity is c – v. A GO signal also moves toward the right starship in the +x direction at velocity c. The right starship moves in the opposite direction with velocity v, toward the oncoming GO signal, thus the closing velocity is c + v. In this frame, each GO signal must travel the length-contracted distance (L / 2) / γ, hence the times at which the GO signals reach the starships are:

GO arrival time at left ship: $T_L = (L / 2\gamma) \;/\; (c - v)$

GO arrival time at right ship: $T_R = (L / 2\gamma) \;/\; (c + v)$

$$T_L - T_R = (L / 2\gamma)\; [\,(c + v) - (c - v)\,] / (c - v)(c + v)$$

$$T_L - T_R = (L / 2\gamma)\; [2\,v]\; \gamma^2 / c^2$$

$$T_L - T_R = L\,v\,\gamma \,/\, c^2$$

The GO signal reaches the starship on the right before reaching the one on the left. The right starship accelerates first and moves away from the other starship. In this frame, the starships' separation becomes greater than L / γ, the length of the contracted string, and the string breaks, just as it did in the space dock frame.

Some may wonder why the separation between moving starships does not contract, as seen in the space dock frame. It does, but in a strange way. Let's examine this in a simple case. Assume each starship accelerates instantaneously to velocity v when it receives its GO signal.

In a frame moving to the right with velocity v, the right starship accelerates $\Delta T = L\,v\,\gamma / c^2$ before the left starship, making the their separation:

$$L / \gamma + L\,v^2 \gamma / c^2 = (L / \gamma)\,(1 + v^2 \gamma^2 / c^2)$$
$$L / \gamma + L\,v^2 \gamma / c^2 = (L / \gamma)\,(\gamma^2)$$
$$L / \gamma + L\,v^2 \gamma / c^2 = L\,\gamma$$

In the moving frame, the separation after the initial instantaneous acceleration is $L\,\gamma$. As seen in the space dock frame, that separation contracts to L.

Now, let's discuss choosing proper inertial frames. The rest frames of the accelerating starships are not inertial. We were careful above to choose a frame moving at constant velocity v that the starships attain at a single instant of time. We could choose any velocity v and come to the same conclusion. Therefore the conclusion is valid at all moments in time.

§5.7 Javelin & Tube Paradox

This next paradox has many guises, including: the pole and the window; the ladder and the barn; and the train and the tunnel. My favorite version is the Ferrari and the parking spot. They all entail an object of length L trying to fit inside an aperture of length L when their relative velocity is close to the speed of light.

Al and Bert are ready to debate another paradox. Are you?

The twins decide to test relativistic length contraction by throwing a javelin of length L through a stationary glass tube also of length L.

Al takes the tube's viewpoint. He says the moving javelin contracts, becoming shorter than the tube. Al says that, at the right moment, the javelin fits entirely within the tube.

Bert takes the javelin's viewpoint. He says the moving tube contracts, becoming shorter than the javelin. Bert says the javelin is never entirely within the tube, some part always sticks out.

Hence the paradox.

The twins agree to launch the javelin at the relativistic velocity v. Al intends to obtain photographic proof that the javelin's length contracts to L / γ. Before the experiment, he positions two cameras to be activated by a laser triggering system. He aims a laser beam across the entry end of the tube. When the javelin's head interrupts the laser beam, it initiates the triggering system. After appropriate delays, laser beams trigger the cameras to image the locations where Einstein predicts the head and the tail of the javelin will be.

The experimental setup is illustrated in Figure 5-5, with the javelin (bold black arrow), about to enter the light gray tube, shown in cross section.

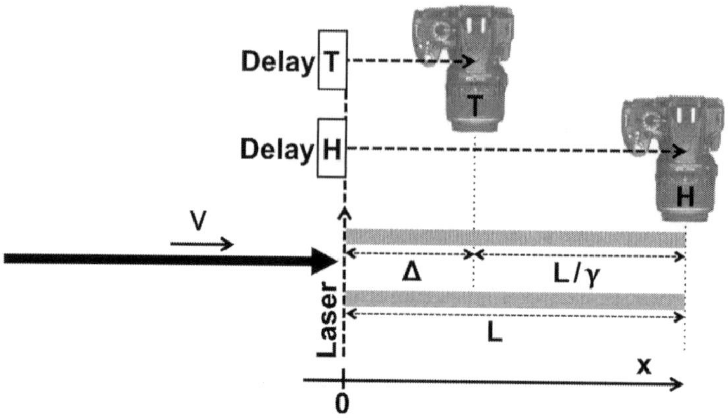

Figure 5-5 Javelin Entering Tube, Triggering Laser Timers

Delay H triggers the photograph of the javelin's head, and Delay T triggers the photograph of its tail.

At a time designated $t = 0$, the javelin breaks the laser beam at the tube entry at $x = 0$. Delay H ensures camera H photographs the javelin's head as it reaches the tube's exit at $x = L$ at time $t = L / v$. Delay T ensures camera T photographs the javelin's tail at exactly the same time $t = L / v$. Einstein says the tail will then be at $x = \Delta$, with $\Delta = L - L / \gamma$. All this is measured in the tube reference frame.

The laser beam feeds both Delay H and Delay T. Both Delays use mirrors to reflect their laser beams back-and-forth parallel to the x-axis for the required time intervals, thereafter directing their beams to their assigned cameras. The cameras shoot when their laser beams turn off, some time after the javelin blocks the laser beam at the tube entry. All vertical light path distances in Figure 5-5 are negligibly small — since this is a thought experiment, we may imagine ideal equipment. Al adjusts the mirror separations in Delay H and Delay T so that both cameras capture the action at $t = L / v$, when the javelin's head arrives at $x = L$ after traveling distance L at velocity v. This is also when the contracted javelin's tail arrives at $x = \Delta$. All this is measured in the tube frame.

After the javelin exits the tube, Al shows Bert the camera images that indeed show the javelin's head at $x = L$ and its tail at $x = \Delta$, just as Einstein predicts for a length-contracted javelin. However, Bert says Al's cameras did not shoot simultaneously. He says camera T shot much later than camera H, giving the javelin's tail enough time to reach camera T by the time it took a picture. By then, Bert says, the javelin's head was well beyond the exit of the contracted tube.

Who is right? What do you think? Ready for the answer?

The key issue here is the absence of universal simultaneity: the two separated cameras H and T cannot shoot at the same time in both a stationary and a moving frame.

Whatever clocks Al employs to provide Delay H and Delay T will appear to run slower in Bert's moving frame. We proved this for the photon clock in Figure 2-3, where the photon's velocity was perpendicular to the relative velocity between the reference frames. We also proved this for the Michelson-Morley experiment in Figure 2-5, where the photon's velocity was parallel to the relative velocity between the frames.

Would you like to try to prove that Delay H and Delay T are both longer in the javelin's frame by the factor γ? I provide that calculation at the end of this section.

Here is the rest of the analysis for the javelin and the tube.

In the **tube** reference frame:

Camera H shoots at:	$t = L / v$
Laser transit time, $x = 0$ to $x = L$:	L / c
Delay H:	$L / v - L / c = L (c - v) / v c$
Camera T shoots at:	$t = L / v$
Laser transit time, $x = 0$ to $x = \Delta$:	Δ / c
Delay T:	$L / v - \Delta / c$

In the **javelin** reference frame:

Delay H:	$\gamma L (c - v) / v c$
Laser transit time, $x = 0$ to $x = L / \gamma$:	$L / \gamma (c + v)$
Delay T:	$\gamma (L / v - \Delta / c)$
Laser transit time, $x = 0$ to $x = \Delta / \gamma$:	$\Delta / \gamma (c + v)$

Note that in the javelin frame, the delay times are increased (multiplied by γ), and the lengths L and Δ are contracted (divided by γ). Also in this frame, the cameras are moving at velocity v toward the oncoming laser beams; hence the closing speed is $(c + v)$. In the tube frame, the laser-camera closing speed is c.

In the **javelin** frame, camera H shoots at:

$$t_H = \text{Delay H} + \text{Laser transit time}$$
$$t_H = \{\gamma L (c - v) / v c\} + \{L / (c + v) \gamma\}$$
$$t_H = L \{\gamma^2 (c - v)(c + v) + v c\} / [v c (c + v) \gamma]$$
$$t_H = L \{c^2 + v c\} / [v c (c + v) \gamma]$$
$$t_H = L / [v \gamma]$$

Recall some equations we will need:

$$\Delta = L - L / \gamma$$
$$L = L / \gamma + \Delta$$
$$\gamma L = L + \gamma \Delta$$
$$1 / \gamma^2 = (1 - v^2 / c^2) = (c^2 - v^2) / c^2$$
$$1 / \gamma^2 = (c - v)(c + v) / c^2$$
$$(c - v) = c^2 / \gamma^2 (c + v)$$

In the **javelin** frame, camera T shoots at:

$$t_T = \text{Delay T} + \text{Laser transit time}$$
$$t_T = \{\gamma L / v - \gamma \Delta / c\} + \{\Delta / \gamma (c + v)\}$$
$$t_T = L / v + \gamma \Delta / v - \gamma \Delta / c + \Delta / \gamma (c + v)$$
$$t_T = L / v + \{(c - v) / c + v / \gamma^2 (c + v)\} \gamma \Delta / v$$
$$t_T = L / v + \{c / \gamma^2 (c + v) + v / \gamma^2 (c + v)\} \gamma \Delta / v$$
$$t_T = L / v + \{c / (c + v) + v / (c + v)\} \Delta / \gamma v$$
$$t_T = \{L + \Delta / \gamma\} / v$$

Therefore, from the javelin's viewpoint, Bert says that, when each camera shoots, the javelin's head and tail are at:

Head:

$$x(t_H) = x(0) + v\, t_H = 0 + v (L / v \gamma)$$
$$x(t_H) = L / \gamma$$

Tail:

$$x(t_T) = x(0) + v\, t_T$$
$$x(t_T) = - L + v (1 / v) \{L + (\Delta / \gamma)\}$$
$$x(t_T) = \Delta / \gamma$$

These exactly match the locations of the cameras, properly length-contracted, as viewed in the javelin's frame.

Thus both Al and Bert agree that camera H's image shows the javelin's head and camera T's image shows its tail, although they disagree about when the cameras shot.

Al "knows" the cameras shot at the same time; he says the photographs prove the moving javelin was shorter than the stationary tube.

Bert says the photographs are correct, but the cameras shot at *different* times; he "knows" that, adjusting for this time difference, the photographs prove the moving tube was shorter than the stationary javelin.

Al and Bert are both right, but each is right only in their own reference frame.

Time is relative, but a photographic image is absolute; all observers must agree about what the image shows.

Here is the calculation for the time dilation of Delay H and Delay T, as seen in the javelin's frame that moves with relative velocity v. Al's camera trigger delay results from bouncing a laser beam back-and-forth between mirrors separated by distance A, as shown in Figure 5-6. Again, we assume the vertical distances are negligible.

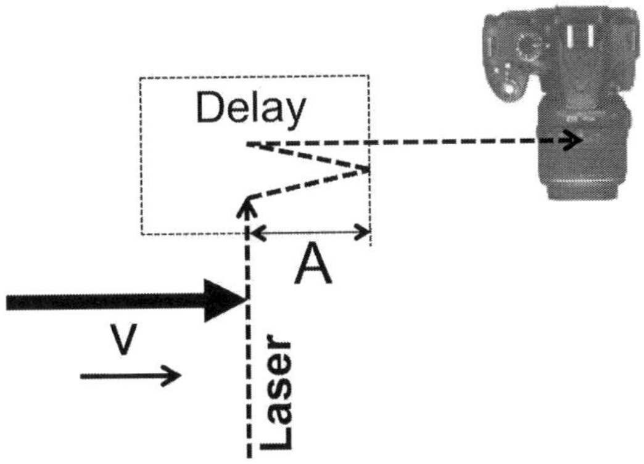

Figure 5-6 Detail of Camera Delay

In the stationary frame of the laser, delay, and camera, the delay time equals:

$$t_0 = 2A / c$$

In the javelin's moving frame, the right-going leg of the delay path has closing velocity c + v, since the mirrors are moving left at velocity v. Hence the time along this leg is:

$$A \;/\; \gamma\;(c + v)$$

Similarly, the left-going leg has closing velocity c – v, since the mirror is moving away from the laser beam. Hence the time along this leg is:

$$A \;/\; \gamma\;(c - v)$$

The total time, the sum of both legs, is:

$$t = (A / \gamma)\;\{1 / (c + v) \;+\; 1 / (c - v)\}$$

$$t = (A / \gamma)\;\{c - v \;+\; c + v\} / (c^2 - v^2)$$

$$t = (A / \gamma)\;\{2c\}\;(\gamma^2 / c^2)$$

$$t = 2A\gamma / c = \gamma\, t_0$$

§5.8 CMB Redshift

This section is not about a paradox or a puzzle. Rather, I wish to debunk an egregious falsehood.

Unfortunately, some "physicists" with dubious agendas have promoted to the general public a major misconception regarding the Cosmic Microwave Background ("CMB") radiation and its redshift. The CMB is the first light released by our universe — the afterglow of the primordial fireball born in the Big Bang.

The early universe was too hot for protons and electrons to bind together to form neutral atoms. These unbound charged particles absorbed and reemitted photons with extreme efficiency, thereby making the universe effectively opaque to light. When the universe was about 379,000 years old, and its temperature had dropped sufficiently, the charged particles combined to form neutral atoms, making the universe transparent to photons with a broad range of frequencies. The light that was then released is the CMB radiation that still persists today.

Redshift is a lengthening of the wavelength of light that shifts its spectrum to lower frequencies. For visible light this means shifting the color toward the red end of the spectrum; hence the term *redshift*. Redshift can occur for several reasons, but the CMB is redshifted due to the expansion of the universe. In fact, the CMB redshift provides some of the most compelling evidence for cosmic expansion and the Big Bang theory. All of this is discussed thoroughly in my ebook ***Our Universe 3: CMB, Inflation & Dark Matter*** and my print book ***Our Universe: The Known & The Unknown***.

This is all well established science.

Some falsely claim the CMB cannot be redshift due to cosmic expansion, because that would violate energy conservation. Hence, they claim, the Big Bang theory must be wrong, and that the universe is only thousands of years old, not billions of years old. These people evidently do not understand, or are choosing to ignore, the most basic concepts of relativity. Their mistake is not recognizing that some physical quantities are relative.

Energy is a *relative* quantity — the result of any energy measurement depends on the reference frame in which that energy is measured. Observers in different frames will measure different values. Energy is always conserved (its total amount does not change) as demonstrated by measurements made on a consistent basis in the *same reference frame*. A simple example is provided at the end of this section.

The energy of CMB photons is the same today as it was 13+ billion years ago, provided both measurements are made in the reference frame of the gases that emitted these photons. That energy corresponds to a temperature of 3000 K.

The energy of CMB photons is the same today as it was 13+ billion years ago, provided both measurements are made in the reference frame in which Earth now resides. That energy corresponds to a temperature of 2.725 K.

On any consistent basis, in any fixed reference frame, the energy of the CMB has been conserved.

In fact, every type of redshift conserves energy.

Here is the simple example I promised. Imagine you are driving on a highway going north, with traffic flowing smoothly at 60 mph in both directions. (This excludes the greater Los Angeles area.) In your reference frame, the northbound cars seem to be moving slowly and hence have low kinetic energy, but the southbound cars appear to speeding by with high kinetic energy.

Now, imagine you pull over and park on the shoulder of the highway. From your new reference frame, the northbound cars are suddenly moving much faster and the southbound cars are moving much slower than before. Has the energy of every other car changed? Has energy conservation been violated? Of course not. The only thing that changed is the reference frame in which you observed the other cars.

Chapter 6
Special Relativity Review

§6.1 Principles of Special Relativity

The *principle of relativity*, first suggested by Galileo, states that the same laws of nature apply to all observers moving at constant velocities, regardless of those velocities. Absolute velocity has no physical significance; only relative velocities are meaningful. Having a constant velocity and being in an inertial frame are equivalent descriptions of non-accelerating motion.

Before Einstein, physicists believed that light was a wave traveling through a medium, as does every other type of wave. They dubbed light's mediam the *luminiferous ether*. If light did have a medium, Maxwell's equations of electromagnetism would be valid only in the rest frame of its medium, thereby defining an absolute velocity, contrary to the principle of relativity.

The Michelson-Morley experiment proved light's speed is the same in all directions, regardless of Earth's changing motion through the cosmos. Unless Earth is unique in the universe, this means *ether does not exist*. Einstein showed relativity and Maxwell's equations are compatible only if *light always moves at the same speed c*, and has no medium.

Einstein did not say: "everything is relative." He said the *laws of nature are universal*, the same for all observers, but *measurements of certain quantities are relative*, definable only with respect to each observer's state of motion.

§6.2 Key Equations of Special Relativity

The *Lorentz transformation* to a reference frame (T, X, Y, Z) moving in the +x-direction with velocity u relative to a "stationary" frame (t, x, y, z) is:

$$X = \gamma (x - u t)$$

$$T = \gamma (t - u x / c^2)$$

$$Y = y$$

$$Z = z$$

Where the relativistic factor γ is: $\gamma = 1 / \sqrt{(1 - u^2 / c^2)}$

Observers in frames moving at different velocities measure different values for time, length, mass, and other quantities that depend on these, such as energy and momentum. In what follows, t_0, L_0, and m_0, are rest frame measurements made by a *co-moving observer*, one moving with the objects being measured. Whereas t, L, and m are the corresponding time interval, length, and mass observed in a frame with a different velocity, one that we call "stationary". To a stationary observer, moving clocks seem to run slower by the relativistic factor γ. Events of fixed duration t_0, such as one Earth-day, seem to take a longer time t in a frame moving relative to Earth.

To a stationary observer, observing phenonema in a moving frame:

actions take more time:	$t = \gamma t_0$	(clocks run slower)
lengths seem contracted:	$L = L_0 / \gamma$	
masses seem increased:	$m = \gamma m_0$	

Relativity is symmetric: all the effects that we see in a moving frame, their observers see in ours.

Sum of *collinear* velocities u & v: $w = (u + v) / (1 + u v / c^2)$

For any object with any velocity: $E^2 = c^2 p^2 + c^4 m_0^2$

$$E = m c^2 = m_0 \gamma c^2$$

Most commonly, physicists use m to denote an object's rest mass, and write energy and momentum as:

$$E = \gamma m c^2$$

$$p = \gamma m v$$

§6.3 Spacetime & 4-Vectors

Space and time are not distinct unrelated quantities, as Newton believed — they are two aspects of one united 4-D entity that Hermann Minkowski named *spacetime*.

In 4-dimensional spacetime, four quantities are required to uniquely specify an *event*: (t, x, y, z).

4-vectors are normal 3-vectors in 3-D with one additional component that makes the 4-vector *Lorentz compliant*, transforming according to the Lorentz transformation. Some important 4-vectors are:

Position: $\boldsymbol{r} = (c\,t,\ x,\ y,\ z)$

Separation: $\boldsymbol{dr} = (c\,dt,\ dx,\ dy,\ dz)$

Momentum: $\boldsymbol{p} = (E / c,\ p_x,\ p_y,\ p_z)$; $\quad \boldsymbol{p}_\mu \boldsymbol{p}_\mu = -\ m_0^2 c^2$

Velocity: $\boldsymbol{v} = (\gamma c,\ \gamma v_x,\ \gamma v_y,\ \gamma v_z)$; $\quad \boldsymbol{v}_\mu \boldsymbol{v}_\mu = -\ c^2$

As defined below, the *dot product* of any two 4-vectors A_μ and B_μ is *invariant*, it has the same value in all inertial frames (we sum over all values of the repeated index μ):

$$A_\mu\ B_\mu = -A_t B_t + A_x B_x + A_y B_y + A_z B_z$$

The *invariant interval* ds^2 and the *proper time* $d\tau$ between two spacetime events are given by:

$$ds^2 = -c^2 dt^2 + dx^2 + dy^2 + dz^2$$

$$c^2 d\tau^2 = -ds^2$$

An ideal clock moving at constant velocity between two spacetime events measures proper time $d\tau$. The interval between two events falls into one of three categories that is the same in all reference frames:

Time-like for $ds^2 < 0$

Space-like for $ds^2 > 0$

Photon-like for $ds^2 = 0$

Objects moving at $v < c$ have time-like intervals everywhere along their paths. Objects with zero mass, and only those, move at $v = c$, along paths whose intervals are all zero. Space-like separated events cannot be causally related — neither can affect the other. Moving along a space-like path requires $v > c$, which is impossible.

Relativity ensures *causality*. For any two time-like separated events, one precedes the other in all frames. The earlier event can affect the later event, but the later event cannot affect the earlier.

The earlier event is in the *past light cone* of the later event, which is in the *future light cone* of the earlier event.

§6.4 Philosophical Insights

There are no true paradoxes in special relativity. But you may enjoy debunking purported paradoxes by identifying common mistakes, including: forgetting that some physical quantities are relative, applying special relativity to non-inertial frames, and assuming simultaneity where none exists.

The development of special relativity, which superseded Newton's long-cherished laws of motion, taught scientists three vital lessons. Firstly, we must realize that even our best theories may be proven inadequate in the face of more and better observations. Secondly, whether or not we like strange new ideas is irrelevant — all that matters is whether or not they agree with experiment. Thirdly, we can learn much from the symmetry of natural laws, and the transformations that leave their form unchanged.

The following seven phenomena of special relativity distinguish it from Newtonian physics, and all are mathematically equivalent, meaning assuming any one, proves all the others. Einstein assumed the first, which has the strongest physical motivation, and from that he derived the other six.

1. Light's speed c is constant
2. Length contraction: $L = L_0 / \gamma$
3. Time Dilation: $t = \gamma\, t_0$ (moving clocks run slower; actions take longer)
4. Mass increase: $m = \gamma\, m_0$
5. Sum w of collinear velocities u and v: $w = (u + v) \,/\, (1 + u\,v / c^2)$
6. Mass is a form of Energy: $E = m\,c^2$
7. Lorentz transformation

§6.5 Illustrative Examples

In this section we examine the power of the Lorentz transformation to solve some simple problems. In relativity, even simple problems can be tricky.

Define a stationary reference frame with coordinates (t, x, y, z) that contains a clock and a ruler. Also define a second reference frame moving with velocity v in the +x-direction with coordinates (T, X, Y, Z). Let the coordinates have the same origin, so that t = x = y = z = 0 is the same event as T = X = Y = Z = 0. This is illustrated in Figure 6-1, with the axes slightly displaced for clarity.

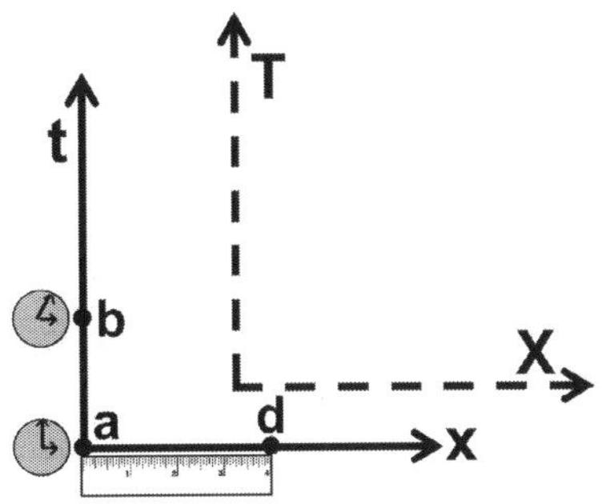

Figure 6-1 Events a, b, and d Viewed in Two Frames

A clock is shown in the xt frame at two different times: $t = 0$ and $t = t_0$. The ruler is shown at $t = 0$, with its ends at $x = 0$ and $x = L_0$. In the stationary frame, events a, b, and d have coordinates:

$$a = (0,\ 0,\ 0,\ 0)$$

$$b = (t_0,\ 0,\ 0,\ 0)$$

$$d = (0,\ L_0,\ 0,\ 0)$$

Let's Lorentz transform the coordinates of events a, b, d to obtain coordinates of the corresponding events A, B, D in the XT frame.

$$X = \gamma (x - v t)$$

$$T = \gamma (t - v x / c^2)$$

$$A = (0, 0, 0, 0)$$

$$B = (\gamma t_0, -\gamma v t_0, 0, 0)$$

$$D = (-\gamma v L_0 / c^2, \gamma L_0, 0, 0)$$

In the XT frame, the length of ruler AD seems to be:

$$D_X - A_X = \gamma L_0$$

In XT, the ruler's length seems *longer* than L_0, the ruler's length ($d_X - a_X$) in the xt frame. Does this mean an XT observer sees a moving ruler *expand*?

What happened to length contraction?

The problem here is that we made an improper comparison.

In the xt frame, a and d are simultaneous, but in the XT frame, A and D have different time coordinates. It makes no sense to calculate the length of a moving ruler by subtracting its end locations at two different times. To properly compare ruler lengths in two different frames, both ruler end positions must be measured *at the same time* t in the xt frame, and *at the same time* T in the XT frame.

We know that the left end of the ruler is at A at T = 0. Let's find out where the right end is at T = 0.

Since D_t is negative, let's wait a time interval $-D_t > 0$, and then measure the location of the right end. During time $-D_t$, an observer in XT sees the ruler move a distance ΔX, given by:

$$\Delta X = (-D_t)(-v) = (+\gamma v L_0 / c^2)(-v) = -\gamma v^2 L_0 / c^2$$

Note that ΔX is negative: in the XT frame, the ruler is moving toward –X at velocity v. The new X position is:

$$D_X + \Delta X = \gamma L_0 - \gamma v^2 L_0 / c^2$$

$$D_X + \Delta X = \gamma L_0 (1 - v^2 / c^2)$$

$$D_X + \Delta X = \gamma L_0 / \gamma^2 = L_0 / \gamma$$

When both ends of the ruler are measured at the same time in the XT frame, its length is indeed *contracted*, as special relativity claims.

Let's now examine the time difference of events A and B:

$$B_t - A_t = \gamma t_0$$

This is *greater* than t_0, the time difference ($b_t - a_t$) in the xt frame. Does this mean an XT observer sees a moving clock run *faster*? What happened to time dilation?

This perhaps unexpected result is again due an improper comparison. In the stationary frame, a and b have the same x-coordinate, but in the moving frame, A and B have different X-coordinates.

To properly compare the passage of time in these two frames, we must compare clock readings *at the same location*. Even if two clocks are counting time at exactly the same rate, we will observe a difference in their readings if one is near and the other is far away. This is due to the finite speed of light. Compared to your wristwatch, a clock on the moon appears to run 1.3 seconds late, even if both keep exactly the same time. It makes no sense to measure your heart rate by subtracting your wristwatch reading at one beat from the lunar clock reading at the next beat.

To compare clock readings at the same location, we define an event g in the xt frame with coordinates:

$$g = (t_0, v t_0, 0, 0)$$

In the XT frame, the corresponding event G has coordinates:

$$G = (\gamma\, t_0 - v^2\, \gamma\, t_0 / c^2, \quad \gamma\, v\, t_0 - \gamma\, v\, t_0, \quad 0, \quad 0)$$

$$G = (\gamma\, t_0\, [1 - v^2 / c^2], \quad 0, \quad 0, \quad 0)$$

$$G = (\, t_0 / \gamma, \quad 0, \quad 0, \quad 0)$$

We see that event G has the same spatial coordinates as A: X = Y = Z = 0. We can now compare the time difference t_0 / γ between events G and A with the time difference t_0 between events g and a. The XT observer sees the moving clock running *slower* by the factor γ, as special relativity claims.

When in doubt, using the Lorentz transformations is the safest approach to properly compare events in different reference frames.

§6.5 Doppler Effect

The most famous relativistic effect on light is the *Doppler effect*. This arises whenever the source of light and its observer are in relative motion.

Let's first consider a stationary atom emitting light of frequency Ω. Now let that atom move toward us at velocity v. Due to relativistic time dilation (see Section §2.3), we see everything in a moving frame running slower, including clocks and atomic oscillations. We therefore say the emitted light has a lower frequency ω^* *at the source* given by:

$$\omega^* = \Omega\, \sqrt{(1 - \beta^2)}$$

Here, we define $\beta = v/c$. But now another factor enters the picture. As these light waves expand outward toward us, they become *compressed*.

Think of it this way: at some time t_1, the crest of a light wave passes us, then another at t_2. If the waves were emitted at the same location, the time interval $t_2 - t_1$ would equal $2\pi / \omega^*$, the period of one full wave oscillation. However, since the source is moving toward us, the second crest is emitted at a point closer to us than the emission point of the first crest. How much closer? By $2\pi\, v / \omega^*$, the source velocity multiplied by one oscillation period, as seen by us. That closer distance means the second wave crest arrives earlier by time $2\pi\, \beta / \omega^*$, the distance change divided by c. We therefore observe a time period between crests of:

Wave period without compression: $2\pi / \omega^*$

Wave period with compression: $(1 - \beta)\; 2\pi / \omega^*$

This means the frequency ω that we actually observe is higher than ω^*, according to:

$$\omega = \omega^* \;/\; (1 - \beta)$$

Substituting ω^* from above yields:

$$\omega = \{\, \Omega \sqrt{(1 - \beta^2)}\, \} \;/\; (1 - \beta)$$

$$\omega = \Omega \sqrt{\{\, (1 - \beta)\, (1 + \beta) \;/\; (1 - \beta)^2\, \}}$$

$$\omega = \Omega \sqrt{\{\, (1 + \beta) \;/\; (1 - \beta)\, \}}$$

For a source moving towards us, the Doppler effect increases light's frequency ($\omega > \Omega$ for $\beta > 0$). This is called *blueshift*, since visible light is shifted toward blue, the higher frequency end of the visible spectrum. For a source moving away from us, the Doppler effect decreases light's frequency ($\omega < \Omega$ for $\beta < 0$). This is called *redshift*, since visible light is shifted toward red, the lower frequency end of the visible spectrum.

Cosmological redshift is **not** due to the Doppler effect. The universe's expanding space stretches the wavelength of all light waves. Light in all directions from every light source is redshifted, with the degree of shifting increasing almost linearly with distance from the source. Almost all remote cosmic bodies are moving away from us, due to the expansion of space — we see their light redshifted.

For more on cosmological expansion and redshift see my ebook ***Our Universe 2: Redshift, Expansion & Dark Energy*** or my print book ***Our Universe: The Known & The Unknown***.

§6.6 Stellar Aberration

Aberration of starlight — the slight motion of stars in the heavens over the course of Earth's yearly orbit of the Sun was observed in the 17th Century. This is remarkable because, for Earth-based observers, stellar aberration is never more than 20 arcseconds (0.006° or 0.1 milliradians).

To understand the cause of stellar aberration, consider a familiar analogy: holding an umbrella in the rain. When rain falls vertically, a stationary person holds their umbrella directly overhead. But for a person walking briskly, call her Alice, the rain's motion is not entirely vertical, but has a horizontal component as well. Alice must tilt her umbrella forward to keep her face dry. Another person walking the opposite way, call him Bob, also tilts his umbrella forward, but Alice's "forward" and Bob's "forward" are in opposite directions.

Stellar aberration is sketched in Figure 37-9. In this image, light from a distant star moves vertically downward as Earth orbits the Sun in a horizontal plane.

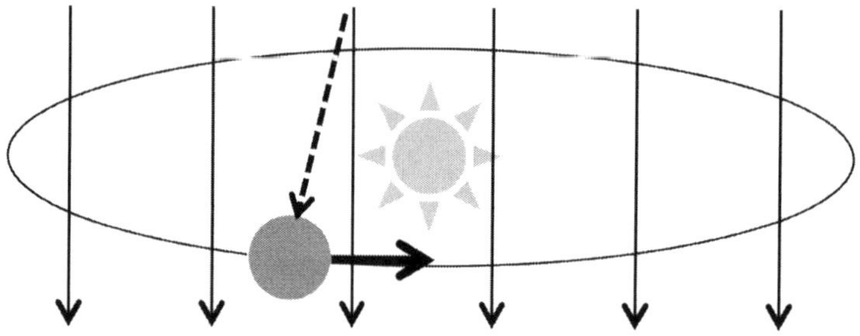

Figure 37-9 Stellar Aberration due to Earth's Motion

In the Sun's frame, the starlight is vertical (solid arrows). But in Earth's frame, the starlight is tilted (dashed arrow). Six months later, when Earth moves in the opposite direction, the tilt is also in the opposite direction. This proves aberration is not in our stars.

To derive the equation for aberration, let v be Earth's horizontal velocity, which is about 67,000 mph, or 107,000 km/hr, or 1.0×10^{-4} c. During a brief time interval Δt, Earth moves horizontally v Δt, while starlight moves downward c Δt. In the Sun's frame, the ratio of those two terms equals the tangent of the tilt angle Θ:

$$\tan\Theta = (v\,\Delta t) \,/\, (c\,\Delta t) = v / c = \beta$$

However, in Earth's moving reference frame, relativistic length contraction changes the apparent horizontal motion by the factor $\gamma = \sqrt{(1-\beta^2)}$. Do we multiply or divide by γ?

Section §2.5 explains in detail why the horizontal motion appears *longer* in Earth's moving frame. (We measure v Δt with length-contracted rulers.) For starlight incident normal to Earth's orbit, the tilt angle θ in Earth's frame is:

$$\tan\theta = \beta / \sqrt{(1-\beta^2)}$$

$$(1-\beta^2)\tan^2\theta = \beta^2$$

$$\tan^2\theta = \beta^2 (1 + \tan^2\theta)$$

$$\tan^2\theta = \beta^2 / \cos^2\theta$$

$$\sin^2\theta = \beta^2$$

$$\sin\theta = \beta$$

$$\theta = 0.1 \text{ milliradian}$$

Chapter 7
Magnetism as a Relativistic Effect

In Section §1.3, we found that the effects of special relativity are extremely modest at normal velocities, those much below the speed of light. For example, even at 1000 mph, time is dilated (slowed down) by only one part per trillion. It seems special relativity is *generally* only important at extremely high speeds.

There is, however, a remarkable exception to *generally* that we explore here.

One of many "gee whiz" experiments that first-year physics students enjoy is running electric currents through two parallel wires. When the currents are switched on, the wires jump together if the currents in both wires flow in the same direction. The wires jump apart if the currents are in opposite directions. Why? Normally, students are told this is due to *magnetism*. That is not wrong, but a more profound answer is: *special relativity*.

At the atomic scale, an electric current is comprised of an immense number of electrons "hopping" from atom to atom in a consistent direction. Any individual electron "hops" extremely rapidly but very infrequently. The "average velocity" of current-conducting electrons is a macro-world concept that does not fully describe the reality of the micro-world. Nonetheless, this average velocity v does usefully describe the current flow observed in the macro-world. The value of v is exceedingly small — typically about 1 foot per hour (30 cm per hour), or about 1/3 of a trillionth of the speed of light. At that velocity:

$$\gamma = 1.000{,}000,000{,}000,000{,}000,000{,}000,05$$

Is it possible that such an itsy bitsy effect can move large objects like wires? YES

Let's start with two parallel wires, called wire A and wire B, that carry equal currents in the same direction. In an electrical conductor like copper, the outermost electrons of each atom are loosely bound to the nucleus and can easily flow from one atom to the next. If an atom loses an electron, it becomes a positively charged *ion*. In the laboratory reference frame, where we are stationary, we observe electrons moving with velocity v through both stationary wires with stationary ions. We also note that, like nearly every macroscopic object, each wire has zero net electric charge — the number of electrons per inch of wire equals the number of ions per inch. So far, there is no apparent reason for a force between the wires — but we're not done yet.

Now, consider how this appears in the reference frame of a current-carrying electron in wire A. Since velocity is relative, this electron "sees" itself and every other conduction electron as stationary, but observes both wires and their ions moving with velocity –v (same speed as above but in the opposite direction). Special relativity says lengths shorten along the direction of motion. Hence the electron sees less space between the moving ions, yielding a higher number of ions per inch than we observe in the lab frame.

The electron also "sees" a lower number of electrons per inch than we observe in the lab frame, which is a bit trickier. To understand why, recall that in the lab, the electrons are moving and thus their spacing appears smaller to us than it appears in the electron's frame. Turning that around, the electron spacing appears larger in the electron's frame than we observe in the lab.

Hence, the electron "sees" more ions and fewer electrons per inch than we observe in our stationary frame. The wires that appear uncharged in the lab appear positively charged to the conduction electrons. The excess positive charge in wire A does not affect the electrons in wire A since the same excess charge exists in every direction and those forces cancel. But the excess positive charge in wire B attracts the electrons in wire A, and *vice versa*, forcing the wires together.

You might ask: does the excess positive charge in each wire force them apart? No, because this excess is only observed in the reference frame of the electrons. In the reference frame of the ions (the lab reference frame), the net charge of each wire is zero.

Thus, the wires pull one another together. Even though the amount of relativistic shortening is incredibly small, there are so many electrons and ions per inch that the total force is substantial — enough to snap the wires together. This should give everyone of us a heightened appreciation for the immense number of atoms in every macroscopic object, even an inch of wire.

If currents in the two wires flow in opposite directions, the electrons in wire A still see an increased number of ions per inch in wire B. But now they also see the electrons in wire B moving at –2v, faster than the ions moving at –v. The observed increase in electrons per inch is nearly twice the increase in ions per inch, making wire B's apparent net charge negative, pushing the wires apart.

What is normally called a magnetic force is at a fundamental level the result of length contraction due to special relativity, even at an exceedingly low velocity.

We discussed above all the essential physics of the origin of forces between conducting wires. The rest of this chapter provides the complete mathematical proofs that you may enjoy or elect to skip.

Consider an infinitely long, cylindrical, current-carrying wire. Let's take the realistic view that current results from the flow of electrons, while the positive nuclei are stationary: their velocity $v_p = 0$. For negative electrons moving to the right with velocity v_n, electric current J flows to the left. In a real conductor, only a few electrons move; those that remain attached to their atoms do not contribute to current flow. For our purposes, we can ignore stationary electrons and neutral atoms, and consider only the conduction electrons and the positive ions they leave behind. We define the wire's positive and negative charge densities as ϱ_p and ϱ_n, respectively.

Let's view all this in two reference frames: frame S in which the wire is stationary and its electrons are moving to the right at velocity v_n; and frame S* that is moving to the right at velocity v_n relative to S. In S* the electrons are stationary and the wire is moving to the left. Figure 7-1 shows the wire as viewed in both frames. The figure also shows a separate charge Q at distance r above the wire's central axis. Q has horizontal velocity v_Q in frame S and v^*_Q in frame S*.

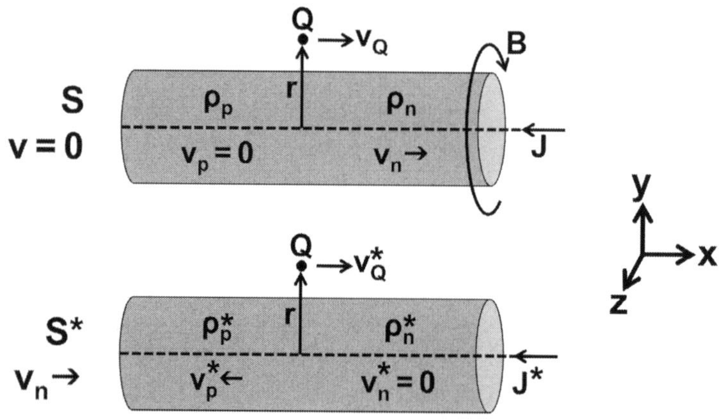

Figure 7-1 Wire & Charge Q in Stationary Frame S and Moving Frame S*

We define W to be the wire's cross sectional area, and positive velocity to be toward the right, toward +x. Therefore $v_n > 0$, and velocities are transformed from S to S* by subtracting v_n.

In S, the wire's magnetic field exerts a vertical force on charge Q that has magnitude $F = Q\ v_Q\ B$. According to the right hand rule, the direction of force F is:

If $(Q\ v_Q) > 0$: F is upward, toward +y, away from the wire

If $(Q\ v_Q) < 0$: F is downward, toward –y, toward the wire

The equations for the magnetic force F acting on charge Q are:

$$B = J \,/\, 2\pi \varepsilon_0 c^2 r$$

$$J = \varrho_n v_n W$$

$$\boldsymbol{F} = Q \boldsymbol{v} \times \boldsymbol{B}$$

$$F_y = Q\,(v_z B_x - v_x B_z) = -Q\, v_Q\, B$$

$$F_y = -Q\, v_Q\, \varrho_n\, v_n\, W \,/\, 2\pi \varepsilon_0 c^2 r$$

In S, the wire's negative charges move to the right with velocity v_n, while its positive charges are at rest with $v_p = 0$. Charge Q moves to the right with velocity v_Q. The wire's positive and negative charge densities are ϱ_p and ϱ_n.

In S*, the wire's negative charges are at rest, $v^*_n = 0$, while its positive charges move with velocity $v^*_p = -v_n$ to the left. Here, charge Q has velocity $v^*_Q = v_Q - v_n$, and the wire's positive and negative charge densities are ϱ^*_p and ϱ^*_n.

Since lengths along the direction of motion contract by γ, the volume containing any group of particles is also reduced by the factor γ. Hence charge densities increase by γ. The relationships between the charge densities in the two frames are:

Positive charge density is higher in S*: $\varrho^*_p = \gamma_n \varrho_p$

Negative charge density is higher in S: $\varrho_n = \gamma_n \varrho^*_n$

Where: $\gamma_n = 1 / \sqrt{(1 - v_n^2 / c^2)}$

The **key point** of this analysis is the asymmetry in charge density changes. The positive charges are stationary in S and moving in S*, therefore their density is greater in S*: ϱ^*_p equals ϱ_p multiplied by γ_n that is greater than 1. The negative charges are stationary in S* and moving in S, therefore their density is greater in S: ϱ_n equals ϱ^*_n multiplied by γ_n.

In S, the wire is neutral, as is almost every macroscopic object. This means $\varrho_p + \varrho_n = 0$. But in S*, the wire has a net positive charge, because:

$$\varrho^*_p + \varrho^*_n = \varrho_p \gamma_n + \varrho_n / \gamma_n$$

Using $\varrho_n = -\varrho_p$:

$$\varrho^*_p + \varrho^*_n = \varrho_p \gamma_n - \varrho_p / \gamma_n$$

$$\varrho^*_p + \varrho^*_n = \varrho_p \gamma_n \{1 - \gamma_n^{-2}\}$$

$$\varrho^*_p + \varrho^*_n = \varrho_p \gamma_n \{1 - (1 - v_n^2 / c^2)\}$$

$$\varrho^*_p + \varrho^*_n = \varrho_p \gamma_n \{v_n^2 / c^2\}$$

Now consider the special case of $v_Q = v_n$. In the S frame, the neutral wire's magnetic field exerts a force F on Q that we derived above. With $v_Q = v_n$, F is:

$$F_y = -Q\, W\, \varrho_n\, v_n^2 \,/\, 2\pi \varepsilon_0 c^2 r$$

$$F_y = +Q\, W\, \varrho_p\, v_n^2 \,/\, 2\pi \varepsilon_0 c^2 r$$

In the S* frame, charge Q is stationary: $v^*_Q = 0$. Since Q's velocity is zero, the wire's magnetic field cannot exert a force on Q. But there must be a force between Q and the wire in every reference frame.

To understand why there must be a force on charge Q in every reference frame, recall what we learned about special relativity. The distance between two objects is a *relative* quantity: its measured value L may be different in different inertial reference frames. But the essential characteristic of how that distance changes is *invariant*. If L increases (or decreases) in one frame, L* must increase (or decrease) in every other frame; the amount of change may be different, but the existence of change and its polarity is cannot be different.

In the limiting case that the objects collide in one frame (L → 0), they must collide in every frame.

Hence, while the wire cannot exert a magnetic force on charge Q in frame S*, where Q is stationary, the wire must exert some other force on Q. The answer to this apparent contradiction lies hidden inside the wire. While the wire is neutral in S, we showed above that it is positively charged in S*. That positive charge exerts a electric force on Q. Let's calculate this force. The electric field $\mathcal{E}$ at a distance r from the center of a wire with charge λ per unit length is (see my book ***Feynman Simplified Part 2***, Section §5.4):

$$\mathcal{E} = \lambda \;/\; 2\pi\,\varepsilon_0\, r$$

For our wire, (charge per unit length) equals (charge per unit volume) multiplied by (the unchanged cross sectional area W). In S*:

$$\lambda^* = (\varrho^*_p + \varrho^*_n)\; W$$

$$\lambda^* = W\; \varrho_p\; \gamma_n\; \{v_n^2 / c^2\}$$

The electric force exerted on Q by the charged wire in S* is:

$$F^*_y = Q\; \mathcal{E}^* = \lambda^* \;/\; 2\pi\,\varepsilon_0\, r$$

$$F^*_y = Q\; W\; \varrho_p\; \gamma_n\; \{v_n^2 / c^2\} \;/\; 2\pi\,\varepsilon_0\, r$$

$$F^*_y = Q\; W\; \varrho_p\; \gamma_n\; v_n^2 \;/\; 2\pi\,\varepsilon_0\, c^2\, r$$

The final step is transforming forces from S to S*. The force on charge Q is vertical, toward +y, orthogonal to the direction of motion. Any force in the y-direction can be written in terms of the rate of change of y-momentum p_y as:

In S: $$F_y = \Delta p_y \;/\; \Delta t$$

In S*: $$F^*_y = \Delta p_y^* \;/\; \Delta t^*$$

Relativistically, y-momentum is unchanged by a Lorentz transformation along the x-axis. But, time does change. For charge Q, S* is its rest frame and Δt* is its *proper time*. In Q's rest frame, clocks in S seem to run slower, which means physical processes seem to last longer. (Unstable particles take more clock time to decay in frames in which they are moving relative to the clock— see Section §2.4.) This means Δt is greater than Δt*, according to:

$$\Delta t = \Delta t^*\; \gamma_n$$

Combining these equations, we find:

$$\Delta p_y = \Delta t\;\; F_y$$

$$\Delta p_y = \Delta t\;\; \{Q\; W\; \varrho_p\; v_n^2 \;/\; 2\pi\,\varepsilon_0\, c^2\, r\}$$

$$\Delta p_y^* = \Delta t^*\;\; F^*_y$$

$$\Delta p_y^* = \Delta t^*\, \{Q\; W\; \varrho_p\; \gamma_n\; v_n^2 \;/\; 2\pi\,\varepsilon_0\, c^2\, r\}$$

$$\Delta p_y^* = \Delta t\;\; \{Q\; W\; \varrho_p\; v_n^2 \;/\; 2\pi\,\varepsilon_0\, c^2\, r\}$$

We see that Δp_y has the same value in both S and S* reference frames, as required by special relativity. This means the "magnetic" force in frame S exactly equals the Lorentz transformed electric force in frame S*, the rest frame of charge Q.

One can interpret this to mean that "magnetism" is a relativistic effect, an artifact of not choosing the simplest reference frame: the rest frame of the charge on which forces act.

Chapter 8
Particle Physics: Decays

Starting in this chapter, I will employ natural units in which the speed of light c equals 1. In major results, I will restore the c's for your convenience. Since most advanced physics books use natural units, you should become comfortable with this practice.

Anytime you wish to find out where all the c's should go in an equation, I recommend using *dimensional analysis*, which is thoroughly discussed in Appendix 2.

High-energy physics is the study of the most fundamental particles, forces, and laws of nature. The highest-energy particles have the shortest wavelengths, enabling them to probe the smallest structures, as explained in my ebook ***Quantum Mechanics 2: Reality, Uncertainty & Schrődinger's Cat*** or my print book ***Quantum Mechanics Strange Theory of the Micro-World***. Particles obtain high energy by moving at nearly the speed of light, thus special relativity is an essential tool in high-energy particle physics.

This chapter explores particle decays. Decay is a process in which an isolated particle spontaneously changes into something else. Of the 17 types of particles, most do decay; the others are called *stable*.

§8.1 Elementary Particles

To begin our exploration of particle physics, let's examine the roster of elementary particles, the most fundamental components of nature. We believe these particles have no internal parts, and are not comprised of anything "smaller". Figure 8-1 shows the 17 types of elementary particles now known.

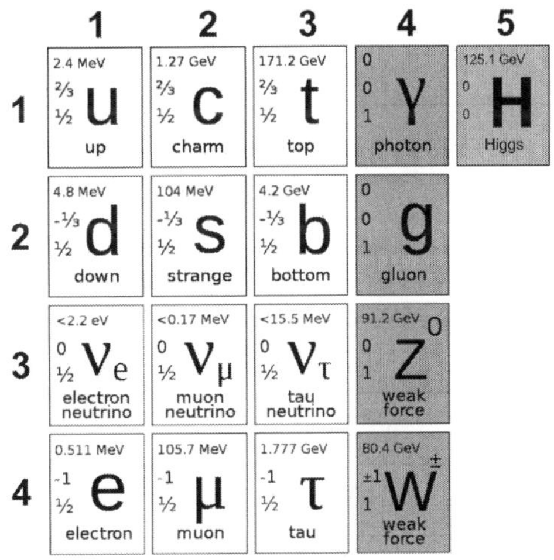

Figure 8-1 The 17 Elementary Particles

Each box in Figure 8-1 describes one particle type, starting with the large one-letter symbol above the particle's name. The three numbers on the left side of each box are: the *mass*, *electric charge*, and *spin* of that particle. Electric charges are stated in units of the proton charge. The small numbers in the figure may be hard to read, so I will repeat them in the text.

The *proton* and *neutron* are not elementary, but are extremely important. A proton is comprised of two *up quarks* and one *down quark*; it has charge +1, mass 938.3 MeV, spin 1/2, and is stable (its lifetime is measured to be at least $1.9\times10^{+34}$ years). A neutron is comprised of two *down quarks* and one *up quark*; it has zero charge, mass 939.6 MeV, spin 1/2, and its mean lifetime is 882 seconds, about 15 minutes.

Particle masses are measured in units of eV/c^2, with 1 eV (1 electron-volt) being the energy an electron gains traversing a one-volt potential. Particle physicists typically use units in which c = 1, and quote masses in eV, or MeV for million eV, or GeV for billion eV.

From left to right across each row of Figure 8-1, the masses are:

Row 1:	u	c	t	γ	H
	2.4 MeV	1.27 GeV	171.2 GeV	0	125.1 GeV
Row 2:	d	s	b	g	
	4.8 MeV	104 MeV	4.2 GeV	0	
Row 3:	ν_e	ν_μ	ν_τ	Z	
	neutrino masses are discussed below			91.2 GeV	
Row 4:	e	μ	τ	$W^\pm$	
	0.511 MeV	106 MeV	1777 MeV	80.4 GeV	

Elementary particles are divided into two very distinct groups: the five bosons that occupy columns 4 and 5 of Figure 8-1, and the twelve fermions that occupy columns 1, 2, and 3.

Every elementary particle of each type is *intrinsically absolutely identical*. Different electrons can have different *extrinsic* properties — properties that depend on our choice of reference frames and coordinate systems — such as energy, linear momentum, angular momentum, and spin orientation. But all electrons have exactly the same intrinsic properties: mass, electric charge, spin, and other properties that define quantum states.

Electrons are intrinsically perfectly identical beyond any macroscopic analog. Quantum mechanics explains that if this were not precisely true, all atoms would immediately collapse.

Each of the 16 other types of elementary particles is equally intrinsically identical. It is not just that physicists cannot find any differences. We know of no natural process that reveals even the slightest difference between particles of the same type.

Fermions are the particles of matter. All elementary fermions have spin 1/2, and are further divided into two groups: six quarks and six leptons.

Six *Quarks* occupy rows 1 and 2 of columns 1, 2, and 3 in Figure 8-1. Row 1 contains the up, charm, and top quarks that all have electric charge +2/3. Row 2 contains the down, strange, and bottom quarks that all have charge –1/3. Quarks interact strongly, bonding with one another to form composite particles. The quark names are entirely fanciful. In the micro-world, "up" and "down" are meaningless. Also, no quark is more strange or less charming than the others.

Six *Leptons* occupy rows 3 and 4 of columns 1, 2, and 3 in Figure 8-1. Row 4 contains the electron, muon, and tau that all have charge –1. Row 3 contains the electron-neutrino, muon-neutrino, and tau-neutrino that all have zero charge. Leptons interact weakly and do not form composite particles.

Fermion generations: the three generations of fermions (other than neutrinos) are distinguished by their masses. In each row of fermions, the third generation particles (column 3) are substantially more massive than the second generation (column 2) that are substantially more massive than the first generation (column 1). Third generation particles *decay* (transform) into second generation particles that decay into first generation particles. These decays happen rapidly, with lifetimes ranging from millionths of a second to trillionths of a trillionth of a second. First generation particles are stable (they appear to be eternal), because there exist no lighter particles of the right type into which they can decay.

Neutrinos are the most elusive particles. The masses of the three types of neutrinos are poorly determined. We do know that the three masses are all nonzero, all very small, and all different. Analyses of particle reactions limit neutrinos masses to under a few eV. Surprisingly, cosmological measurements provide a much more stringent limit: the sum of the three masses is less than 0.1 eV. In some sense, it is easier to measure the mass of the universe and subtract everything else than to measure the neutrino masses directly.

The four *Bosons* in column 4 in Figure 8-1 all have spin 1. These are the exchange particles of forces: the photon for electromagnetism, gluons for the strong force, and Z and $W^{\pm}$ for the weak force. Estimated lifetimes are 3.16×10^{-25} sec for the $W^{\pm}$, and 2.64×10^{-25} sec for the Z. In isolation, photons and gluons have infinite lifetimes; both are massless and therefore cannot decay to anything else. Both can be emitted and absorbed by other particles.

The *Higgs Boson*, alone in column 5 of Figure 8-1, has charge 0, spin 0, and a predicted lifetime of 1.56×10^{-22} sec. Hyped as the "God Particle", the Higgs boson is thought to be the origin of the masses of the other elementary particles. Particles that interact strongly with Higgs are very massive; those interacting weakly with Higgs have small masses. Photons and gluons are massless because they do not interact at all with Higgs. The composite particle masses have a quite different source. For example, only about 1% of a proton's mass, the masses of its three quarks, is Higgs related; the other 99% is the kinetic and interaction energy of its quarks. Thus, about 99% of the mass of all normal matter is not due to Higgs, making it seem a bit less Godly.

Unstable particles, those that decay into other particles, do not have definite masses. For particle lifetime Δt, the uncertainty principle stipulates a mass uncertainty — a *mass width* — Δm, given by:

$$\Delta m = \hbar \ / \ \Delta t$$

$$\Delta m = (6.58\times10^{-16}\text{ eV-sec}) \ / \ \Delta t$$

Antiparticles: each type of fermion has a corresponding antiparticle, not shown in Figure 8-1. Particles and their antiparticles have the same mass and spin, but all their other quantum properties have opposite polarities. For example, the electric charge of the antielectron is +1. Since their quantum properties cancel one another, a particle / antiparticle pair can be produced from pure energy, energy without mass. Conversely, when a particle and its antiparticle collide, they may *annihilate* — both vanish completely, their masses convert into pure energy, and no trace remains that either entity ever existed. The resulting energy may become a pair of photons, or new particle / antiparticle pairs.

Bosons are their own antiparticles, or if you prefer, the particle / antiparticle distinction does not apply to bosons.

§8.2 Power of Relativity

Relativistic kinematics imposes many constraints on particle decays. As we will soon discover, regardless of its energy, a particle of rest mass m:

- Cannot decay if m = 0
- Cannot decay to only one other particle
- Cannot decay to a group of particles whose combined masses exceed m

Additionally, all particle decays must conserve energy, momentum, charge, and certain other quantum numbers that we will discuss later. Conserving a quantity means the total amount of that quantity is the same after the decay as it was before the decay.

Let's discover why all this is true.

There is no doubt that physics can be complex, particularly mathematically. One of the survival skills of successful physicists is finding the simplest way to look at complex problems. Einstein excelled at making problems as simple as possible. But as he said, do not make them *too* simple — so simple that you lose the key physics.

One of the most powerful tools of relativity theory is the Lorentz transformation.

Einstein said the laws of nature are universal, the same in all reference frames; it is certain measurements that are relative, having different values in different reference frames. As discussed in Section §3.4, observers in every reference frame must always agree on the outcome of any physical process, such as particle O spontaneously changing into particles A, B, and C. Different observers may measure different values of energy and direction for each particle, but they will agree that the process occurred, and they will agree on the identity of each particle — those properties are not relative.

If nature's laws allow a certain physical process to occur in one reference frame, it is allowed in all frames. Conversely, if nature's laws forbid a physical process in one frame, it is forbidden in all frames. If we can solve the mathematics of a physical process in any frame, we can use the Lorentz transform to express that solution in any other frame.

Therefore, we might as well solve problems in the frames in which they are simplest.

In particle reactions, it is often simplest to consider problems in their center-of-mass ("CM") frame, the reference frame in which total momentum is zero. A more appropriate name would be the center-of-momentum frame; both terms are commonly used and mean exactly the same thing.

In Section §4.1, we discovered that any particle's energy and 3-vector momentum form a covariant 4-vector (E, ***p***) that changes from one frame to another according to the Lorentz transform. Let's see how this works by starting in the particle's rest frame where (E, ***p***) = (m, 0, 0, 0), transforming to a frame moving with velocity v in the +x direction, and then transforming back to the original frame.

Transform to +v frame:
$$E = \gamma (m - 0) = \gamma m$$
$$p_x = \gamma (0 - v m) = -\gamma m v$$

Transform back to rest frame:
$$\gamma (E + v p_x) = \gamma^2 m - \gamma^2 m v^2 = m$$
$$\gamma (p_x + v E) = -\gamma^2 m v + \gamma^2 m v = 0$$

As shown, in Section §4.1, the square of any 4-vector is an *invariant scalar*, a number that is the same in all reference frames, and the square of any particle's 4-momentum (E, ***p***) = $-m^2$.

In the CM frame, the total energy before any decay is m, the decaying particle's rest mass. After the decay, the total energy is also m, which must be at least as large as M, the sum of the masses of all the decay products. If M > m, energy cannot be conserved, even if each decay product has zero kinetic energy in the CM frame.

If M = m, energy could be conserved if every decay product is at rest in the CM frame. While that is mathematically possible, quantum mechanics dictates that the probability of this is zero, because there is no *phase space* for the decay products to occupy. If a decay product had exactly zero kinetic energy, it would not move, its position uncertainty and momentum uncertainty would both be zero, violating the uncertainty principle.

Hence, the mass of the decaying particle must exceed the sum of the masses of its decay products. For example, consider this reaction:

$$p^+ \rightarrow n^0 + e^+ + \nu_e$$

Here, a proton decays into a neutron, an antielectron, and an electron-neutrino. Electric charge and all quantum numbers are conserved in this reaction. But since the neutron's mass exceeds the proton's mass, an isolated proton cannot decay through the above reaction (or indeed through any other known process). However, this reaction does occur within nuclei, for instance in the decay of carbon-11 and oxygen-15, where additional energy is provided by a change in the state of other particles.

One might think: "If the proton moves fast enough, it would have enough kinetic energy to exceed the decay product masses, so it could decay." But special relativity shows that is impossible. What counts is not the proton's total energy in some frame, but its total energy in its own CM frame, which is always m. If a particle moves rapidly, so must its decay products, thus increasing their energy as well.

§8.3 No Decays to Only 1 Particle

In our reference frame, call it the "Lab" frame, the velocity v of any massive particle must be less than 1 (recall in our units c = 1). This ensures the relativistic factor γ and all components of the Lorentz transform are finite. We can therefore transform from the Lab frame to the particle's own CM frame (where v = 0). Here, the particle's momentum p = 0 and its total energy E = m. These conditions must also be true after the decay, because nature requires energy and momentum conservation.

If particle A could decay to only one other particle B, then B would also have p = 0 and E = m, in the CM frame. B and A would have the same mass, electric charge, and all other quantum numbers that define a particle's identity. Hence B and A are identical, indistinguishable particles — there is no physical reality to the statement "A decayed to A".

§8.4 Decays to 2 Particles

We next consider one particle decaying to two other particles. Label each particle's mass, energy, and 3-vector momentum: m_j, E_j, and $\boldsymbol{p}_j$, for j = 0 (for the decaying particle) and j = 1 and 2 (for the decay products). The total 4-vector momenta before and after the decay are:

Before decay: $(m_0, p_0 = 0)$

After decay: $(E_1 + E_2,\ \boldsymbol{p}_1 + \boldsymbol{p}_2)$

These two 4-vectors must be equal; hence their squares must be equal:

$$(m_0, 0) \bullet (m_0, 0) = -m_0^2$$

$$(E_1 + E_2,\ \boldsymbol{p}_1 + \boldsymbol{p}_2) \bullet (E_1 + E_2,\ \boldsymbol{p}_1 + \boldsymbol{p}_2) = -(E_1 + E_2)^2 + (\boldsymbol{p}_1 + \boldsymbol{p}_2)^2$$

Equating the squares: $$m_0^2 = E_1^2 + E_2^2 + 2\,E_1 E_2 - p_1^2 - p_2^2 - 2\,\boldsymbol{p}_1 \bullet \boldsymbol{p}_2$$

Eqn. #1: $$m_0^2 = m_1^2 + m_2^2 + \{\, 2\,E_1 E_2 - 2\,\boldsymbol{p}_1\ \boldsymbol{p}_2 \cos\theta \,\}$$

Here, θ is the angle between $\boldsymbol{p}_1$ and $\boldsymbol{p}_2$.

Recall that (for c = 1): $$E^2 = p^2 + m^2$$

This holds for any particle, and ensures: E >= p. A particle's energy E equals its momentum p only if the particle has zero mass. For any massive particle, its energy exceeds its momentum. Therefore the expression in { }'s in Eqn. #1 must be positive. It could only be zero if $m_1 = m_2 = \theta = 0$, otherwise it is greater than zero.

Clearly, we could generalize Eqn. #1 to any number of decay products.

In the CM frame, total momentum 3-vector is zero by definition, both before and after the decay. So, $\boldsymbol{p}_1 + \boldsymbol{p}_2 = 0$. Let's define:

$$\boldsymbol{p} = \boldsymbol{p}_1 = -\boldsymbol{p}_2$$

In the CM frame, Eqn. #1 is:

$$m_0^2 = m_1^2 + m_2^2 + 2\sqrt{\{(p^2 + m_1^2)(p^2 + m_2^2)\}} + 2\,p^2$$

Unfortunately this does not reduce to anything simple, but at least the equation has only one solution.

Figure 8-2 shows a particle decaying into two particles in both the CM frame and the Lab frame. The decaying particle is the black dot on the horizontal midline, and the two decay products are shown at the tips of the angled arrows.

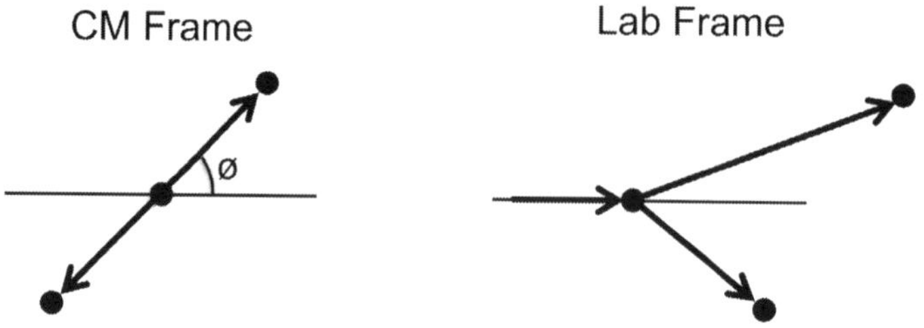

Figure 8-2 Decay to 2 Particles in CM and Lab Frames

In the CM frame, the two decay product momenta p are opposite in direction but equal in magnitude, making their sum zero. The magnitude of p in the CM frame is uniquely determined by the three masses. The only remaining variable, or as we say *degree of freedom*, is the angle ø in the CM frame, as shown above. Given ø, everything else is uniquely determined.

§8.5 Photons Cannot Decay

We saw above that in any two-particle decay, the measured energy and momentum of each decay product and Eqn. #1 determine m_0, the mass of the decaying particle. If the decaying particle is a photon, m_0 must be 0. But, as we proved in the prior Section, m_0 is zero only when both decay products have zero mass and their momenta are parallel (θ, the angle between them, is zero). So far, kinematics allows photons to decay only into other photons moving in exactly the same direction, provided the total energy of the decay products equals the original photon's energy. Such decays would change the spectrum of light from celestial sources but not the total energy.

The prior paragraph discussed photon decays into two particles, but we can generalize that to any larger number of particles. Consider a photon decaying to particles A, B, and C. This is mathematically equivalent to a photon decaying to C and D, followed immediately by D decaying to A and B. From the prior logic, C and D must be collinear photons, as must A and B.

But relativity forbids even that. At the speed of light, the relativistic γ factor equals infinity, and time stops. A photon never "experiences" the passage of time. Photons may be created, travel across the entire universe, and be captured by our telescopes, but all these events would be simultaneous for them. Viewed in any specific reference frame, isolated photons can never change, because change requires time. Hence they can never decay.

When a photon hits something else, the 4-vector sum (E, p) of photon and "else" has $E^2 - p^2 = m^2 > 0$ if "else" has mass. The CM m^2 must still be > 0 even if "else" is a photon, because two photons cannot collide if they are moving in exactly the same direction. The γ factor of photon + "else" is therefore finite, allowing time to flow, and interactions to occur.

§8.6 Finding Invisible Particles

High-energy particle collisions can create particles that physicists have never seen before — even particles that no one ever expected. In these collisions, nature generously produces every type of particle that exists, some less frequently than others, but constrained only by the collision's CM energy.

Even if we have no idea what to look for, new particles fly out of these collisions, waiting to be discovered. But, there is a major problem: many types of particles, particularly the ones that have not yet been discovered, are invisible and impossible to detect directly. We must devise techniques to detect such particles indirectly.

One major cause of invisibility is lack of electric charge. The strong and weak nuclear forces only extend over subatomic distances; since all our particle detectors are made of atoms, they are too big to directly detect the actions of those forces. Gravity is too weak to reveal the actions of individual particles. That leaves the electric force as the primary means of probing particle-scale reactions. While we have some success detecting reactions caused by neutral particles, we are far more successful detecting charged particles.

Another major cause of invisibility is extremely short particle lifetimes. In this case, Eqn. #1 can provide the definitive evidence we seek.

Let's imagine that, after one specific collision, we observe two particles emanating from a collision — both particles emerge at the same instance, and both trajectories align with the collision point. By various means, we can often determine each particle's mass m, momentum ***p***, energy E, and charge. (A particle's mass and charge almost always uniquely identifies its particle type.)

We can then employ Eqn. #1 to calculate m_0, the effective mass of the pair of detected particles in the pair's CM frame. If the two detected particles are truly the decay products of a particle X, m_0 is the mass of X. If, after examining thousands or millions of collisions, we identify many instances of the same two types of particles having the same effective CM mass m_0, we can be confident they are the decay products of an X particle.

In principle, this technique works with any number of detected particles, not just two, but it becomes more complicated as that number increases.

§8.7 Decays to 3 Particles

As we discovered above, a decay into two particles has only one degree of freedom, the angle ø in the CM. For decays into three particles, there are two degrees of freedom. Repeating prior logic, O decaying to A, B, and C, is equivalent to O decaying to C and D followed immediately by D decaying to A and B. There is one degree of freedom in the decay of O and another in the decay of D.

Physicists analyze decays into three particles with 2-dimensional charts call Dalitz plots, as shown in Figure 8-3. Here, each observed decay is plotted as one small dot with the axes being: the combined mass-squared of A and B horizontally, and the combined mass-squared of A and C vertically. In accordance with Eqn. #1: if A and B are the only decay products of particle X, their combined mass is the mass of X.

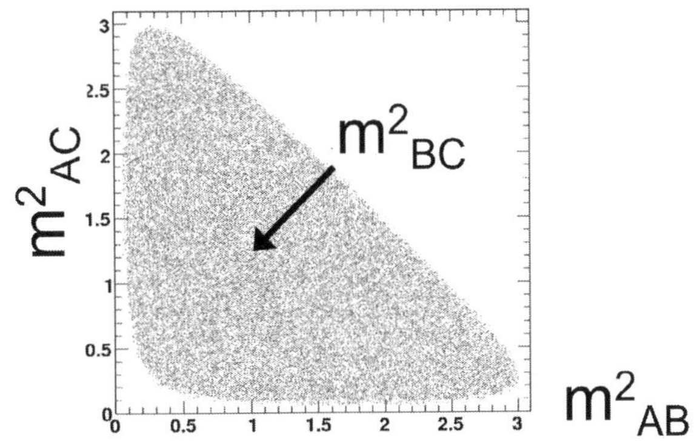

Figure 8-3 Dalitz Plot of Decay to 3 Particles

We describe the Dalitz plot in Figure 8-3 as "flat"; there is no apparent preference for any specific mass value. It seems the degrees of freedom in these decays are uniformly distributed.

Conversely, the Dalitz plot on the left side of Figure 8-4 has a dramatic vertical stripe; something interesting is clearly happening. The histogram on the right shows the same data plotted according to the number of decays with various values of the square of the combined AB mass.

These data indicate that what we observe as O decaying to A, B, C, is actually O decaying to D and C, followed rapidly by D decaying to A and B. If the lifetime of particle D is short enough, D leaves no detectable track. The only evidence that D existed is in the AB mass plot. Literally hundreds of different types of particles, including the Higgs boson, have been discovered using such mass plots.

If particle D decays through the strong force, its mean lifetime might be trillionths of a trillionth of a second. Even with the largest time dilation factor of any man-made particle, $\gamma = 40{,}000$, D's track would be less than one atom long — impossible to detect.

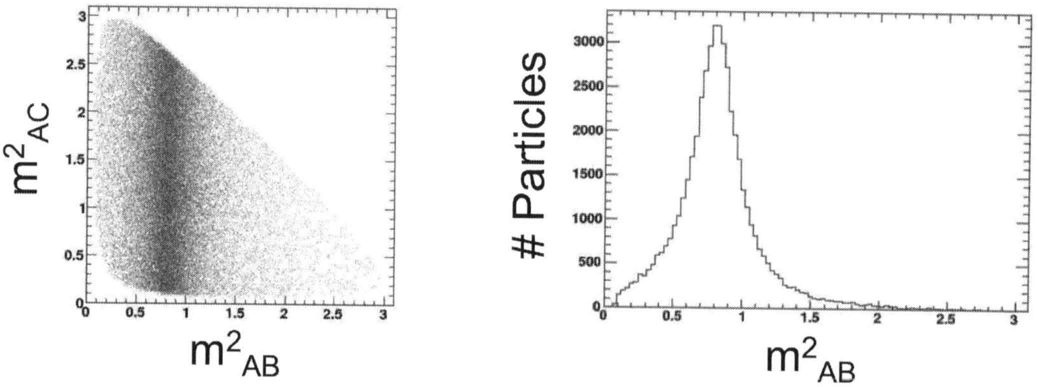

Figure 8-4 Dalitz Plot (left), Histogram of Particle Distribution (right)

As discussed in Section §8.1, the uncertainty principle of quantum mechanics states that the mass of an extremely short-lived particle does not have a precisely determined value: for lifetime Δt, the width of the mass distribution Δm is:

$$\Delta m = \hbar \ / \ \Delta t$$

$$\Delta m = (6.58\times10^{-16}\ \text{eV-sec}) \ / \ \Delta t$$

For lifetime $\Delta t = 3\times10^{-24}$ sec, $\Delta m = 219$ MeV, nearly a quarter of a proton's mass. If we produce a million D particles, they would not all have the same mass; their masses would be distributed across a broad peak, as in the right side of Figure 8-4. From the width of the mass peak, we can calculate the decaying particle's lifetime.

Dalitz plots also identify the *spin* of short-lived, undetected particles. Figure 8-5 shows the Dalitz plots of unseen particles, with spin 0, 1, and 2, that decay into three observed particles.

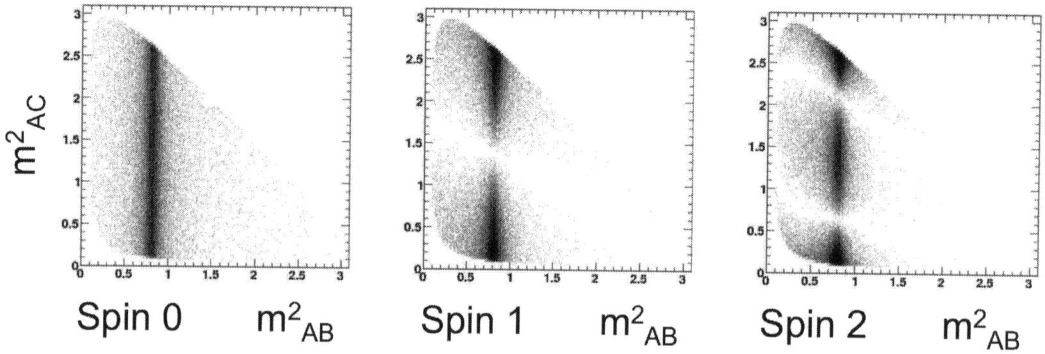

Figure 8-5 Dalitz Plots of Decays of Spins

For more about particles, spin, the uncertainty principle, and quantum mechanics see my ebooks ***Quantum Mechanics 2: Reality, Uncertainty & Schrodinger's Cat*** and ***Quantum Mechanics 4: Identical Particles*** or my print book ***Quantum Mechanics Strange Theory of the Micro-World***.

Chapter 9
Particle Physics: Collisions

In this chapter, we use relativity to explore high-energy particle collisions.

§9.1 Particle-Target Collisions

A particle beam is a group of identical particles that all have almost the same energy and are moving in almost the same direction. Accelerators produce *primary beams* containing immense numbers of particles, usually protons or electrons, with extremely limited variations in energy and direction.

The first several generations of particle accelerators produced a single primary beam of high-energy particles. When such a beam hits a target, a stationary block of matter, some of the beam particles scatter while others produce a spray of secondary particles. Experiments on both the scattered primary particles, and on the more exotic secondaries, produced ground-breaking discoveries for many decades.

Physicists can also produce *secondary beams* from the spray of particles produced by primary-on-target collisions. The spray can be filtered to select certain types of particles for a variety of subsequent experiments. We can, for example, obtain secondary beams of unstable particles with moderate lifetimes, such as "long-lived" kaons. These kaons have a mean lifetime of 51 nanoseconds, and at $v = 0.995$ c, they can travel hundreds of meters before decaying. Secondary beams typically contain far fewer particles and have greater variation in energy and direction than do primary beams.

In the sketch below, a high-energy particle, with energy E, momentum p, and mass M, enters from the left and strikes a single particle of mass m in a stationary target. Four particles exit the collision, moving to the right. The total momentum 4-vector is $(E + m, \boldsymbol{p})$ before the collision, since the stationary target particle initially had zero kinetic energy and momentum.

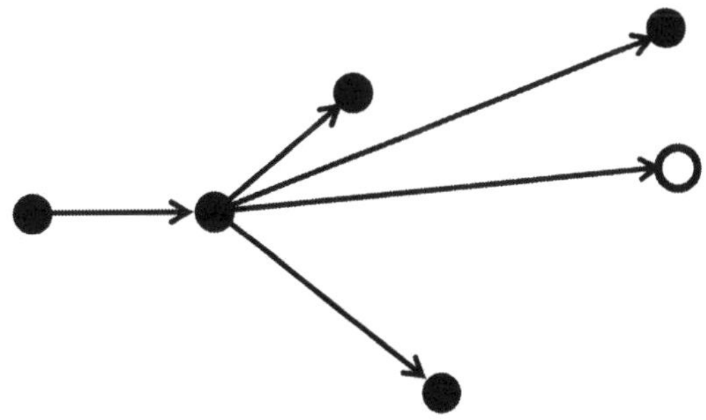

Figure 9-1 High Energy Particle Hits Stationary Target Particle

Each particle is represented by a solid black dot, except for one that has a white center. We will return to that shortly. Due to energy and momentum conservation, the sum of the 4-vectors of all exiting particles must equal $(E + m, \boldsymbol{p})$. Now, let's compute $-A^2$, the square of the pre-collision 4-vector:

$$-A^2 = -(E+m)^2 + p^2$$

$$A^2 = E^2 + 2\,m\,E + m^2 - p^2$$

Eqn. #2:

$$A^2 = M^2 + 2\,m\,E + m^2$$

If M = m:

$$A = \sqrt{\{2m\,(E+m)\}}$$

Recall that the square of any 4-vector is an invariant scalar, the same number in all reference frames. Hence, in the center-of-momentum frame, CM, the sum of the 4-vectors of the exiting particle is (A, 0).

Remember the sole white-centered particle in Figure 9-1? Let's imagine the black dots are protons and the white-centered dot is an antiproton, and discover what is required to produce it.

Protons are the least massive of a group of strongly interacting particles called *baryons*. Each type of particle has a quantum property call *baryon number*. Protons have baryon number +1 and antiprotons have baryon number –1. Exhaustive, precise experiments show that strong force interactions conserve the total baryon number.

Heavier baryons decay to lighter ones that decay to even lighter ones. But that decay sequence ends at protons. Protons cannot decay via the strong force, because no lighter baryon exists. This is why our universe has so many protons. Some theorists speculate that protons should decay via other forces according to their models, but extensive searches have failed to find any evidence of proton decay. Experiments have established that the proton mean lifetime is at least 10^{34} years, about a trillion, trillion times longer than the current age of the universe. While experiments have not proven protons never decay — 10^{34} years is a bit short of eternity — don't hold your breath.

In Figure 9-1, the baryon number before the above collision is +2, the incident proton plus the target proton. Since we want one antiproton with baryon number –1 after the collision, there must be exactly three other baryons after the collision to maintain the net baryon number at +2. Thus, there must be a minimum of four exiting particles: three protons and one antiproton. Many other combinations with more post-collision particles would also yield baryon number 2, but all those would increase the total mass of exiting particles that increases the required primary beam energy.

The minimum value of A required to produce an antiproton is 4m, 4 times the proton mass. (Particles and their antiparticles have identical masses.) The corresponding minimum incident energy E is:

$$2\,m\,(E + m) = 16\,m^2$$

$$E = 7m$$

The minimum kinetic energy of a proton required to produce antiprotons is 7 – 1 = 6 times the mass of a proton. The Bevatron at U. C. Berkeley was designed for just this purpose, with a beam energy of 6.6 times the proton mass, providing a 10% margin.

§9.2 Omega Minus Discovery

Another momentous new particle discovery was the Ω^-. Nobel Laureates Murray Gell-Mann and Yuval Ne'eman both predicted a very strange particle, one with a strangeness quantum number of –3. (Physicists quantify everything, including strangeness.) They based their prediction on the belief that the then known strongly interacting particles are not fundamental, but are comprised of "smaller" particles called *quarks*. Discovering the Ω^- convinced most physicists that the quark model is indeed correct.

Figure 9-2 is an image of an Ω^- produced by the collision of a K^- and a proton in a hydrogen bubble chamber, a primary particle physics research instrument of the mid 20th Century.

A bubble chamber contains a large pool of pressurized liquid at a temperature near its boiling point. Typical conditions are liquid hydrogen at 27 K at a pressure of 5 atmospheres (500,000 pascal or 74 psi). As charged particles traverse the liquid, they ionize atoms along their path, depositing energy that warms the nearby liquid very slightly, thereby creating bubble nucleation sites.

Immediately after a particle beam transits the chamber, a high-speed piston suddenly decreases the liquid's pressure, dropping its boiling point. Macroscopic bubbles then form at the nucleation sites, creating a visible trail of bubbles along the path of every charged particle that recently passed through the liquid. Electrically neutral particles do not ionize the liquid, and hence leave no tracks in bubble chambers. Neutral particles are invisible in most other particle detectors as well.

Bubble chambers also incorporate strong magnetic fields, causing positive particles to turn in one direction, while negative particles turn in the opposite direction. The turning direction identifies a particle's charge, while the turning radius quantifies its momentum.

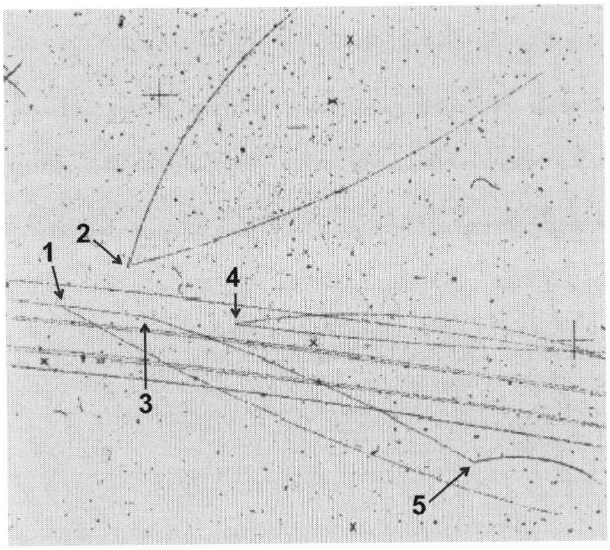

Figure 9-2 Bubble Chamber Image of Omega-Minus

The reactions occurring at each indicated point above (beyond the tip of each arrow) are:

1. $K^- + p^+ \rightarrow \Omega^- + K^0 + K^+$
2. $K^0 \rightarrow \pi^- + \pi^+$
3. $\Omega^- \rightarrow \Lambda^0 + K^-$
4. $\Lambda^0 \rightarrow p^+ + \pi^-$
5. $K^- \rightarrow \pi^- + \pi^0$

In Figure 9-2, a secondary beam of negative K mesons with charge –1 enters from the left, leaving 8 long, slightly curved tracks, some of which are nearly on top of one another. One of these K mesons (kaons) hits a stationary proton at point 1, producing the Ω^-, a neutral kaon, and a kaon with charge +1. The unseen K^0 (neutral particles leave no tracks) decays to two pions at point 2, producing a large "V" shape. At point 3, the Ω^- decays to a Λ^0 and a negative kaon, which decay at points 4 and 5 respectively.

The masses of the particles exiting point 1 total:

$$m_\Omega + m_{K0} + m_{K+} = 2664 \text{ MeV}$$

From Eqn. #2, the minimum kaon beam energy to drive this reaction is:

$$\sqrt{[}\,(2664^2 - m_K^2 - m_P^2)\ /\ 2\, m_P] = 3184 \text{ MeV}$$

The actual beam energy was 4200 MeV.

§9.3 Particle-Particle Collisions

The newest particle accelerators produce two primary particle beams that collide with one another. Getting two beams to collide is much more difficult than hitting a large stationary target with one beam. This challenge has been compared to hitting one needle with another needle thrown from two miles away — with both needles moving at nearly the speed of light.

The motivation for attacking this great challenge is that colliding beams provide much more energy for the production of new particles. According to Eqn. #2, the available CM energy in a high-energy-beam-hits-target collision is: $\sqrt{(2\ m\ E)}$, where E is the beam energy and m is the proton mass. By comparison, the available CM energy in a beam-hits-beam collision is: 2 E.

For E = 1000 m, two colliding beams provide 44 times more CM energy than a beam-target collision.

§9.4 Tevatron at Fermi Lab

The Tevatron at the Fermi National Laboratory near Chicago, Illinois, features a main ring whose circumference is 7 km (4.3 miles). The main ring is in the foreground of Figure 9-3. At the left and slightly above center, is the 12-story, somewhat triangular, main office building. The Tevatron achieved an energy of 1 TeV (Trillion electron Volts) in each beam; hence its name. The total CM energy corresponded to the mass of 2100 protons.

The Tevatron was the last U.S. high-energy particle accelerator. It was shutdown in 2011.

Figure 9-3 Tevatron at Fermi National Lab near Chicago

The Tevatron produced two high-energy beams — a clockwise-circulating proton beam, and a counterclockwise-circulating antiproton beam. This worked well because particles and their antiparticles bend in opposite directions in an accelerator's magnets. An array of particle detectors surrounded the two locations where the beams collided.

Since particles and antiparticles have opposite quantum numbers, all pre-collision quantum numbers netted to zero, eliminating the need for extra exiting particles just to conserve quantum numbers. Hence, all the CM energy was available for new discoveries. Figure 9-4 shows a beam of high-energy protons entering from the left and colliding with an equally high-energy beam of antiprotons entering from the right, producing four exiting particles.

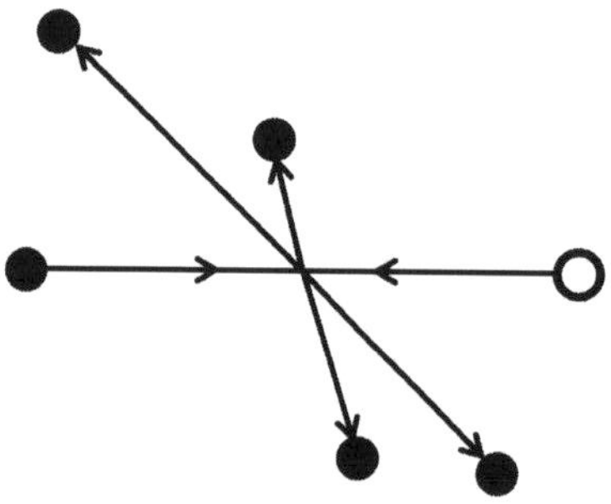

Figure 9-4 Proton-Antiproton Collision in CM

In 1995, the Tevatron discovered the top quark, the last fundamental fermion in the Standard Model of Particle Physics, as well as several other composite particles.

On July 2, 2012, two Tevatron teams announced the possible existence of a previously-unidentified particle of mass 115 to 135 GeV. They had analyzed 500 trillion events taken over a 10-year span, and found a 2.6 standard deviation excess of events above background, corresponding to a 1-in-550 chance that this signal was a statistical fluke.

Two days later, on July 4, 2012, two LHC teams announced their discovery of the Higgs Boson, with measured masses of 125.3 ± 0.4 GeV (by CMS) and 126 ± 0.4 GeV (by ATLAS).

§9.5 LHC at CERN

The Large Hadronic Collider, LHC, is the culmination of nearly 100 years of accelerator development, and is the world's only remaining truly high-energy particle accelerator. The LHC is part of CERN, the European nuclear physics research consortium located near Geneva, Switzerland.

Figure 9-5 LHC at Cern with Swiss Alps in Background

Figure 9-5 shows an outline of the LHC main ring superimposed on a photograph of its beautiful setting. The main ring has a circumference of 27 km (17 miles), and is as much as 175 m (570 feet) below ground. The ring spans the border between Switzerland and France. Two counter-rotating proton beams collide at seven points along the main ring, at the indicated circles.

The LHC has so far achieved an energy of 6.5 TeV in each beam. At that energy, literally hundreds of particles are produced in each collision. Collisions are no longer analyzed photographically, as in the Ω^- discovery. Instead, all particle detectors have high-speed electronic readouts. At its maximum design performance, the LHC is expected to produce 10 million gigabytes of data annually that will be analyzed by 170 computing facilities in 36 countries.

Figure 9-6 is an image of ATLAS, one of the LHC's primary detectors for analyzing particles produced by the colliding beams. If you look carefully, you might see a man standing directly below the main beam line, on the lowest level of the detector platform.

Figure 9-6 ATLAS Detector at LHC

Chapter 10
Newtonian Gravity

Gravity is replete with contradictions.

Gravity seems so simple: everything falls. My 18-month-old granddaughter succinctly explained: "Gravity go down." Gravity certainly seems simpler and much more familiar than nature's other forces: electromagnetism, the strong nuclear force, and the weak force. Yet in many ways, gravity's effects are more surprising and more complex than those of the other forces. Some of those surprising effects are discussed in Chapter 14.

Gravity also seems incredibly feeble. The gravitational attraction between a proton and an electron is 2000 trillion, trillion, trillion times weaker than their electrostatic attraction. But while the other forces fade for ever-larger objects, gravity relentlessly accumulates. Each of the 10^{57} particles in the Sun attracts each of the 10^{51} particles in the Earth. Despite gravity's seemingly feeble fundamental strength, the total force between Sun and Earth is $10^{-39} \times 10^{57} \times 10^{51} = 10^{+69}$ — a fabulously enormous number.

For cosmic bodies, gravity is the most powerful force of nature, dominating all others. Gravity controls the fate of the universe and everything in it.

And while everyone refers to "the force of gravity", Einstein explained that gravity is not a force at all, but rather the effect of geometry.

What then is gravity?

Let's follow the historical advances in mankind's understanding of gravity. It started in the heavens, and ultimately came down to Earth.

Even before the beginning of recorded history, the motions of celestial bodies have intrigued mankind. Of paramount interest was the Sun, which dictated the cycle of life. Many pre-historic civilizations built great monuments that seem designed to track the Sun's annually recurring movements, perhaps to predict animal migrations and the best times to sow and harvest.

Next in importance were the planets, those few bright spots that deviated from the otherwise simple rotation of the myriad stars. The word planet derives from a Greek word for "wandering star." The odd movements of some planets puzzled the Greeks, who believed that all heavenly body would surely follow the most "perfect" paths: circles.

Greco-Roman astronomer Claudius Ptolemy (100-168 A.D.) developed a *geocentric* model of the universe based on circular motion. His epic work ***Almagest*** is the only substantial surviving treatise of ancient astronomy. Ptolemy described the universe as a set of concentric shells that rotate around Earth. He explained the retrograde motion of some planets in terms of epicycles, complex nested combinations of circular motions that sometimes turn backwards. He also said the radius of the Sun's shell is 1210 times Earth's radius, while the radius of the shell of the stars is 20,000 Earth-radii. Actually, the distance to the Sun is 23,500 Earth-radii, and the distances to stars that are visible by eye are between 2.4×10^{17} and 2.4×10^{20} Earth-radii.

Nicholas Copernicus (1473-1543) published the first *heliocentric* model of our Solar System, with the Sun at the center, and the planets, including Earth, revolving around it. Copernicus argued that planetary motions are simpler if one assumes they orbit the Sun rather than the Earth, as had always been assumed before.

The next step was making the Copernican model quantitative.

The mathematician Johannes Kepler (1571-1630) developed his famous laws of planetary motion through detailed analysis of the observations of Tycho Brahe (1546-1601). Brahe compiled the most thorough and precise measurements ever made in naked-eye astronomy, which Kepler acquired upon Brahe's death.

§10.1 Kepler's Laws

Kepler's three laws of planetary motion are:

1. Planets orbit in *ellipses*, with the Sun at one *focal point*.
2. A planet sweeps out *equal areas* in *equal times*.
3. A planet's orbital *period squared* is proportional to its *semi-major axis cubed*.

Figure 10-1 shows an ellipse, an elongated circle, that has two *focal points* (the central black dots). Each point on the curve has the same total distance to the two foci. In the figure, two lines connect the two foci to one point on thc curve. The sum of the lengths of the two lines is the same for every point on the ellipse. An easy way to draw an ellipse is to place tacks at the two foci, tie opposite ends of a string to each focal point, stretch the string with the tip of a pencil, and trace a path completely surrounding the two foci, while keeping the string taut.

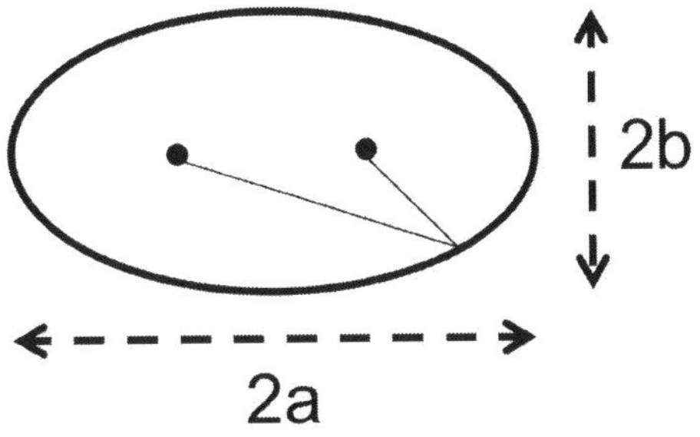

Figure 10-1 Axes of an Ellipse

Ellipses have two orthogonal axes: the major axis is the ellipse's largest extent (the horizontal distance labeled 2a), while the minor axis is its smallest extent (labeled 2b). Normally, we speak of an ellipse's *semi-major axis* a, and *semi-minor axis* b, each one half of the corresponding full axis. Ellipses have a special optical property: if the curve is a mirror, light emitted at one focal point comes to a perfect focus at the other focal point. This is because, by the definition of an ellipse, the total path length from one focal point to the curve and back to the other focal point is the same for all light rays.

This also works for sound waves. In fact, in an elliptical room, a person whispering at one focal point is clearly heard by a person at the other focal point. It is said politicians have eavesdropped on opponents by cleverly arranging desks in an elliptical room.

The *eccentricity* of an ellipse is defined as the distance between foci divided by the major axis. If we bring the foci together, the eccentricity becomes zero and the ellipse becomes a circle of radius $r = a = b$. The eccentricities of planets in our Solar System range from 0.0068 (Venus) to 0.2056 (Mercury), with Earth at 0.0167 and *dwarf planet* Pluto at 0.2488.

As a planet orbits the Sun, its distance of closest approach is called its *perihelion* and its farthest distance is its *aphelion*, from Greek words meaning near to and far from Helios, the Sun God. A planet's perihelion plus its aphelion always equals 2a, its major axis.

Kepler's second law is illustrated in Figure 10-2. This shows that the orbit of a planet is an ellipse with the Sun (black dot) at one focal point. The Sun is *not* at the center of the ellipse. As the planet orbits, its radial line to the Sun sweeps out a two-dimensional region. Two such regions are shown below in gray. The left region was swept during time interval Δt when the planet was near the Sun, and the right region was swept during the same interval Δt when it was far away.

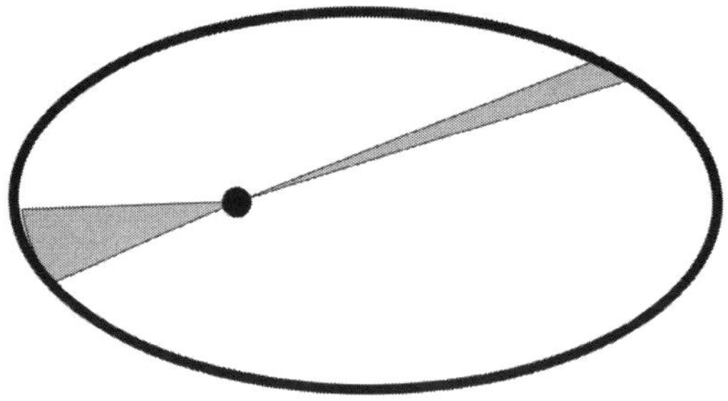

Figure 10-2 Kepler's Second Law

Kepler's second law states all such regions are equal in area. This is because planets move faster when they are nearer the center of gravity. The region on the left is wider but has shorter sides than the region on the right. While the shapes of swept regions vary, Kepler's second law says a planet sweeps equal areas in equal times everywhere along its orbit.

Kepler's third law relates the orbits of the planets to one another. It says the square of a planet's period T (the time required to complete one orbit) is proportional to the cube of its semi-major axis. This is written:

$$T^2 \sim a^3$$

We will shortly discover that the proportionality constant is M, the mass of the gravitational source. We often call Kepler's third law the *123-law*, a moniker that helps us remember the exponents:

$$M^1\ T^2 = a^3$$

This equation is valid for any stable, two-body, gravitational system if: M is measured in multiples of the Sun's mass (M = 17 means a mass 17 times that of our Sun); T is measured in Earth years; and a, the semi-major axis, is measured in AU (Earth's average distance to the Sun). For Earth the 123-law is easy: $1^1 \times 1^2 = 1^3$. The values for the Sun's other planets are:

Planet	T (year)	a (AU)	T^2	a^3
Mercury	0.241	0.387	0.058	0.058
Venus	0.615	0.723	0.378	0.378
Mars	1.881	1.523	3.537	3.537
Jupiter	11.86	5.20	140.7	140.9
Saturn	29.46	9.54	868	867
Uranus	84.01	19.2	7058	7078
Neptune	164.8	30.1	27156	27128

The *123-law* is slightly inaccurate for our Solar System because it does not account for the gravitational attractions of the planets on one another. But it still works quite well.

Kepler's second law (equal areas in equal times) applies to each planet individually; it does not say that different planets sweep out the same areas per unit time. With the 123-law, we can show this for the simple case of circular orbits. During one complete orbit of period T and radius R:

Area A swept in one orbit:

$$A = \pi R^2$$

$$A = \pi (M\,T^2)^{2/3}$$

$$A / T = \pi\ M^{2/3}\ T^{1/3}$$

If all planets swept the same area in the same time, the right side of the last equation would be the same for every planet, which it clearly is not.

Kepler's laws apply not just to our Solar System — their scope is universal.

Figure 10-3 shows a *binary star* system comprised of two nearby stars, Sirius A and Sirius B, that are bound by their mutual gravitational attraction. Sirius A is about twice as massive as Sirius B, which is about our Sun's mass. Sirius A is the large black dot at the origin of the coordinate axes. The many small black dots, some with dates, show the relative positions of Sirius B, as it traces its 50-year long orbit. While we see the orbit at a tilt, this ellipse would make Kepler proud.

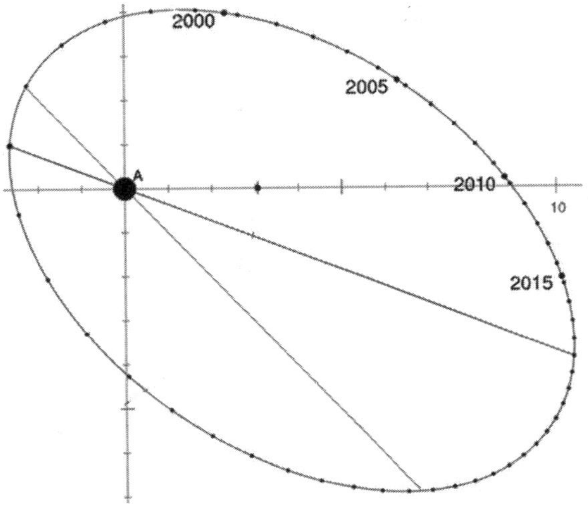

Figure 10-3 Orbit of Sirius B around Sirius A

Kepler's 123-law is used by astronomers to compute the masses of distant stars, galaxies, black holes, and more. When astronomers observe a smaller object orbiting a much larger body, they use the 123-law to compute the mass M of the larger body from the period T and semi-major axis a of the smaller object. It's as easy as 123, thanks to Johannes Kepler's 400-year-old equation.

§10.2 Galileo Gets Ball Rolling

The ancient Greeks were among humanity's first scientists. Their pursuit of knowledge was driven by philosophical discourse and weight of authority. Aristotle was the ultimate authority on science. The Greeks believed objects fell because their "natural condition" was to be down, as close to the Earth as possible. They reasoned this natural condition must be stronger in heavier objects, thus it made sense that heavier objects fell faster.

The Greeks eschewed experimentation: they did not try dropping objects of different weights to find out if what "made sense" was actually true. They felt the real world was impure, corrupted by spiteful demigods and untrustworthy mortals. Another factor discouraging experimentation was that science was then the sole province of the elite, who felt experimentation had the scent of labor and was thus beneath them. They believed logic was the more certain path to understanding the ultimate, hidden truths that existed only in the ideal heavenly realm of their gods. Their approach dominated science for two millennia.

Italian Galileo Galilei (1564-1642) is generally considered the first modern scientist. He was the first to employ systematic experimentation, and the first to apply modern scientific methods to the study of gravity. Galileo devised tests that discriminated between alternative theories. He also built special devices to enhance measurement precision, such as calibrated inclined planes that slowed a ball's descent.

Galileo discovered the principle of relativity and the law of inertia. He also discovered that, ignoring air resistance, all bodies fall at the same rate. His discoveries invalidated what natural philosophers had proclaimed for millennia, and his ideas played a critical role in the development of Newton's laws.

Galileo, a contemporary of Kepler, was the first to observe the heavens with a telescope. He made dramatic discoveries. He found that the Moon is not just a patch of light, but is a world of its own, with mountains and valleys like Earth. He also discovered Jupiter's four largest moons — the Galilean Moons — the first objects ever seen that definitively orbit something other than Earth.

§10.3 Conditions for Circular Orbit

Our next step is to understand the conditions required to keep an object in circular motion. This is really all about geometry. The left side of Figure 10-4 shows an object (black dot near the letter "d") following a circular path of radius R (the dashed arc). The object starts at the top at time t = 0, and initially moves downward with speed v. Shortly later, at t = Δt, the object has moved a distance s = v Δt. The angle between the object's radial lines at t = 0 and t = Δt is θ. (A radial line connects an object to a circle's center.)

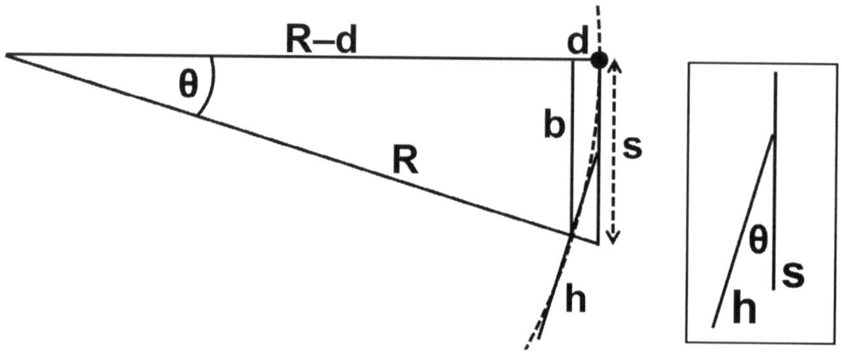

Figure 10-4 Circular Motion with Magnified Image in Box

The distance the object moves horizontally to the left during Δt is d. The line h is tangent to the circular arc where it crosses the object's radial line at t = Δt.

The key element here is θ, which is also the angle between s and h, because s is perpendicular to the radial line at t = 0 and h is perpendicular to the radial line at t = Δt. This angle is shown magnified in the box at the right. Without the action of a force, the object would continue moving in its original downward direction. To follow the circular path, the direction of motion must be rotated by angle θ during time Δt. The required velocity change, Δv, is:

$$\Delta v = v (\tan\theta)$$

$$\Delta v = v (s / R) = v^2 (\Delta t / R)$$

In the limit that Δt goes to zero, the velocity change — the required acceleration a — is:

$$a = \Delta v/\Delta t = v^2 / R$$

Before continuing, let's obtain the equation relating R, b, and d in Figure 10-4.

$$b^2 + (R - d)^2 = R^2$$

$$b^2 + R^2 - 2\,d\,R + d^2 = R^2$$

$$b^2 = d\,(2\,R - d)$$

Our key result is: as viewed from above, an object moving at speed v around a circle of radius r, must experience a *centripetal* acceleration (an acceleration toward the center) of:

For a circular orbit: $$a = v^2 / r$$

§10.4 Newton's Universal Gravity

England's most famous scientist, Sir Isaac Newton (1642-1727), was the next to advance our understanding of gravity.

Newton's *Principia Mathematica*, published in 1687, is thought by many to be the most influential book in the history of science.

Newton was prolific as a physicist, mathematician, alchemist, and theologian.

His other scientific discoveries include his laws of motion, the reflecting telescope, and the theory of color. Yet, he is said to have spent more time on his unorthodox biblical interpretations than on science.

As England's Master of the Mint for 28 years, he reformed the currency, routed out counterfeiters, and greatly increased England's wealth and fiscal stability. For this, and not for his outstanding contributions to science, Queen Ann knighted him in 1705.

After his death, a postmortem found very high levels of mercury in Newton's body, no doubt due to his extensive work in alchemy. This might explain his "erratic" behavior in later life.

Newton and Gottfried Wilhelm Leibniz, a German mathematician and philosopher, independently developed an entirely new branch of mathematics: *calculus*. Calculus enabled Newton to formulate physical laws by relating forces to small changes, and relating sequences of small changes to global motion. This was the dawn of a new age of analytical science.

With Kepler's third law, the acceleration equation of Section §10.3, and his own laws of motion, Newton deduced the law of gravity.

For simplicity, let's assume a circular planetary orbit. Kepler's third law then says that a planet's period squared is proportional to the cube of its orbital radius. With k being the proportionality constant, Kepler's 123-law states:

$$k\,T^2 = r^3$$

Let's next substitute orbital period T in the prior equation with the circle's circumference divided by the planet's velocity v:

$$T = 2\pi r / v$$

$$k\,(4\pi^2 r^2 / v^2) = r^3$$

$$v^2 = 4\pi^2 k / r$$

$$a = v^2 / r = 4\pi^2 k / r^2$$

We just proved that gravitational acceleration equals a constant divided by the square of the distance between a planet and the Sun. This is the most famous part of Newton's force law: its inverse squared dependence.

Since all of the Sun's planets must obey the same law, they all must have the same constant k, and their gravitational acceleration must depend only on distance. Because $F = m\,a$, we can also deduce that the force law must have the form:

$$\text{Sun's Force} = (\text{mass of planet})\ Z\ /\ r^2$$

The unknown factor Z can only depend on intrinsic properties of the Sun and the gravitational force, such as its strength; Z cannot depend on the properties of planets or anything else.

Newton achieved all this because of his genius as a mathematician. The final step required his genius as a physicist. Newton postulated that mass is the source of the gravitational force, and furthermore that all masses exert a gravitational force according to one common law.

Newton declared there is a *universal law* of gravity: the same law applies to planets orbiting the Sun, to moons orbiting planets, to the trajectory of cannon balls, and to the descent of apples from trees.

Never before had science had a universal law of anything. With Newton's theory of gravity, the human mind achieved a cosmic scope for the very first time.

Newton's universal law of gravity is:

$$F = G\ M\ m\ /\ r^2$$

Here, M is the mass of one body, m is the mass of another body, F is the force that each body exerts on the other, G is Newton's gravitational constant (6.674×10^{-11} $m^3\ kg^{-1}\ s^{-2}$), and r is the distance between the centers of the two bodies.

The discoveries of Galileo and Newton proved that planets do not need to be *pushed* to continue orbiting (inertia takes care of that), but they do need to be *pulled* toward the Sun to keep revolving around it (else they would fly off like bolas launched by a gaucho).

The force that keeps a planet in orbit points toward the Sun, not along the planet's orbit.

Let's discover how Newton explained our Moon's orbit, as illustrated in Figure 10-5.

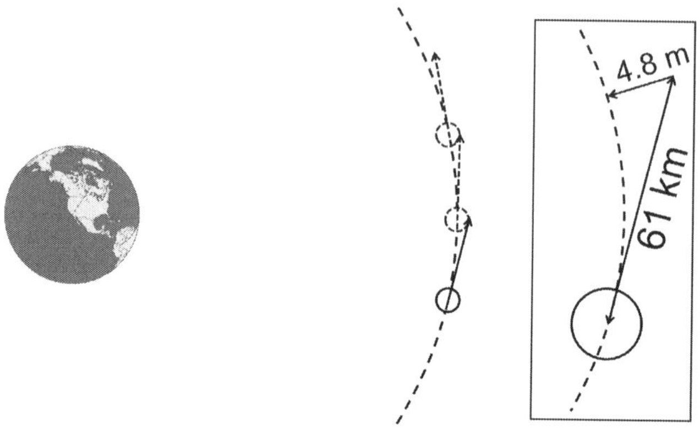

Figure 10-5 Why Moon Orbits Earth

In Figure 10-5, the solid circle and solid arrow indicate the Moon's position and velocity at one particular moment, call it t = 0. The dotted circles and arrows indicate the Moon's position and velocity at later moments, at t = 1 and t = 2. The boxed image to the right is a magnified view of the Moon at t = 0.

A stable orbit requires a balance of gravity and velocity.

Without gravity, the Moon would forever move in a straight line in the direction of its current velocity, the solid arrow in Figure 10-5. It would leave Earth and never look back. But because of gravity, at every moment the Moon also falls toward the Earth, just as apples fall from trees.

The Moon's fall toward Earth is indicated by the short solid arrow in the boxed image. If the Moon were ever brought to a stop for an instant, it would thereafter begin falling straight toward Earth. Its impact would vaporize our oceans and melt the planet's surface.

Fortunately, what actually happens is the sum of both arrows — the Moon falls toward Earth, and at the same time, also moves forward with its current velocity. Every one minute, the Moon moves forward 61 km (38 miles) and drops 4.8m (16 feet) toward Earth. A 16-foot drop seems tiny compared with 38 miles, but it is exactly the right amount to move the Moon from its straight-line-to-nowhere trajectory into its orbit around Earth.

For comparison, every one minute, the Earth moves forward 1788 km (1111 miles) along its current velocity and falls 21 m (70 feet) toward the Sun.

By staring at apples, Newton understood the motion of the Earth and Moon. (Ain't physics cool.)

§10.5 Orbits of Two Moving Bodies

When physicists examine the orbits of celestial bodies, the most common situations involve smaller bodies orbiting much more massive bodies, such as Earth orbiting the Sun, the Moon orbiting Earth, or the Sun orbiting the center of the Milky Way Galaxy.

There are, however, some very interesting situations in which the orbiting bodies have comparable masses. In Chapter 34, we will explore gravitational waves produced by two massive bodies orbiting one another in close proximity.

Let's therefore consider the general two-body case: two objects of masses m_1 and m_2 orbiting one another, with no restrictions on the magnitudes of either mass.

In Newtonian physics, the gravitational forces and accelerations between an object of mass m_1 at position $\boldsymbol{r_1}$ and an object of mass m_2 at position $\boldsymbol{r_2}$ are related by:

Body 2 acting on body 1: $F_{12} = m_1 \boldsymbol{a_1} = G\ m_1\ m_2 \ / \ (\boldsymbol{r_2} - \boldsymbol{r_1})^2$

Body 1 acting on body 2: $F_{21} = m_2\ \boldsymbol{a_2} = G\ m_1\ m_2 \ / \ (\boldsymbol{r_1} - \boldsymbol{r_2})^2$

Newton's third law: $F_{21} = -\ F_{12}$

$m_2\ \boldsymbol{a_2} = -\ m_1\ \boldsymbol{a_1}$

$\boldsymbol{a_2} = -\ m_1\ \boldsymbol{a_1} \ / \ m_2$

Define *relative acceleration*: $\boldsymbol{a} = \boldsymbol{a_1} - \boldsymbol{a_2} = \boldsymbol{a_1}\ (\ 1 + m_1 \ / \ m_2\ )$

$\boldsymbol{a} = \boldsymbol{a_1}\ m_1\ (\ m_2 + m_1\ ) \ / \ m_1\ m_2$

$\boldsymbol{a} = \boldsymbol{a_1}\ m_1 \ / \ \mu$

Define *reduced mass*: $\mu = m_1\ m_2 \ / \ (\ m_1 + m_2\ )$

Define total mass M: $M = m_1 + m_2$

Define separation r: $r^2 = (\boldsymbol{r_2} - \boldsymbol{r_1})^2$

$F_{12} = \mu\ \boldsymbol{a} = G\ \mu\ M \ / \ r^2$

The prior equation describes a body of mass μ orbiting a stationary body of mass M in a circular orbit of radius r.

We have thus reduced a two-moving-bodies problem to a much simpler one-moving-body problem.

§10.6 Success & Limitations of Newtonian Gravity

Newton said that gravity is caused by mass alone, and that only mass responds to gravity. Since we now know that light is made of photons that have zero mass, Newton's laws say light is not affected by gravity. That is incorrect.

Newton also said changes in the positions of massive objects are felt instantaneously throughout the universe. For example, as the Sun orbits the galaxy, Earth immediately feels the Sun's pull from its new position. For that to be true, whatever "causes" gravity must travel with infinite velocity, a decidedly unappealing concept. And by the way, what is it exactly that actually "causes" gravity? What is the mechanism by which the Sun reaches out and pulls on Earth across 93 million miles of empty space? No one knew, not even Newton. This mysterious, unseen mechanism is called an *action-at-a-distance*.

Not understanding gravity's mechanism troubled physicists, including Newton. While Newton had explained remarkably well what gravity *does*, he did not explain *how* gravity works. Nevertheless, Newton's theory was a tremendous achievement that remained the flagship of science for more than 200 years.

Newton's equations of gravity are still used in almost all calculations of objects moving within our Solar System, where its predictions are valid to five decimal digits (1 part in 100,000) or better. But when we seek greater precision, or deal with objects much more massive than the Sun, we must turn to Einstein.

Rather than saying Newton's theory is *wrong*, it is might be more precise to say that it is not as comprehensive and does not provide as profound an understanding as does Einstein's theory.

Chapter 11
Einstein's Theory of Gravity

Chapter 10 explored Newton's ground-breaking theory of universal gravity.

Here, we will discover how Einstein's theory rejects all of that.

Einstein said gravity is not a force after all, but the result of the "curved" geometry of our universe. He said all forms of energy, not just mass, cause gravity and that gravity affects all forms of energy, not just mass. Since Einstein previously said mass and energy are equivalent, we should have seen that last part coming. In Einstein's gravity, all action is local, there is no action-at-a-distance. The Sun curves geometry *where it is*, and Earth responds to the geometry *where it is*. Geometry is the mechanism that links the Sun and Earth — curving the geometry in one location affects the geometry everywhere (like sitting on one side of a bed sheet).

Clearly, the concepts of Albert Einstein (1879-1955) and Sir Isaac Newton (1642-1727) are very different, as summarized here.

Contrasting views of gravity from the two greatest physicists in history

Newton's Gravity

- Gravity is a Force
- Its Mechanism is Unknown
- Action at a Distance
- Acts at Infinite Speed
- Gravity is tied to Mass
- Space and Time are:
  — Two Distinct Entities
  — Absolute
  — Unchanging

Einstein's Gravity

- Gravity is Geometry
- Its Mechanism is Curvature
- All Action is Local
- Acts at Light Speed
- Gravity is tied to Energy
- Spacetime is:
  — One United Entity
  — Relative
  — Dynamic

Einstein believed that if we were able to visualized the full multi-dimensional geometry of our universe, we would see that everything moves along the straightest possible path, and no one would ever have bothered to invent the word "gravity".

Without spacetime curvature — if the geometry of the universe were Euclidean, rather than curved — there would be no gravity and all objects would move in straight lines.

Figure 11-1 Earth Moving in Space Curved by the Sun

Figure 11-1 is an *embedding diagram* that shows only two of our universe's four dimensions (three dimensions of space plus time). No one I know can draw, or even imagine, all four dimensions at once. We can only draw part of the full geometry, and hope that conveys the essential concepts.

In Figure 11-1, our universe is represented by only the deformed two-dimensional surface, crisscrossed by white lines. To make them easier to see, the Sun and Earth are shown as balls lying on the surface that is our universe, but it would be more precise to represent them as flat disks entirely within this two-dimensional surface. Anything outside that surface is not part of our universe.

Figure 11-1 shows why Earth orbits the Sun, according to Einstein. The Sun curves the geometry of spacetime, much as a bowling ball placed on a taut bedsheet would depress and stretch the fabric. Like the bedsheet, spacetime curves most near the Sun, but it is also distorted at great distances from the source of curvature; in fact, curvature extends, at an ever-decreasing degree, to the farthest reaches of spacetime (or bedsheet). Earth simply follows the straightest possible path in this curved geometry. It rolls along the curved surface of spacetime, like a marble rolling around the inside of a salad bowl.

Figure 11-2 also illustrates Einstein's concept.

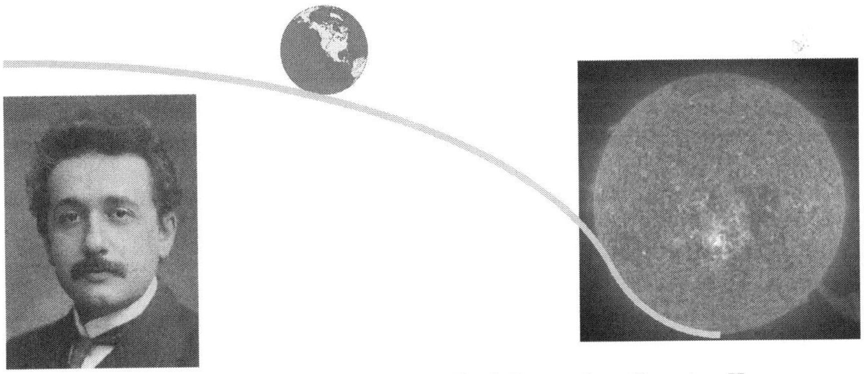

Figure 11-2 Sun Curves Space There, Earth Responds to Curvature Here

Here, spacetime is represented by the one-dimensional curved gray line — the other three dimensions of spacetime are not shown. Again, Sun and Earth are shown as balls sitting on spacetime, but they should properly be represented by disks completely contained within spacetime (the gray line). The Sun curves spacetime where it is, which curves spacetime to an ever-decreasing extent all the way to the end of the universe (if that exists). Earth is affected only by the curvature of spacetime where it is. Thus, Einstein said, there is no *action-at-a-distance* — all actions are *local*.

We might think of our universe bending in a fifth dimension, corresponding to the vertical direction in Figure 11-1. That larger space, with five or more dimensions, is called *hyperspace*.

Digesting the concept of spacetime curvature is the greatest challenge in understanding general relativity. Because no one can truly visualize more than three dimensions, even those proficient with the mathematics of general relativity are not entirely comfortable with spacetime curvature. Our intuition is too deeply rooted in 3-D Euclidean geometry.

Euclidean geometry is what we learned in high school: parallel lines never cross, the interior angles of every triangle sum to 180 degrees, the circumference of a circle is π times the diameter, and all that. Those marvelous theorems are true only on a *flat* surface, like an idealized sheet of paper. These theorems are not true on a curved surface, like a sphere. On Earth's surface, for example, the meridians are parallel at the equator, but intersect at both poles.

As we progress through this book, I will endeavor to lessen your natural discomfort with "spacetime curvature" by providing more examples, illustrations, and explanations.

In Chapter 15, we will begin quantitatively exploring the geometry of curved spacetime and its implications for gravity.

Einstein also said that changes in gravity are really changes in geometry. He called these changes *gravitational waves* that ripple through spacetime like ripples on a pond's surface, except gravitational waves travel at the speed of light.

If the Sun were to vanish this instant, we would continue to see its light and feel its gravity for another 500 seconds, because that is how long it takes light and gravity to travel 150 million km (93 million miles). After 500 seconds, the Sun's light and its gravity would vanish together. Not to worry, there is no known mechanism by which the Sun could instantly vanish.

Einstein also completely changed our understanding of space and time.

Newton thought there could never be any disagreements about the length of a mile or the duration of a second because space and time were absolute and fixed — the same for everyone, everywhere, and always. He also believed space and time were two completely unrelated entities.

Einstein showed that space and time are relative — different observers measure different values for distance and time. He also showed that space and time are intimately united as one entity: *spacetime*. They are really two sides of one coin. When one changes reference frames, some of what was space can become time, and some of what was time can become space. In Einstein's theory of general relativity, spacetime is dynamic — the geometry of spacetime is constantly changing as mass and energy move.

Einstein explained why, ignoring air resistance, all free-moving objects fall at the same rate, as Galileo had discovered. This is because such objects travel through the same curved, four-dimensional geometry, along the same curved paths — paths that have nothing to do with their masses. In fact, light, which has zero mass, also moves along these same curved paths.

Therefore, Einstein said, light bends as it passes near a massive body like the Sun, as our next chapter explores.

Chapter 12
Gravity Bends Starlight

As we discussed in the last chapter, in Newton's theory, gravity is produced by and has effect on mass alone. Thus, light, which has zero mass, should not be affected by gravity. But, in Einstein's theory, gravity (more precisely curved spacetime) is produced by and has effect on all forms of energy, including light. This distinction provides an opportunity to establish which theory best describes nature.

§12.1 Gravity Acts on Light

Consider the situation shown in Figure 12-1.

Normally, the Sun is not close to our line of sight to star A; light from star A then travels to Earth along a straight line, and we observe the star in its actual location.

However, when the Sun moves into our line of sight, light from star A bends as it passes very close to the Sun. At that time, looking back along the light ray that reaches us, the light appears to come from location A* instead of A, and we see star A appearing to be closer to star B than it actually is.

According to Einstein, starlight that passes very close to the Sun, just grazing its surface, bends by 1/2000th of one degree. Light passing twice as far from the Sun's center bends half as much. We will calculate this bending angle in Section §24.4.

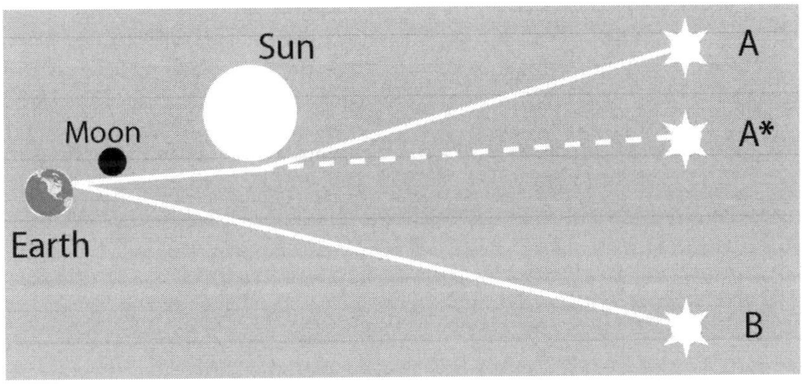

Figure 12-1 Light from Star A Bends due to Sun's Mass

§12.2 Eddington Validates General Relativity

When Einstein published general relativity in 1915, he predicted the bending angle of starlight passing the Sun. Even though the angle seems quite small, it was within the measurement capabilities of skillful and meticulous astronomers. There was, however, one major challenge: the measurement could be done only during a total solar eclipse. Stars near the Sun are visible only when the Moon blocks the Sun's intense light.

Total solar eclipses are spectacular as I can personally attest. Unfortunately, they are infrequent, brief, and localized. They occur somewhere on Earth about once every 18 months, often in inaccessible places. Totality lasts only minutes, and covers only a tiny sliver of Earth's surface. At any specific location, a total solar eclipse occurs only once every 370 years.

Astronomers have only a brief moment to capture an eclipse, and little chance to overcome mishaps or bad luck. If the sky is cloudy, the opportunity is lost.

The solar eclipse of 1919 raced across the South Atlantic. British astronomer Sir Arthur Eddington made an extraordinary effort to get the most from this less than ideal opportunity to test Einstein's revolutionary theory.

Eddington was a Quaker and a conscientious objector who refused military conscription, which was not well accepted in England during World War I. Only with the aid of powerful friends was he able to avoid prison and continue his research.

Eddington led eclipse expeditions to both sides of the South Atlantic, to Brazil and to the island of Principe near Africa. During the eclipse, his teams feverishly took picture after picture, through partially cloudy skies, hoping to get at least one good image.

Eddington succeeded, but just barely.

One of Eddington's photographs convincingly showed a star whose position had shifted from its normal location, by just the slight amount Einstein had predicted. Figure 12-2 shows Eddington and his famous photograph proving the bending of starlight (arrow points to the shifted star).

Figure 12-2 Eddington and Eclipse Photograph Confirming Bending of Starlight.

§12.3 Who's Third?

Standing under an immense portrait of Sir Isaac Newton, Eddington formally presented his findings to a meeting of Britain's Royal Society. His efforts had produced one definitive photograph. Eddington declared that Einstein's prediction of the bending of starlight was confirmed, and that Newton's theory of gravity must give way to Einstein's theory of general relativity.

After his presentation, so the story goes, someone asked Eddington if it were true that only three people in the world understood general relativity. After a lengthy pause, he was asked why he refused to answer the question. Eddington replied: "I'm just trying to think who the third person is."

It is a tribute to international scientific cooperation that near the end of the catastrophe that was World War I, a British scientist devoted so much effort to test a radical alternative to Newton's theory of gravity that was proposed by a German scientist.

Immediately following Eddington announcement, the success of Einstein's theory was proclaimed on the front page of major newspapers worldwide. He was heralded as the greatest genius of his time, and almost instantly became an international celebrity, the Elvis Presley of his day. Einstein became more famous than any scientist has ever been, before or since.

Once again, a better understanding of gravity ushered in a new age of science.

§12.4 Right Moon

It is also worth noting how lucky we are that Eddington's experiment was even possible. As seen from Earth, our Sun's angular diameter (how much of the sky it covers) is about one-half degree (0.5338°). By pure coincidence, the angular diameter of the Moon, as seen from Earth, is also about one-half degree (0.5286°). The Moon's elliptical orbit causes its distance from Earth and its observed angular diameter to vary by ± 12%. Earth's orbit around the Sun is more circular; the Sun's distance and its observed angular diameter vary by ± 1.7%. Even with these variations, the Sun and our Moon are very closely matched in apparent size as seen from Earth.

Earth's scientists can therefore observe total solar eclipses with unobstructed views of stars appearing near the Sun, making Eddington's experiment possible.

Astronomers have found nearly 200 moons in our Solar System. From my own analysis of the available data, no other moon provides solar eclipses on its host planet that are nearly as useful as does Earth's moon.

Without this remarkable coincidence, general relativity would have been substantially harder to prove, thereby significantly retarding the advancement of science.

Chapter 13
Equivalence Principle

Einstein's theory of special relativity applies only to those *special* situations in which everything moves with a constant velocity — with no accelerations, and therefore no forces. Despite the great success of this theory, there was clearly a need to extend it to include forces and, most importantly, to incorporate gravity.

Einstein succeeded in extending special relativity, publishing his theory of general relativity in 1915. General relativity applies *generally* — to all states of motion, with or without forces, and does incorporate gravity.

We believe general relativity is valid in all but two circumstances. It fails only where all our current physical theories fail: (1) at the center of black holes, and (2) at the instant the universe began. In these two circumstances, the mathematics on which all our physical theories depend may not be applicable. There, space and time may not be continuous and smooth, as calculus requires.

If we exclude, for now, discussion of these two extremes, all available evidence shows Einstein's theory of general relativity is valid for any and all circumstances.

§13.1 Equivalence Principle

Legend has it that in 1907, a clerk in the Swiss patent office gazed out his window and saw workmen on the roof of a tall building on the opposite side of Bern's main square. The patent clerk wondered what a person might feel during their descent, if they fell from that great height. (Let's avoid considering how the descent ends.) Einstein concluded a free-falling person would feel no gravity at all.

Einstein also conceived his famous elevator thought experiment. My version is: a physicist in an elevator at the 100th floor is serenely reading a physics book, when suddenly a cable breaks, dropping the elevator. While this poor soul haplessly falls toward an unfortunate sudden stop, he feels and sees no evidence of gravity. Naturally, our physicist is eager to verify this with several (quick) experiments. He drops his keys, and observes them hang motionless in midair. Then he places his keys on the ceiling, and sees that they remain there. And then he... (oops). Einstein again concluded that a free-falling person feels no gravity at all.

Einstein called this realization his "happiest moment" (not the "oops" part). To Einstein, this insight meant free fall is the simplest and most natural state of motion. It is the floor that makes us deal with gravity — by preventing us from falling, the floor makes physics seem more complicated than it really is.

Einstein extended this realization, forming the fundamental principle of general relativity. Just as the constancy of the speed of light is the fundamental principle of special relativity, the equivalence principle plays that role in general relativity.

Nearly everyone today has seen "weightless" astronauts floating in air within the International Space Station. To us, it is no great surprise that a person in free fall feels no gravity. But in 1907, Einstein's realization was pure genius and completely revolutionary. It led him to proclaim:

<u>The Equivalence Principle</u>

The effects of a uniform gravitational field and

the effects of a uniform acceleration are *indistinguishable*.

Let's discover what the equivalence principle means and why it is so important.

Firstly, a gravitational field from a massive body, such as Earth, is never perfectly uniform — its acceleration is different in different places. The acceleration points "down", but "down" points in different directions in London, Vancouver, and Sydney. The strength of gravity also varies depending on the distance to Earth's center, being slightly less atop Mt. Everest than on the Dead Sea.

The change in Earth's gravitational acceleration over a distance r is generally proportional to r^2 / R^2, where R is Earth's radius. Therefore, gravitational non-uniformity decreases very rapidly as r becomes much less than R.

Locally, within a small enough region ($r \ll R$), the acceleration of Earth's gravity is approximately uniform. *Locally*, the equivalence principle can be employed, and gravity can be replaced by a uniform, constant acceleration without altering the motions of any objects, the outcome of any interactions, or the results of any measurements.

Figure 13-1 illustrates the equivalence principle using the example of a cannon firing a cannonball in four different scenarios. Each image shows the cannonball at four equally spaced moments in time. In each image, the horizontal dotted line indicates the height of the cannon barrel at the instant it fires.

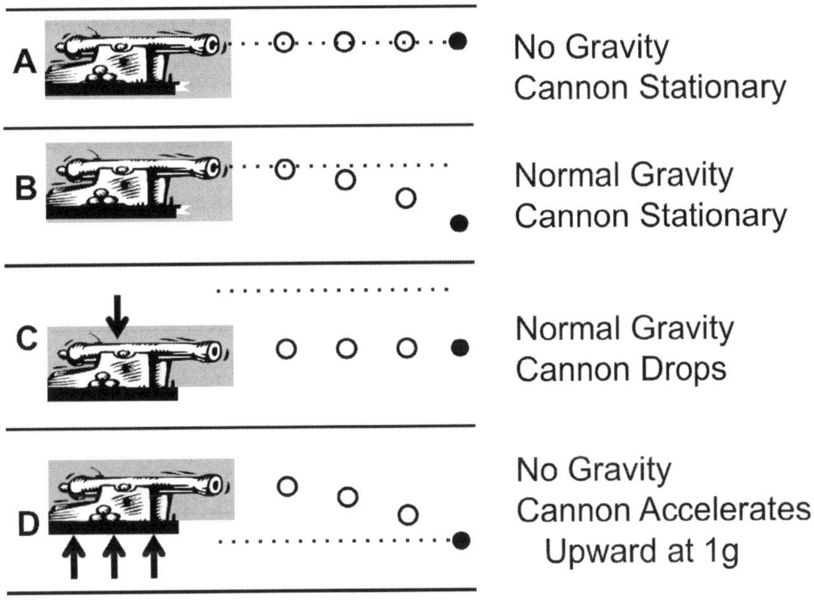

Figure 13-1 Equivalence Principle: Cannon Fires in Four Scenarios

Image A assumes zero gravity; the cannonball travels on a straight horizontal line at constant velocity. Image B shows the cannonball in normal gravity; its trajectory is a parabola, the combination of moving horizontally at constant velocity and dropping at an accelerating velocity.

Next, imagine that at the instant the cannon fires, it drops and falls freely in normal gravity. Image C shows this from the "cannon's point of view": since the cannon and the cannonball fall at the same rate, the cannon "sees" the cannonball moving entirely horizontally at a constant velocity.

Lastly, we switch gravity off and strap jets to the cannon's base. At the instant the cannon fires, we launch it upward with an acceleration of 1 g. Without gravity, the cannonball moves entirely horizontally at a constant velocity. But from the "cannon's point of view", shown in image D, the upward accelerating cannon "sees" the cannonball moving downward with acceleration –1 g.

So, after the smoke clears, what does all this mean?

The cannonball trajectories in images A and C are the same: from the cannon's viewpoint, the cannonball moves entirely horizontally at constant velocity. Thus, physics is the same in a free-falling reference frame (C) as it is in a zero-gravity reference frame (A).

The cannonball trajectories in images B and D are also the same: from the cannon's viewpoint, the cannonball falls downward along a parabola. Thus, physics is the same in a zero-gravity reference frame accelerating upward at 1 g (D) as it is in a stationary, normal-gravity reference frame (B).

All this demonstrates the equivalence principle: motion in uniform gravity is indistinguishable from motion with uniform acceleration. Within a closed elevator, it is impossible to know whether you are in a free-falling elevator in normal gravity, or in a stationary elevator in zero gravity. It is similarly impossible to know whether you are in a normal elevator in normal gravity, or in an upward-accelerating elevator in zero gravity.

The equivalence principle is a powerful tool; it can cut through the myriad mathematical Gordian Knots of general relativity. It says, in any geometry however complex, locally there are free-falling reference frames in which all the effects of gravity vanish. The paths of these zero-gravity reference frames are called *geodesics*. Along any geodesic, at least *locally*, nature's laws are completely described by special relativity in flat spacetime.

Einstein's remarkable insight greatly simplified the development of general relativity. With the equivalence principle, it took him "only" another eight years to complete his theory.

§13.2 Praise for General Relativity

Einstein's theory of general relativity is the bedrock of our understanding of the universe. All modern cosmology begins with general relativity. It opens our eyes to the expansion of the universe, black holes, exploding stars, and many other cosmic phenomena. General relativity is as close to being the work of just one single person as any concept in science has ever been.

General relativity is considered by almost all physicists to be the most profound and most beautiful physical theory.

Brian Greene, author of ***The Elegant Universe***, eloquently called it: "The choreography of the cosmic dance of space, time, matter and energy." Paul Dirac said general relativity was probably the greatest scientific discovery ever made. Max Born, J. J. Thomson, and Sir Arthur Eddington all went even further and said general relativity was perhaps the greatest achievement of human thought. All these statements are from experts on good science. Thomson discovered the electron and received the 1906 Nobel Prize. Dirac received the 1933 Nobel Prize and Born received the 1954 Nobel Prize, for their contributions to the development of quantum mechanics. Eddington received the Royal Society Royal Medal, the Royal Astronomical Society Gold Medal, and a knighthood for his contributions to astronomy and astrophysics.

In my opinion, general relativity is a pinnacle achievement of human culture to be cherished along with our greatest achievements in music, literature, art, medicine, and other fields of intellectual endeavor.

Einstein was once asked why he had been able to make great discoveries that others had not. He replied, "It's not that I'm so smart, it's just that I stay with problems longer." Einstein was certainly wrong about the first part, but right about the second. He worked on some problems, almost continuously, for a decade or more. Such intensive thought is truly remarkable, particularly in our age, when for many, instant gratification can't come soon enough.

Conceptually, general relativity is extraordinarily elegant, but its mathematics is extremely difficult. In nearly 100 years, less than a handful of exact solutions to Einstein's equations of general relativity have been discovered.

In later chapters, we will discuss a simple approach to understanding that math. We will also explore the two most important solutions of Einstein's equations of general relativity, and find out what these tell us about the expansion of the universe and the curved spacetime near a massive body, such a black hole.

Chapter 14
Gravity's 3 R's

Gravity seems simple. But, there are actually four different gravitational effects that vary quite differently with distance. These effects are:

- Normal acceleration of gravity
- Tidal forces of gravity
- Time dilation due to gravity
- Gravitational waves

We explore gravitational waves in chapters 33 and 34. In this chapter, we explore the other three effects.

§14.1 Natural Units

Throughout this chapter, I will simplify the equations by employing natural units, in which c = G = 1, where c is the speed of light and G is Newton's constant. How can we get rid of constants that seem to appear in every equation? By realizing why they are there in the first place.

Most physical constants have units: c is often said to be 3×10^8 meters per second. But why use meters? Why use seconds? Only because someone else used them before, and enshrined them as "standards". Using conventional units, terms, and definitions facilitates communication with other people, which is a good thing. But we should also realize that meters and seconds have no significance in nature — they are just arbitrary choices made by human beings. If we made a different arbitrary choice, c would have a different value, such as 200 trillion fathoms per fortnight.

The fact that these constants have bizarre numerical values is nature's subtle hint that we did not make the most *natural* choice.

We can, for example, choose to measure time in years and distance in light-years (1 light-year is 9.46 trillion km or 5.88 trillion miles). Then c is 1 light-year / 1 year = 1. Similarly, the acceleration a of gravity at a distance r from a body of mass M is usually written: $a = G\,M / r^2$. If we use meters, kilograms, and seconds, G is the unappealing number 6.674×10^{-11} m^3 kg^{-1} s^{-2}. We may have already chosen units for distance and time, but we still have options for measuring mass. Choosing an appropriate unit of mass makes G = 1.

If we go *au naturel*, there is no need to remember a lot of bizarre numbers, our equations will be less cluttered, and we will be able to focus more attention on what they really mean.

§14.2 Mass Matters, Not Size

The normal force of gravity between two bodies, A and B, is proportional to the product of the masses of each body divided by the square of r, the distance between their centers, as in:

$$F = M_A\,M_B / r^2$$

This equation assumes the sizes of A and B are both very much smaller than r.

Newton proved a marvelous theorem: for two spherically symmetric bodies, the gravitational force that each exerts upon the other is independent of how their mass is distributed. My proof of this theorem is in Section §14.6.

We may therefore assume that any such body has all its mass concentrated at a single point. Since large macroscopic objects are almost always spherically symmetric, or nearly so, this theorem applies to nearly every cosmic body.

Newton's theorem makes calculating the gravitational attraction between A and B quite simple. If both bodies are viewed as single points, every particle in A is at the same distance, and in the same direction, from every particle in B. We need only calculate one value of $\boldsymbol{r}$, the displacement vector from the center of A to the center of the B. We then multiply the mass of A by the mass of B and divide by r^2.

Without this theorem, we would have to calculate the force contribution for every combination of one particle in A and another particle in B. For each particle pair, we would have to calculate their displacement vector $\boldsymbol{r}$, take the product of those two particle masses and divide by r^2. Finally, we would have to vectorially sum the contributions of every particle pair, with each contribution pointing in a different direction. For two small planets, each with 10^{50} particles, and 10^{100} combinations of particle pairs, calculating their gravitational attraction would be impossible, as a practical matter.

Newton's theorem means an object's size is irrelevant and so is its density; only its total mass counts. It turns out that the same theorem is true in general relativity: we may compute the overall acceleration of gravity assuming all of a symmetric body's mass acts at its center.

§14.3 Inside a Shell

Let's discover something remarkable about gravity inside a thin, spherically symmetric shell of any material (like a balloon). Pick any point P inside a hollow shell, and place at P the tips of two slender cones, whose axes are exactly anti-parallel, as shown in Figure 14-1. Extend the cones until they intersect the shell, so that each cone encircles a small patch of the shell.

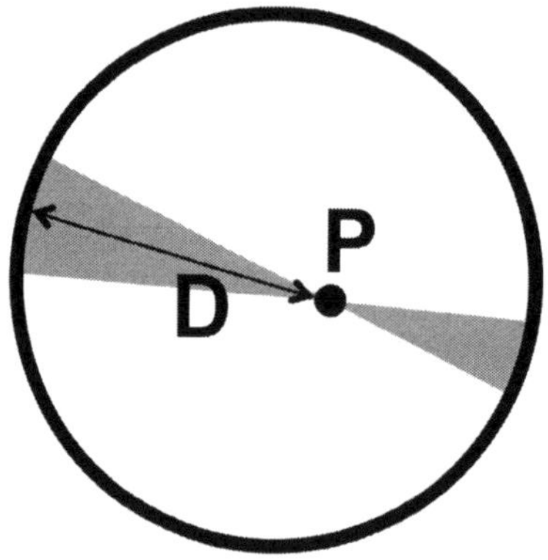

Figure 14-1 Gravity Inside a Shell

Let D be the distance from point P to the left patch. Recall that the shell, unlike the figure, is really three-dimensional. The area, and hence mass, of the left patch is proportional to D^2.

Since the force of gravity is proportional to mass divided by D^2, the distance D cancels — the force from the left patch is independent of D. The same applies to the right patch, so both patches attract equally, and they exactly cancel one another because their forces are anti-parallel. That is true for any pairs of cones oriented in any direction. Summing over cones pointed in all directions, we cover the entire shell and find zero gravitational force at point P. But since P is any arbitrary point within the shell, the gravitational force is zero at every point inside any spherically symmetric shell of matter.

A thick, symmetric shell is just the sum of many thin concentric shells, so the theorem applies to those as well, even if the thin shells have different mass densities.

G. D. Birkhoff proved these special properties of symmetric masses are also valid in general relativity.

Let's next use this result to delve into a symmetric massive body.

Newton's theorem, discussed in Section §14.1, is valid for the gravitational force between symmetric objects that do not overlap, or for the gravitational acceleration *outside* a symmetric object. But interesting things happens when we go inside.

Imagine following Jules Verne on a ***Voyage to the Center of the Earth***.

As we descend and reach a distance r from Earth's center, we can imagine dividing Earth into two parts: the solid ball beneath us comprising all of Earth's mass that lies at radius < r; and a hollow shell above us comprising all of its mass that lies at radius > r.

We just proved that the hollow shell surrounding us exerts no gravitational force on us, regardless of how thick and massive it may be. However, we are still attracted by the force of gravity of the ball beneath us. But its mass is less than Earth's total mass.

If Earth had a uniform density throughout (it actually does not), the mass attracting us at distance r from the center would be proportional to r^3, while its force strength is proportional to $1/r^2$ as usual. The net effect is that the attractive force is proportional to r, which drops to zero at the center.

At Earth's center, the pressure is immense, but there is no gravity. This is also true for the Sun and any other symmetric massive object.

We can also prove this with a symmetry argument. At its center, Earth looks the same in every direction. There is no special direction in which gravity should point, so it must be zero.

§14.4 Tidal Force of Gravity

Newton's law of gravity explained a long-mysterious phenomenon: Earth's tides.

Many had suggested tides were due to water being "pulled" by the Moon. But if the oceans' water was pulled to Earth's Moon-facing side, there would be only one high tide each day as the world turns under its distorted oceans. Why then are there two high tides each day?

Newton's law provides the explanation: Earth's tides are due to the *non-uniformity* of the Moon's gravitational field — due to its pull being stronger at some places on Earth's surface than at others.

Figure 14-2 illustrates the Earth-Moon system, with the Moon on the right and Earth's oceans shown as a gray ellipse. For clarity, everything in this image is greatly exaggerated.

First note that water on Earth's Moon-facing side is somewhat closer to the Moon, and is therefore pulled more strongly by it, than water on Earth's opposite side. This is represented by the longer arrow on Earth's right compared with Earth's left.

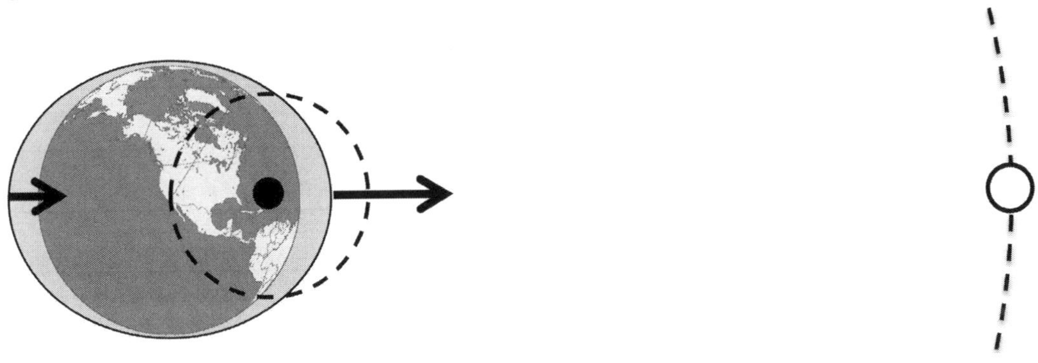

Figure 14-2 Earth, Moon & Tides

While water flows easily, the bulk of our planet is rigid rock; it moves as a single body. The pull from the Moon, averaged over all Earth's rock, is a force whose strength is about the average of the two arrows shown above. That force (not shown in the figure to reduce clutter) effectively acts at Earth's center, pulling Earth toward the Moon.

To sum up: the Moon's pull is greatest on the Moon-facing water, moderate on the rocky Earth, and least on the water on the opposite side. Tides occur because Moon-facing water is pulled away from the rocky planet that is itself pulled away from the water on the opposite side. Thus there are two water bulges, one bulging toward the Moon and one bulging in the opposite direction.

As the Earth turns, coastal sites experience two high tides and two low tides every 24 hours and 50 minutes. If the Moon were stationary relative to Earth, the tidal cycle would repeat every 24 hours, exactly matching Earth's rotation. But since the Moon moves as Earth turns (both in the same direction), each day, it takes an extra 50 minutes for Earth to rotate enough to catch up to the Moon.

Now look at the last element in Figure 14-2: a black dot inside a dashed circle overlaying Earth. Just as Earth exerts a gravitational force on the Moon, the Moon exerts an equal gravitational force on Earth (Newton's third law). Earth must orbit in order to avoid falling directly onto the Moon.

Because Earth is 81 times more massive than our Moon, Earth does not actually go around the Moon, but rather both bodies orbit their common *center of mass* that is located at the black dot in Figure 14-2. The center of mass is on the line connecting the two bodies' centers, but 81 times closer to Earth's center than to the Moon's center. The average distance between centers is 384,400 km, so the center of mass is 4671 km from Earth's center, about 73% of Earth's radius. The dashed circle indicates the orbit of Earth's center that is always at the point farthest from the Moon.

Similarly, the planets, particularly Jupiter, exert their gravity on the Sun, causing it to orbit around the common center of mass of the Solar System that lies just outside the Sun's surface.

So far, our description has been qualitative. Now let's calculate the Moon's forces that cause tides on Earth.

The moon, at the far right in Figure 14-3, exerts a gravitational force F_N on water on the near side of Earth, force F_M on Earth's rocky middle, and force F_F on water on the far side of Earth. Let r be the distance between the centers of Earth and the Moon, and let e be Earth's radius. Note that, at our North Pole, the Moon's gravitational force is pointed slightly downward; the downward component, call it F_S, is given by: $F_S = F_M \sin\phi = F_M\, e / r$, assuming $e \ll r$.

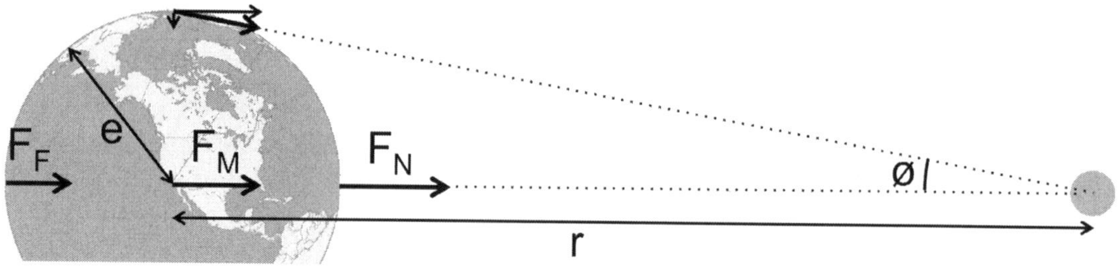

Figure 14-3 Moon's Forces at Various Places on Earth

At the nearest point on Earth, the distance to the Moon's center is r – e. At Earth's farthest point, the distance is r + e.

Let's calculate the Moon's gravitational force on a tiny object of mass m, at the near, middle, and far distances (recall G = 1):

Nearest to Moon: $F_N = m\ M\ /\ (r - e)^2$

Middle of Earth: $F_M = m\ M\ /\ r^2$

Farthest from Moon: $F_F = m\ M\ /\ (r + e)^2$

Next, let's calculate the excess force on the near side relative to the middle:

Near side excess force:

$$F_N - F_M = m\ M\ \{ (r - e)^{-2} - r^{-2} \}$$

$$F_N - F_M = m\ M\ \{ r^2 - (r - e)^2 \}\ /\ \{ r^2 (r - e)^2 \}$$

$$F_N - F_M = m\ M\ \{ r^2 - r^2 + 2\,e\,r - e^2 \}\ /\ \{ r^2 (r - e)^2 \}$$

Continuing the calculation of force on Earth's near side minus the force on Earth's middle:

$$F_N - F_M = m\ M\ \{2\,e\,r - e^2\}\ /\ \{r^2\,(r-e)^2\}$$

Since e << r: $F_N - F_M = m\ M\{2\,e\,r\}\ /\ \{r^4\}$

Near side excess force: $F_N - F_M = +\ 2\ m\ M\ e\ /\ r^3$

The excess force on the far side relative to the middle is the same with – e replaced by + e:

Far side excess force: $F_F - F_M = -\ 2\ m\ M\ e\ /\ r^3$

Thus the near side water is pulled toward the Moon, away from Earth by $F_N - F_M$, while Earth is pulled away from the far side water by that same force.

Lastly, let's calculate the sideways forces at Earth top and bottom, where the tides are lowest.

From the equation just above Figure 14-3, the sideways forces are:

$$F_S = F_M \sin\phi = F_M\, e\, /\, r$$

$$F_S = m\ M\ e\ /\ r^3$$

This is exactly one-half of the excess forces on the near and far sides. But there is the same sideways force at every point along a vertical great circle at Earth's middle, all pushing inward, and thereby equaling the horizontal forces in magnitude. The sideways forces push water away from Earth's middle, complementing the near-side and far-side forces that pull water to those two ends. The net effect is two high tides, and two low tides that maintain their alignment with the Moon, as Earth turns under its oceans.

The Sun also contributes to Earth's tides. The Sun is 27 million times more massive than the Moon, but it is also 389 times farther from Earth, on average. Tidal effects are proportional to M/r^3, and 389 cubed equals 59 million. Thus the Sun's tidal effects on Earth are only 46% of the Moon's, on average. When the Sun and Moon align (when the Moon is at maximum zenith at noon or midnight), they act in unison to produce our greatest tides, *spring tides*. When Sun and Moon are separated in our sky by 90 degrees, their tidal effects are in opposition, producing minimal tides, *neap tides*. Since the orbits of Earth and Moon are not perfectly circular, and all heavenly bodies move, the magnitude of Earth's tides constantly change.

§14.5 Gravitational Time Dilation

The third effect of gravity is *time dilation*. In Chapter 17, we will discuss how gravity slows time, making clocks near massive bodies run slower than those more distant.

NIST, the U.S. National Institute of Standards and Technology, designs and maintains the world's most precise clocks. NIST's scientists have compared the readings of two identical atomic clocks placed at different elevations. They found that a clock one foot (30 cm) closer to Earth's center ran slower by about 30 parts per billion, billion, confirming Einstein's prediction of gravitational time dilation.

Time dilation is a critical issue for GPS, the Global Positioning System, which is now commonly used for navigation in a wide variety of vehicles. If GPS were not corrected for the time dilation effects of relativity, it would have a cumulative error of 7 miles (12 km) per day.

Let's gain some insight into why time runs slower closer to a massive body such as Earth.

Imagine a laser atop the *Leaning Tower* of Pisa firing a pulse of light toward the ground. Because uniform gravity is equivalent to acceleration, Einstein said all forms of energy are affected equally by gravity, including light that has zero mass. Anything that falls in a gravitational field gains energy. A rock falling down a mountainside gains kinetic energy as it moves faster and faster, while losing an equal amount of potential energy. A laser beam must also gain energy as it drops in Earth's gravitational field. Light cannot speed up, its speed is always c; instead it gains energy by increasing its frequency (more wave cycles per second).

To be specific, assume our laser atop the *Leaning Tower* delivers a light pulse of duration 1.00 seconds that contains 500 trillion wave cycles, for a frequency of 500 THz ("tera-Hertz"). For this light to have a higher frequency when it reaches the ground, these 500 trillion wave cycles must occur in less time. If the frequency on the ground is 505 THz, the pulse duration must be 0.99 seconds. To measure a shorter duration, a clock on the ground must run slightly slower than another atop the *Leaning Tower*.

(For ease of discussion, I greatly exaggerated the frequency change. From the 183-foot height of the *Leaning Tower* to the ground, the actual change for 500 THz light would be +3Hz.)

This means time runs slower nearer Earth's center, where the gravitational potential is more negative. Gravitational time dilation is proportional to M/r — as is gravitational potential — where M is the gravitating mass and r is the distance to its center.

Gravity manifests itself in these three very different effects — acceleration, tidal forces, and time dilation — each varying with a different power of distance r — the 3 r's of gravity.

§14.6 Proof of Newton's Theorem

For those who love calculus, here is my proof of Newton's theorem that the gravitational acceleration outside a spherically symmetric body depends on the body's total mass, not on its size or density. This theorem allows us to calculate forces assuming all of the body's mass is concentrated at its center. We will see that the theorem holds even if the body's density varies with radius, provided that it is spherically symmetric, not varying from one direction to another.

Let's calculate the force on an object of mass m at point P, a distance D from the center of a ball of radius R and mass M. Our strategy for this calculation is chopping the ball into onion-ring-like pieces, as follows:

1. Divide the ball into a series of thin concentric hollow shells.
2. Divide each shell into a series of thin rings.
3. Compute the contribution of an individual ring.
4. Sum the contributions of all rings within one shell.
5. Sum the contributions of all shells to find the total force at P.

In a spherically symmetric ball, each shell has a uniform mass density μ (mass per unit volume).

Figure 14-4 shows the cross section of one shell represented by the bold circle, and the cross section of a thin ring, the portion of the shell between the pair of dashed vertical lines. The ring has radius y, and all its particles are virtually at the same distance r from P. The shell has radius b and thickness db. The angle between b and the x-axis is θ. The vector ***r*** points from a particle in the ring to point P.

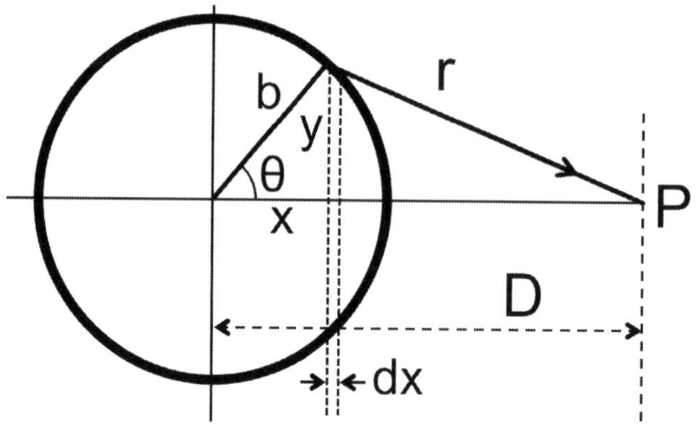

Figure 14-4 One Ring of a Spherical Shell

The ring's horizontal width dx corresponds to an arc length along the shell's circumference of:

$$\text{Arc length} = dx / \sin\theta = dx\ b\ /\ y$$

The parameters of the ring are:

$$\text{Ring circumference} = 2\pi\, y$$

$$\text{Ring surface area} = 2\pi\, y\ (dx\ b / y) = 2\pi\ b\ dx$$

$$\text{Ring volume} = (2\pi\ b\ dx)\ db$$

$$\text{Ring mass} = (2\pi\ b\ dx)\ db\ \mu$$

Symmetry requires the total force at point P to be in the –x direction (there is as much mass at +y as at –y), so all we need sum are the x-components of all force contributions.

Let f be the gravitational force exerted on a tiny body of mass m at point P by the portion of the ring shown in Figure 14-4, which is:

$$f = -\ m\ (2\pi\ \mu\ b\ db\ dx)\ /\ r^2$$

The fraction of f in the x-direction is (the distance between P and the ring along the x-axis) divided by (the total distance r between P and the ring). This is:

$$f\ (D - x)\ /\ r$$

The x-component of the gravitational force from the ring at x is then:

$$f_x = -\,m\ 2\pi\ \mu\ b\ db\ dx\ (D - x)\ /\ r^3$$

This equation has two ring variables, r and x, which is one too many. Noticing that we have two right triangles, b-y-x and r-y-(D–x), we obtain an equation for (D – x) and dx using the Pythagorean theorem.

$$(D - x)^2 + y^2 = r^2$$

$$D^2 - 2\,D\,x + (x^2 + y^2) = r^2$$

Add D^2 to both sides:

$$2D^2 - 2\,D\,x + (b^2) = r^2 + D^2$$

$$2D\ (D - x) = r^2 + D^2 - b^2$$

Now take the x derivative:

$$-\ 2\ D = 2\ r\ dr/dx$$

$$dx = -\ (r\ /\ D)\ dr$$

Use these to eliminate x terms:

$$f_x = -\ m\ 2\pi\ \mu\ b\ db\ (-\,r\,dr\,/\,D)\ (r^2 + D^2 - b^2)\,/\,2\ D\ r^3$$

$$f_x = m\ \pi\ \mu\ b\ (db\ /\ D^2)\ (r^2 + D^2 - b^2)\ dr\ /\ r^2$$

Now integrate over r to sum all rings for a shell of radius b.

$$\int f_x\, dr = m\,\pi\,\mu\, b\ (db / D^2)\ \int \{\, 1 + (D^2 - b^2) / r^2 \}\ dr$$

$$\int f_x\, dr = m\,\pi\,\mu\, b\ (db / D^2)\ \{\, r - (D^2 - b^2) / r \,\}\ \Big|_{rmin}^{rmax}$$

We must evaluate the quantity in { }'s at the maximum and minimum values of r. Note that we can rewrite $(D^2 - b^2)$ as $(D + b)\,(D - b)$.

At its maximum, $r = D + b$:

$$\{\,\} = (D + b) - (D + b)\,(D - b)\,/\,(D + b)$$

$$\{\,\} = (D + b) - (D - b) = +2b$$

At its minimum, $r = D - b$:

$$\{\,\} = (D - b) - (D + b)\,(D - b)\,/\,(D - b)$$

$$\{\,\} = (D - b) - (D + b) = -2b$$

$$\{r - (D^2 - b^2) / r\}\ \Big|_{rmin}^{rmax} = +2b - (-2b) = 4b$$

The total force from a hollow shell of radius b and thickness db is therefore:

$$\int f_x \, dr = m \, \pi \, \mu \, 4 \, b^2 \, db \, / \, D^2$$

The final step is integrating over b, from b = 0 to b = R, summing all shells to obtain total force F.

$$F = (m \, \mu \, / \, D^2) \int_0^R 4\pi \, b^2 \, db$$

$$F = (m \, \mu \, / \, D^2) \, \{ 4\pi \, b^3 \, / \, 3 \} \, \big|_0^R$$

$$F = (m \, \mu \, / \, D^2) \, 4\pi \, R^3 \, / \, 3$$

We recognize $4\pi R^3 / 3$ as the volume of a ball of radius R. Multiplying this volume by density μ yields the ball's mass M. Hence the force of gravity on the mass m at point P at a distance D from a ball of radius R and mass M is:

$$F = M \, m \, / \, D^2$$

The total force from the ball thus depends on its mass M, but not on its radius R. We have therefore shown that the gravitational force *exerted by* a large spherically symmetric body acts as if the body were compressed into a single point. This is true as long as the body's mass is spherically symmetric — its density can vary radially but cannot vary with latitude and longitude.

We can also show that the gravitational force *exerted upon* a large symmetric body acts as if the body were a single point. By Newton's third law, the force exerted by each particle of mass M on the mass m at point P equals the force exerted by m on that same particle of M. Using the same math, the total force F exerted by m on M is independent of M's radius. Hence we may assume M is a single point.

This is true as long as M is rigid and remains spherically symmetric. Earth's tides do depend on Earth's radius, but arise because the oceans are not rigid — as they flow in response to the Moon's gravity, the oceans are not spherically symmetric.

The same conclusions apply to electrostatic forces.

Chapter 15
Spacetime in General Relativity

We now begin exploring curved spacetime, the essence of general relativity.

Because "curved" space and spacetime are such an unintuitive concepts, we will digest them in small morsels. Let's start by carefully reviewing all we have said before about space and time.

Newton viewed space and time as universal reference standards that are unrelated, unchanging, and absolute. The motion and evolution of objects and energy were measured relative to those fixed standards. Any competent person who measured the length of a yardstick or the duration of a day would get the one and only "right" answer.

Einstein said all of that is wrong.

Einstein's theory of special relativity espouses an entirely new concept of space and time. He said space and time are continually changing and are relative — different for different observers. The length of a yardstick or duration of a day does not have just one "right" answer. Each observer's measurements are correct in their own reference frame, while being different from those of other observers in relative motion. Furthermore, Einstein said space and time are intimately related, two different aspects of one combined entity: *spacetime*. Different observers comparing the same two events may measure different combinations of spatial separation versus time change. Einstein showed that the laws of nature are best expressed in 4-dimensional spacetime.

In Newton's vision, space and time are like a Shakespearean stage. They provide unchanging references for the positions and pace of a cosmic drama played out by the actors — mass and energy.

In Einstein's vision, spacetime is like Cirque du Soleil's stage. Spacetime continually changes, influencing actors' motions, and playing an essential role in the cosmic drama.

Einstein's theory of general relativity describes how spacetime changes and constrains the motion and evolution of matter and energy.

§15.1 General Covariance

A foundational principle of general relativity is *general covariance*: the laws of nature must be valid in all coordinate systems and all reference frames. While components of vectors change from one frame to another, the equations that describe nature must not change. Nature is indifferent to our choice of coordinates and reference frames.

While the covariance, or lack thereof, of an equation is obvious in flat 2-dimensional space, covariance becomes more profound and complex in curved 4-dimensional spacetime.

Perhaps surprisingly, quantum mechanics and string theory are *not* covariant; their equations are valid in a spacetime that is static and has zero curvature. While this problem has long-been widely recognized, no one has yet been able to reformulate these theories to make them covariant and coordinate-system-independent.

This is one of the great challenges in merging general relativity and quantum mechanics into a more comprehensive theory of quantum gravity.

§15.2 Metrics Define Geometry

Everything we need to know about spacetime is provided by something quite simple: knowing the distance between nearby points. The distance between all pairs of nearby points completely defines the geometry of spacetime. An equation that provides those distances is called a *metric*.

General relativity provides the foundation for understanding cosmic phenomena: our Solar System, neutron stars, binary pulsars, black holes, galaxies, and our universe as a whole. Different phenomenon may have different impacts on the geometry of spacetime, and thus have different metrics.

The first essential step toward understanding any cosmic phenomenon is discovering its metric.

§15.3 Metrics in Two & Three Dimensions

Simple geometries have the simplest metrics. Let's review Euclidean geometries, starting with two dimensional plane geometry.

We choose a horizontal (x) and a vertical (y) axis, and plot two points, A and B, as shown in Figure 15-1. Metrics focus on distances between nearby points, so we shall employ the customary notation of calculus: "dq" denotes an incremental change in the variable q. However, for geometries with zero curvature, our metric equations are valid for distances of any size. The Pythagorean theorem provides ds, the distance between points A and B, for any pairs of points A and B.

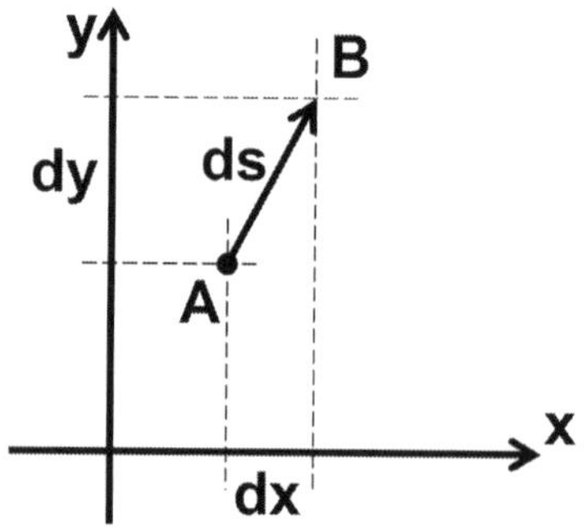

Figure 15.1 Events A and B in (x, y) Coordinates

In Euclidean 2-D (x, y) coordinates, the distance between A and B is given by:

$$ds^2 = dx^2 + dy^2$$

Alternatively, we could choose some other axes (u, v), as shown in Figure 15-2. In two dimensions, any set of (u, v) coordinates is equivalent to rotating (x, y) by an angle ø and adding displacement (u_0, v_0).

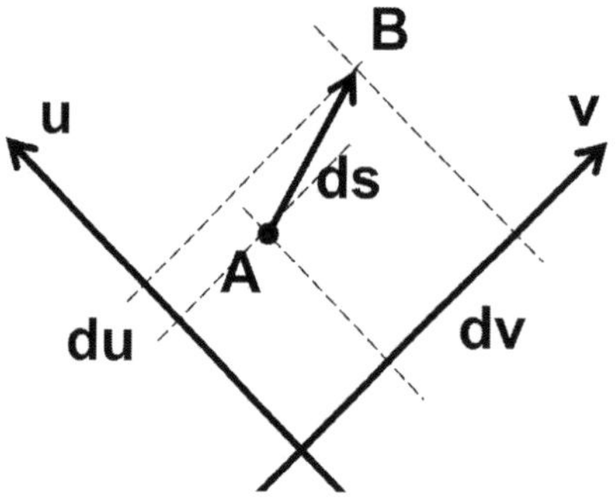

Figure 15-2 Events A and B in (u, v) Coordinates

The distance ds between points A and B, in (u, v) coordinates is:

$$ds^2 = du^2 + dv^2$$

Let's prove that our choice of coordinates has no impact on the distance between A and B. The equations for transforming (x, y) coordinates into (u, v) coordinates are:

$$u = x \cos\phi + y \sin\phi + u_0$$

$$v = y \cos\phi - x \sin\phi + v_0$$

These equations' differentials are:

$$du = dx \cos\phi + dy \sin\phi$$

$$dv = dy \cos\phi - dx \sin\phi$$

The interval ds^2 in (u, v) is:

$$ds^2 = du^2 + dv^2$$

$$ds^2 = dx^2 \cos^2\phi + 2\, dx\, dy \cos\phi \sin\phi + dy^2 \sin^2\phi + dy^2 \cos^2\phi - 2\, dx\, dy \cos\phi \sin\phi + dx^2 \sin^2\phi$$

$$du^2 + dv^2 = ds^2 = dx^2 + dy^2$$

This proves ds is *invariant* under any 2-D coordinate transformation — the AB line length is the same regardless of how we set up our coordinates. While this seems obvious in a simple 2-D example, this critical fact may be less obvious in more complex geometries.

To expand into a 3-D Euclidean space, we need only add a third term, dz^2, to make our distance equation valid for any two points A and B.

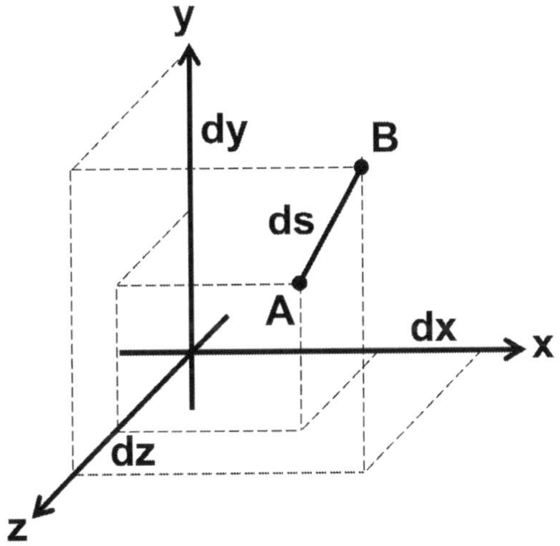

Figure 15-3 Events A and B in (x, y, z) Coordinates

In Euclidean 3-D (x, y, z) coordinates, the distance between A and B is given by:

$$ds^2 = dx^2 + dy^2 + dz^2$$

A summary of metrics in zero-curvature spaces is provided in Section §15.8.

§15.4 Metric in 4-D Spacetime

What about 4-dimensional spacetime? To add more spatial dimensions, we would just need to add mores squares of coordinate differentials. Time, however, is not a spatial coordinate. The requirement of special relativity that the speed of light be the same in all reference frames mandates we *subtract* dt^2.

Figure 15-4 shows two *events* A and B in an initial (x, t) reference frame (left side of figure), and the same events in an (X, T) frame that is moving at high speed in the +x-direction (to the right in this figure). The transformation between these two frames is a Lorentz transformation, often called a "boost". We discussed Lorentz transformations in zero-curvature spacetime in Section §2.1.

Figure 15-4 does not show the y- and z-axes of either reference frame, because nothing changes in those dimensions, and because I cannot draw four dimensions.

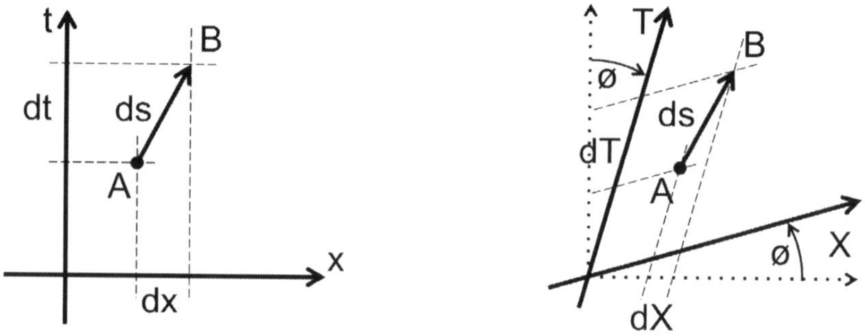

Figure 15-4 Transformation from (x, t) frame to (X, T) frame

In 4-D spacetime, the square of the displacement between two events is called an *interval*. In flat 4-D spacetime, the interval is given by:

$$ds^2 = -\,dt^2 + dx^2 + dy^2 + dz^2$$

$$ds^2 = -\,dT^2 + dX^2 + dY^2 + dZ^2$$

Recall that, in the 2–D rotational transformation in Figure 15-2, both axes rotated by the same amount, and in the same direction. Here, the X and T axes rotate by the same amount, but in opposite directions; both rotate toward the line X = T. With c = 1, the Lorentz transformation from frame (x, t) to a frame (X, T) that is moving in the +x-direction at velocity v is:

$$X = \gamma x - v \gamma t$$

$$T = \gamma t - v \gamma x$$

$$\gamma = 1 \,/\, \sqrt{(1 - v^2)}$$

In the (X, T) frame, ds^2 is:

$$ds^2 = -\,dT^2 + dX^2$$

$$ds^2 = -\,\gamma^2 dt^2 + 2\gamma^2 v\, dx\, dt - \gamma^2 v^2 dx^2 + \gamma^2 dx^2 - 2\gamma^2 v\, dx\, dt + \gamma^2 v^2 dt^2$$

$$ds^2 = -\,\gamma^2 dt^2 (1 - v^2) + \gamma^2 dx^2 (1 - v^2)$$

$$ds^2 = -\,dt^2 + dx^2$$

This proves ds^2 is invariant, the same in all reference frames. Please do not be upset that our coordinate axes are no longer perpendicular. In short order, they will not even be straight — we will be measuring with overcooked spaghetti.

§15.5 Intervals & Causality

In Sections §4.5 and §4.6, we discussed three categories of intervals, and their relationships to causality. Let's briefly review these important concepts. We define:

$$dr^2 = dx^2 + dy^2 + dz^2$$

$$ds^2 = -\,dt^2 + dr^2$$

The interval between two events along the path of a photon is always zero in any reference frame. During a time dt, light moves a distance $dr = 1 \times dt = dt$, so $dr^2 - dt^2$ is always 0 (c=1). Any object whose velocity v is less than 1 has intervals along its path that are all less than zero, because $dr = v\,dt < dt$. If the intervals along a path are greater than zero, $dr = v\,dt > dt$, v would have to exceed the speed of light, which precludes any physical entity or information-carrying signal from traveling that path.

Two events A and B with interval greater than zero are said to be *space-like separated* (their spatial separation dr exceeds their temporal separation dt). For two such events A and B, there exists a reference frame in which they occur at the same time but at different locations.

Two space-like separated events A and B cannot have a causal relationship. Nothing that happens at A can cause an effect at B; and nothing that happens at B can cause an effect at A.

Two events A and B with interval less than zero are *time-like separated* ($dt > dr$). A rocket can travel between them at a constant subluminal velocity v ($v < c = 1$). In the rocket's rest frame, A and B occur at the same spatial coordinates but at different times. If A precedes B in any one reference frame, A will precede B in every reference frame. A is then said to be in the *past-light cone* of B; what happens at A can cause an effect at B. B is then said to be in the *future-light cone* of A; what happens at B cannot cause an effect at A.

Since ds^2 is invariant, events that are time-like separated in one frame are time-like separated in all frames. Similarly, events that are space-like separated in one frame are space-like separated in all frames. Finally, events with interval zero in one frame have interval zero in all frames.

§15.6 What Is Curvature?

Most of us learned Euclidean geometry, also called plane geometry, in high school. We know the Pythagorean theorem, and we know that within a 2-D planar surface:

- Parallel lines **never** cross
- Any triangle's interior angles sum to π radians
- Any circle's (circumference / diameter) = π

None of this is true in curved (non-Euclidean) space.

The simplest non-Euclidean space is a sphere, the 2-D surface of a 3-D ball, such as Earth's surface. On a sphere, the interior angles of every triangle sum to more than π radians (180 degrees), the circumference of every circle is less than π times its diameter, parallel lines can cross, and the Pythagorean theorem is invalid.

Let's see why Euclidean geometry is invalid on Earth's curved surface, which for the sake of simplicity, we will assume is a perfectly spherical surface.

All Earth's meridians (lines of constant longitude) are perpendicular to the equator, and therefore all meridians are parallel to one another there. Yet, they all intersect twice, at both the North and South Poles. Figure 15-5 shows a triangle on Earth's surface that has three 90-degree angles. Here, two meridians meet at a right angle at the North Pole, and each hits the equator at a right angle; the three interior angles sum to $3\pi/2$ radians.

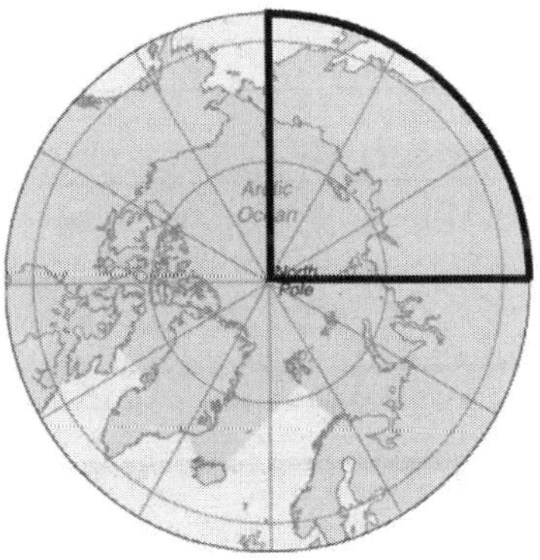

Figure 15-5 Polar View of a Triangle on Earth

Figure 15-6 shows another polar view of Earth. The equator is a circle whose circumference C equals π times a diameter d. In non-curved 3-D space that diameter runs through Earth's core and $C / d = \pi$.

But, in the space of Earth's 2-D curved surface, the equator's diameter must remain with the 2-D space — the extent of its universe. The 2-D diameter runs up along a meridian, across the North Pole, and down the other side, as Figure 15-6 shows. This diameter is precisely half of Earth's circumference, hence $C / d = 2 < \pi$.

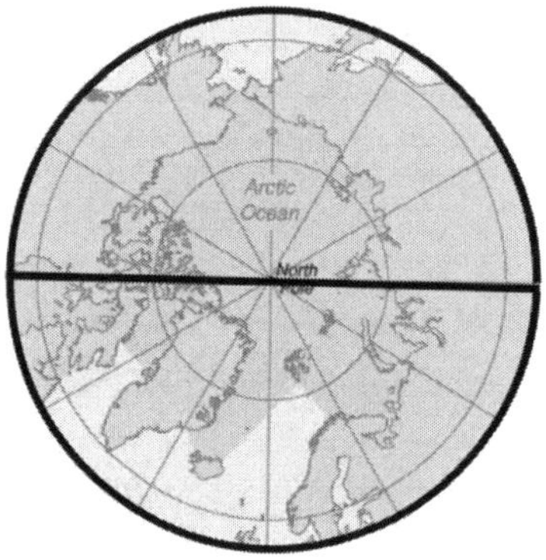

Figure 15-6 Polar View of a Circle around Earth's Equator

Thus, within a spherical 2-D surface:

- Parallel lines **do** cross
- Any triangle's interior angles total **> π** radians
- Any circle's (circumference / diameter) **< π**

In topology, the *radius of curvature* R of any function y(x) is given by:

$$1 / R = d^2y/dx^2 \ / \ ([dy/dx]^2 + 1)^{3/2}$$

A proof of this equation is provided in §15.7.

Topologically, y(x) is a one-dimensional curve, and its *curvature* 1 / R can be positive or negative depending on the polarity of d^2y/dx^2. Two-dimensional surfaces have two radii of curvature. At each point P on a surface, there is one direction along which the radius of curvature has a maximum value; call that radius R_u in the u-direction. While in the orthogonal direction, the radius of curvature has a minimum value; call that radius R_v in the v-direction. In all other directions, the radius of curvature at any point P is between R_u and R_v.

A sphere has the same radius of curvature in every direction at every point. A sphere's radii of curvature are both always inside the surface; hence the product $(R_u R_v)$ is always positive. We therefore say spheres have *positive curvature*.

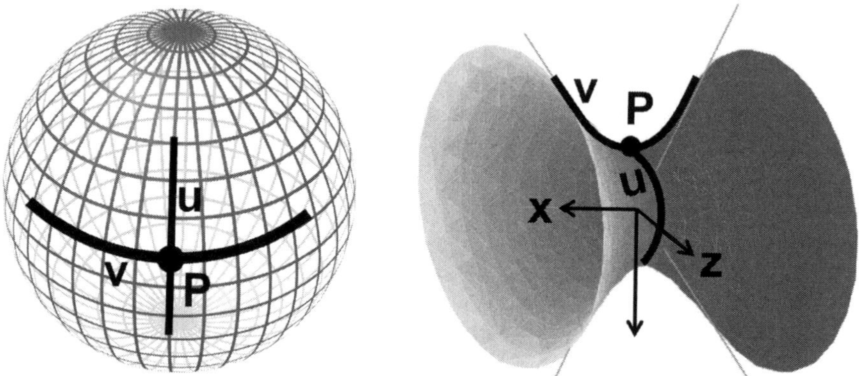

Figure 15-7 Sphere (left) & Hyperboloid (right), latter image by Krishna Vedala

The right hand image in Figure 15-7 shows a hyperboloid with u- and v-curves that intersect at point P and are aligned to the maximum and minimum radii of curvature (u loops around the hyperboloid's neck). This surface is described by:

$$0.25\, y^2 + 0.16\, x^2 - z^2 = 1$$

The center of curvature along u is inside the hyperboloid's surface, while the center of curvature along v is outside — the surface bends inward in the u-direction and outward in the v-direction. This means R_u and R_v have opposite polarity, and ($R_u R_v$) is negative. This means hyperboloids have *negative curvature*. Other familiar surfaces with negative curvature are saddles and potato chips (crisps in the U.K.).

Within a 2-D hyperbolic surface:

- Parallel lines **do** cross
- Any triangle's interior angles total **<** π radians
- Any circle's (circumference / diameter) **>** π

The characteristics of various curvatures are summarized in Section §15.9.

The key point is: the curvature of any space can be determined *within the space itself* from measurements of circles and triangles contained within that space.

Normally, we think of detecting curvature by looking at a surface from the outside. We see a tennis ball is round, and from satellite images we see that Earth is round. But, being able to look from the outside is not the only way to detect curvature. People first discovered that the Earth was round before being able to leave its surface; the ancient Greeks determined Earth's radius from measurements on its surface.

This is very important in cosmology and astrophysics. We want to know the curvature of our universe, and the curvature of space near a black hole. We cannot step outside our universe to see its shape, so it is very fortunate that we can measure its curvature from within.

Imagining that our spacetime bends into another dimension may help visualize its curvature, but this is only a conceptual tool. We really do not know whether or not our spacetime does curve into some exotic dimension, a dimension beyond the four in which we exist (t, x, y, z). The math is *consistent* with spacetime bending in a fifth dimension, but math alone *cannot prove that*.

We know nothing about what might exist in a fifth dimension, but whatever it may be, it lies outside our universe. From a practical viewpoint, we do not need to know the answers to such intriguing metaphysical questions. All we need to know are the properties of circles and triangles everywhere within the 4-D spacetime of our universe.

Everything we need to know about the intricacies of our curved spacetime continuum can be obtained from the distance between points and the properties of triangles and circles.

That's not too scary, is it?

§15.7 Derivation of Radius of Curvature

This Section derives the equation for the radius of curvature R of any function y(x). The equation quoted in Section §15.6 is:

$$1 / R = d^2y/dx^2 \;/\; ([dy/dx]^2 + 1)^{3/2}$$

At any point (x, y) along any physically realistic function y(x), we can approximate y(x) with a Taylor series as:

$$y(x + \Delta x) = y(x) + \Delta x \; dy(x)/dx + \Delta x^2 \; d^2y(x)/dx^2 /2 + \ldots$$

While a Taylor series can continue to infinity, for a small enough value of Δx, we can approximate y(x) with three constants:

y(x), the value of y at x

$s = dy(x)/dx$, the slope of y at x

$a = d^2y(x)/dx^2$, the acceleration of y at x.

We also know that every circle is characterized by three parameters: its radius R, and the X and Y coordinates of its center. With these three parameters, we can match a circle to the three constants listed above. The radius R of the matching circle equals the radius of curvature of function y(x) at x in the limit that Δx goes to zero.

Without loss of generality, let's simplify the math by choosing a coordinate system in which the circle's center is at X = Y = 0. We then have:

$$x^2 + y^2 = R^2$$

$$y = \pm \sqrt{(R^2 - x^2)}$$

The first derivative of y is:

$$s = dy/dx = \pm (R^2 - x^2)^{-1/2} (1/2) (-2x)$$

$$s = \pm (R^2 - x^2)^{-1/2} (-x)$$

The second derivative of y is:

$$a = d^2y/dx^2 = ds/dx$$

$$a = \pm \{ (R^2 - x^2)^{-3/2} (-1/2) (-2x) (-x) + (R^2 - x^2)^{-1/2} (-1) \}$$

$$a = \pm (R^2 - x^2)^{-3/2} (-x^2 - R^2 + x^2)$$

$$a = \pm \{- R^2 / (R^2 - x^2)^{3/2} \}$$

Now use this result to evaluate the denominator of the quoted equation.

$$(1 + s^2)^{3/2} = \{1 + x^2 / (R^2 - x^2) \}^{3/2}$$

$$(1 + s^2)^{3/2} = \{R^2 - x^2 + x^2 \}^{3/2} \;/\; (R^2 - x^2)^{3/2}$$

$$(1 + s^2)^{3/2} = R^3 \;/\; (R^2 - x^2)^{3/2}$$

Finally, evaluate the right side of the quoted equation.

$$a / (1 + s^2)^{3/2} = \pm \{- R^2 / (R^2 - x^2)^{3/2} \} \;/\; \{ R^3 \;/\; (R^2 - x^2)^{3/2} \}$$

$$a / (1 + s^2)^{3/2} = \pm 1 / R$$

$$d^2y/dx^2 \;/\; (1 + [dy/dx]^2)^{3/2} = \pm 1 / R$$

The denominator is always positive, hence the polarity of R is determined by the second derivative of y.

QED

§15.8 Metrics in Zero-Curvature Spaces

In topology, saying a space is "flat" means it is Euclidean — having zero curvature everywhere.

A flat space can have any number of dimensions.

In rectilinear coordinates, the metrics in flat 2–D, in flat 3–D, and in flat 4–D spacetime are:

Flat 2-D: $ds^2 = dx^2 + dy^2$

Flat 3-D: $ds^2 = dx^2 + dy^2 + dz^2$

Flat 4-D spacetime: $ds^2 = -dt^2 + dx^2 + dy^2 + dz^2$

In spherical coordinates, these metrics are:

Flat 2-D: $ds^2 = r^2\,d\theta^2 + r^2 \sin^2\theta\,d\phi^2$

Flat 3-D: $ds^2 = dr^2 + r^2\,d\theta^2 + r^2 \sin^2\theta\,d\phi^2$

Flat 4-D spacetime: $ds^2 = -dt^2 + dr^2 + r^2\,d\theta^2 + r^2 \sin^2\theta\,d\phi^2$

Where:

r is the distance from the coordinate origin
θ is the polar angle ($\theta = 0$ at the North Pole and $\theta = \pi$ at the South Pole)
ϕ is the azimuthal angle (longitude, ranging from 0 to 2π)

§15.9 Summary of Curvature Characteristics

There are three categories of curvature. Plane surfaces have zero curvature. Spheres are defined to have positive curvature. In this context, positive means > 0. That leaves negative curvature, such as the surfaces of a potato chip or a saddle, which curve inwards in one direction and outwards in another.

The three curvature categories, and the properties of triangles and circles in each, are shown below.

	Curvature Categories		
Characteristic	Positive	Zero	Negative
Triangles: Σ Angles	$> \pi$	π radians	$< \pi$
Circle: C / D	$< \pi$	π	$> \pi$
Can Parallel Lines Cross?	yes	no	yes
Examples	sphere	plane	saddle

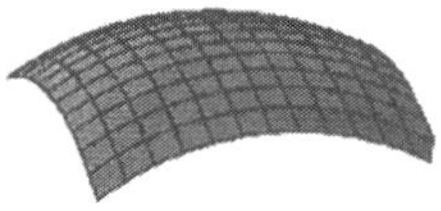
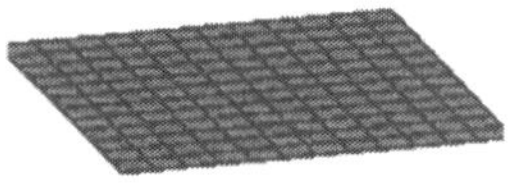
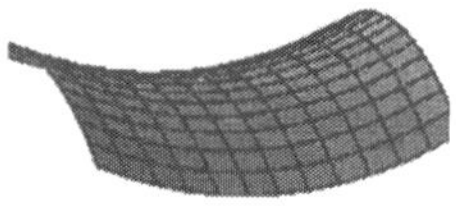

Here, "Σ Angles" denotes the sum of the interior angles of any triangle, "C / D" is the ratio of any circle's circumference C to its diameter D.

General relativity follows the nomenclature of topology by designating "flat" any surface that has zero curvature and is thus Euclidean.

Note that a cylindrical surface is flat. Think of it this way: imagine drawing a triangle and a circle on a flat sheet of paper and then rolling the paper into a cylinder. Rolling does not stretch or tear the paper, nor does it change the angles of the triangle or the circumference or diameter of the circle. Hence, the rolled paper is topologically flat. Euclidean geometry applies equally on a cylindrical surface as it does on a plane surface — both have zero curvature and are flat.

Conversely, a torus, the surface of a donut, is not flat. One might make a torus by taking a cylinder and bending it around and joining its opposite ends, as sketched in Figure 15-8. Because the inside diameter of a torus is necessarily less than its outside diameter, it is very clear that this maneuver requires compressing the inside and the stretching the outside. Pushing some points closer together while pulling others apart changes the geometry, distorting any straight lines, triangles, and circles that were drawn on the original cylinder. The surface is no longer flat.

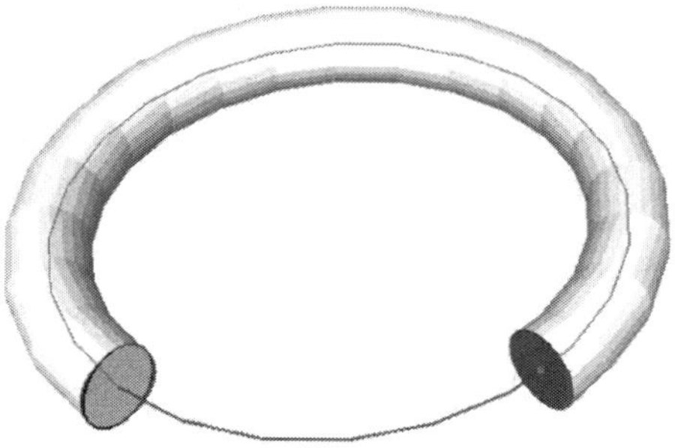

Figure 15-8 Bending a Cylinder into a Torus

Chapter 16
Curvature Controls Motion

Einstein said gravity is not a force between objects, but rather the effect of objects moving through curved spacetime. If spacetime had zero curvature, everything would move in a perfectly straight line and there would be no "gravity".

How does "spacetime curvature" control motion?

§16.1 Terrestrial Motion in Curved Space

It may seem strange, but most of us are already familiar with the shape of space controlling motion.

Consider a flight from Los Angeles to Moscow. The shortest flight path is the arcing black curve shown on a map of Earth in Figure 16-1. This path certainly does not look like a straight line, but it is in fact the straightest possible line across Earth's surface between those two cities.

The flight path looks strange only because this map distorts Earth's actual geometry. On this map, the South Pole has the same width as the equator, which is of course nonsense. This map attempts to use a flat surface to represent a curved surface, which is impossible to do without serious distortion.

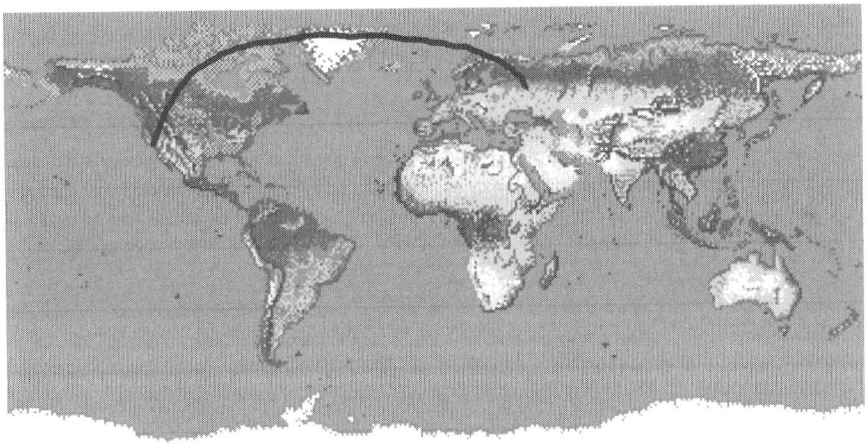

Figure 16-1 Flight Path from Los Angeles to Moscow

When we look at the same flight path in a polar view in Figure 16-2, it makes more sense.

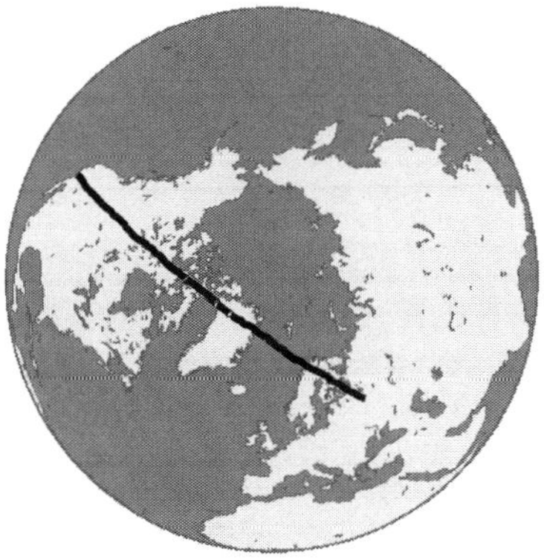

Figure 16-2 Flight Path from Los Angeles to Moscow

The map in Figure 16-2 is a better representation of Earth's far north.

On Earth's surface, the shortest paths are called great circle routes. In the topology of general relativity, the shortest paths are called *geodesics*.

Here's another example: a children's slide. Any object that enters the slide at the top must go round and round in order to go down, because that is the shape of the space.

Figure 16-3 Playground Slide

This same concept, applied to a geometry with many more dimensions than we can perceive, explains why Earth orbits the Sun. For Earth to move forward in time, it must continually circle the Sun because that is the shape of the spacetime of our Solar System, which is dominated by the Sun.

The key point is that motions look simpler when we correctly understand the geometry through which objects move. Newton invented the "Force of Gravity" to explain motions that would need no explanation, if we fully perceived the geometry of the curved spacetime in which we exist.

§16.2 Meet a Friend: 2 M / r

As elegant as the concepts and principles of general relativity are, its mathematics is often extremely difficult. Some equations contain hundreds of terms. When a single equation spans several sheets of paper, it become much more difficult to manage. Relativists therefore go to great lengths to squeeze each equation onto a single page — they use the finest pens, the largest paper, and the densest abbreviations they can.

Fortunately, after mounds of scratch paper filled with intermediate calculations, many important equations finally reduce to something quite simple.

If you can handle 2 M / r, you can understand a great deal about black holes, the expansion of the universe, time dilation, and more. Let's meet a helpful friend: 2 M / r.

Let's consider a rocket preparing to blast off Earth's surface on a mission to deep space. Its first challenge is escaping Earth's gravity. While on the surface, the rocket's gravitational potential energy U is large and negative:

$$U = -M\ m\ /\ r$$

Here, G = 1, M is Earth's mass, m is the rocket's mass, and r is the rocket's distance from Earth's center.

The gravitational potential U, shown as a dotted curve in Figure 16-4, becomes less negative as r increases; U approaches zero as r approaches infinity. To escape Earth's gravitational attraction, the rocket must acquire enough kinetic energy, $T = m\, v^2 / 2$, to bring its total energy, U + T, up to at least zero, the value of U very far from Earth.

The minimum escape velocity, V_{esc}, is the velocity that makes the rocket's total energy equal to zero. Therefore:

$$0 = U + T$$

$$0 = -M\, m / r \quad + \quad m\, V_{esc}^{\,2} / 2$$

$$V_{esc}^{\,2} = 2\, M / r$$

Note that the rocket's mass m is not a factor in V_{esc}; the escape velocity is the same for any mass.

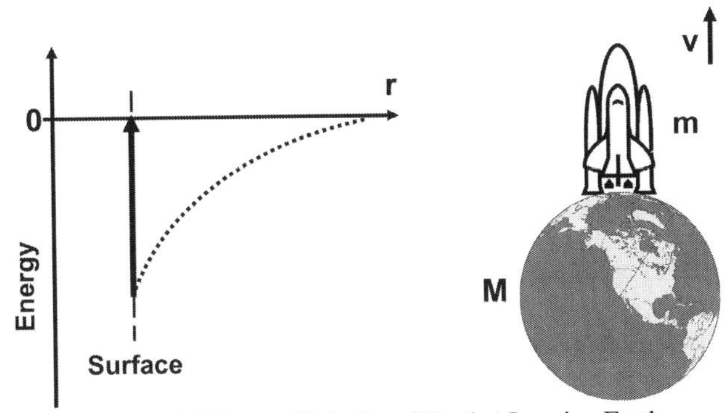

Figure 16-4 Escape Velocity of Rocket Leaving Earth

At sea level on Earth, V_{esc} = 25,300 mph (11.3 km/sec). If Earth were less massive, or if the rocket started farther from Earth's center, V_{esc} would decrease. Recall we are using natural units, with c = G = 1; in other units:

$$V_{esc}^{\,2} = 2\, G\, M / r\, c^2$$

§16.3 Schwarzschild Metric

When Einstein published general relativity in 1915, no one, including Europe's greatest mathematicians, could derive an exact solution of his field equations. Einstein derived some approximate solutions for situations in which gravity was modest, but everyone wanted more.

German physicist and mathematician Karl Schwarzschild (1873-1916) found the first exact solution. Schwarzschild was a child prodigy in mathematics, and later became director of Germany's Astronomical Observatory. He joined the German army in World War I, and became an artillery officer, which was quite extraordinary at that time since Karl was Jewish. Unfortunately, Schwarzschild died on the eastern front — but not before publishing a solution for the metric that will forever bear his name:

$$ds^2 = -\Omega\, dt^2 + dr^2 / \Omega + r^2\, d\theta^2 + r^2 \sin^2\theta\, d\phi^2$$

Where: $\Omega = 1 - 2\, M / r$

With the c's and G's: $\Omega = 1 - 2\, G\, M / r\, c^2$

The Schwarzschild metric is valid *outside* any stationary, *spherically symmetric* body of mass M. The complete derivation of this metric is provided in Appendix 7.

The metric employs spherical coordinates, with r = 0 at the center of M. Polar angle θ ranges from 0 to π radians, and corresponds to latitude in terrestrial coordinates, but with a different range. Azimuthal angle ø ranges from 0 to 2π, and corresponds to longitude, also with a range shift. Finally, r is the *coordinate distance* from the center of M to the point at which the metric is being calculated; r is what the distance would be if there were zero curvature (if Ω equaled 1).

Comparing the Schwarzschild metric with those in Section §15.8, we find curvature changes only the time (dt) and radial distance (dr) terms, which both include the same new factor Ω. The angular terms (dθ and dø) are unaffected by curvature, because we assumed the massive body is spherically symmetric.

The Schwarzschild metric is the most important metric in general relativity.

Normally, r is greater than 2M, and the curvature factor Ω is less than 1. Multiplying the temporal term by Ω decreases the magnitude of the dt term. That is equivalent to making time run slower — this is *gravitational time dilation*. Dividing the radial term by a factor less than 1 increases the magnitude of the dr term. This is equivalent to increasing radial distance — to stretching space radially, as we will discover in the next Section.

§16.4 Curvature of Space

According to Einstein's theory of general relativity, all forms of energy stretch space and dilate time (make clocks run slower).

While no 2-D or 3-D image can accurately portray action in four or more dimensions, Figure 16-5 is an *embedding diagram* that suggests how space stretches around a massive object. The left image shows 2-D space funneling downward into a chasm centered on mass M.

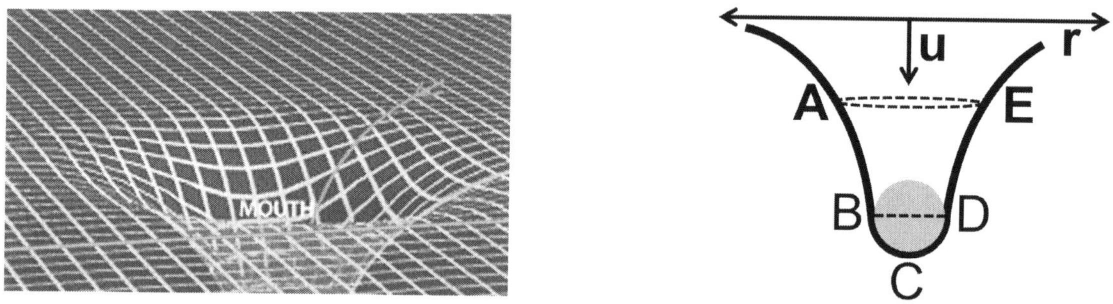

Figure 16-5 Curvature near Massive Body

The right image is a cross section of this curving spacetime (bold black curve) and the gray mass that curves it. Here, r is the radial coordinate, the *coordinate distance* from the center of M, and u is the curvature coordinate. The downward direction u is **not** one of the dimensions of our spacetime; we show it only to suggest curvature. Our entire spacetime is restricted to the 2-D surface represented by the bold curve. The figure assumes the gray body has uniform mass density. For clarity, Figure 16-5 shows the gray mass as a ball filling the "bottom of the well". In reality, that body lies entirely within the curved surface; if shown properly, it would be a gray arc extending from B to C to D.

The dashed circle in the right image, between points A and E, represents the circular orbit of a satellite around the gray mass. Within our 4-D spacetime, the circle's circumference has its "normal" length but its diameter runs from A to B to C to D to E. The stretching of space makes this diameter much greater than the coordinate distance, the "straight" line from A to E that does not lie within our spacetime. Similarly, the gray object's diameter is not the "straight" line from B to D, but rather the length of the arc from B to C to D.

Let's confirm the statement that outside a typical celestial body r >> 2M, where r is the coordinate distance from the center of a body of radius R and mass M. Here are two important examples.

	Earth	Sun
Mass:	4.435 mm	1.4767 km
Radius:	6 371 km	696 392 km
2 M / r:	1.394×10^{-9}	4.221×10^{-6}

It may be startling to see masses expressed as lengths, and such tiny lengths at that. This results from using units in which $c = G = 1$. In more common units, the curvature factor Ω is $(1 - 2\,M\,G\,/\,r\,c^2)$. When we switch to natural units, we rescale masses by the factor G/c^2 that is a very small quantity. The masses of normal cosmic bodies, expressed as distances, are much smaller than their physical sizes. The only exceptions are neutron stars ($2M/r$ can be 0.3) and black holes ($2M/r$ can be 1).

The Schwarzschild metric is an exact solution only outside a symmetric, stationary, massive object. Most celestial bodies are nearly round, however most rotate to some extent. Earth spins once every 24 hours, and our Sun turns once every 28 days (averaged across its surface).

Rotation introduces two effects: a non-spherical mass distribution, and *frame-dragging*.

The first effect is easy to understand. Rotating bodies are never precisely spherically symmetric. Matter along a rotating body's equator has more angular momentum than matter near its poles. This makes the body oblate — its equatorial radius exceeds its polar radius. For example, Earth's polar radius is 6357 km (3950 miles) compared to its equatorial radius of 6378 km (3963 miles), corresponding to an oblateness of 0.3%.

The second effect is pure general relativity: as massive bodies spin, thcy twist spacetime like water swirling down a drain. This is called frame-dragging, requiring the more complex Kerr metric that we will explore in Chapter 29.

Fortunately, many celestial bodies rotate slowly enough that the Schwarzschild metric suffices.

Chapter 17
Gravitational Time Dilation

In the last chapter, we discovered that, in the Schwarzschild metric, the curvature of spacetime affects only time and radial distance. The metric is:

$$ds^2 = -\ \Omega\ dt^2\ +\ dr^2 / \Omega\ +\ r^2\, d\theta^2\ +\ r^2 \sin^2\theta\ d\phi^2$$

Where: $$\Omega = 1 - 2\,M / r$$

Far from black holes and neutron stars, the curvature factor Ω is generally just slightly less than 1.

The metric's radial term, dr^2, is divided by Ω, thereby increasing the magnitude of the radial term, effectively stretching space radially. The curvature factor Ω multiplies the temporal term, dt^2, thereby reducing the effect of time, effectively slowing the passage of time. We say time is *dilated*.

Recall from Section §2.3 that time is also dilated by uniform relative motion, according to special relativity. These two sources of time dilation are independent.

§17.1 Time Dilation in Weak Gravity

Time dilation is generally a more important consequence of spacetime curvature than the stretching of space. This is because most objects have velocities much less than 1, light's speed. Thus changes in their spatial coordinates, such as dr, are much smaller than changes in their time coordinate, dt. Mathematically:

$$dr/dt <= v << 1$$

$$dr << dt$$

Most of the effects we attribute to gravity, such as balls falling from the Leaning Tower of Pisa, are actually primarily due to gravitational time dilation, not to the stretching of space. This clearly does not apply to light.

How large are the effects of time dilation?

Let's consider an ideal clock falling freely in the vicinity of a large body of mass M. "Falling freely" means not subject to any force, subject only to the curvature of spacetime. An inert body orbiting M is at fixed radius r is free-falling. The time measured by a free-falling ideal clock is called *proper time*. In the clock's rest frame, it is stationary and the derivatives of all its spatial coordinates are zero. The equation for the clock's proper time in its own rest frame is:

$$d\tau^2 = -\ ds^2 = +\ dt^2\ (1 - 2\,M / r)$$

$$d\tau = dt\ \sqrt{(1 - 2\,M / r)}$$

In weak gravity, where r >> M, we can employ a useful approximation for the square root.

For x << 1: $$(1 - x)^2 = 1 - 2\,x + x^2 \rightarrow 1 - 2\,x$$

Hence: $$\sqrt{(1 - 2\,x)} \rightarrow 1 - x$$

$$\sqrt{(1 - 2\,M / r)} \rightarrow 1 - M / r$$

$$d\tau = dt\ (1 - M / r)$$

Curvature therefore dilates time by the factor:

$$1 - M / r$$

From Section §16.4, at Earth's surface:

$$M / r = 0.000,000,000,7$$

$$r / M = 1.4 \text{ billion}$$

Our analysis shows that the magnitude of time dilation at Earth's surface due to Earth's mass is only 1 part in 1.4 billion. This means clocks on Earth's surface run slower by 2.25 seconds per century than a zero-gravity clock, a hypothetical clock not subject to gravity from any source.

Tiny as this seems, this is the dominant cause of everything we attribute to Earth's "gravity".

§17.2 Relativity and GPS

The GPS (Global Positioning System) is the most important practical application of general relativity. The number of GPS receivers worldwide in 2021 is estimated to be 6.5 billion, nearly equal to the human population. At least 24 GPS satellites are active, orbiting Earth at an elevation of 20,200 km (12,550 miles). Each GPS satellite is equipped with an atomic clock that tracks time to a precision of seconds per billion years.

Each GPS satellite emits a stream of radio pulses that encode the time and position at which each pulse is emitted. These pulses are detected by GPS receivers in cars, planes, ships, and myriad other applications. GPS receivers compare each pulse's arrival time with its coded emission time. Multiplying that time difference by the speed of light yields the satellite-to-receiver separation. By repeating this procedure on pulses from several satellites, the receiver can calculate its location.

The crucial factor that makes GPS work is a proper understanding of Time.

Recall from the prior Section that clocks on Earth's surface run slower than zero-gravity clocks by 2.25 seconds per century. Since GPS satellites orbit 26,600 km from Earth's center, their orbital radius r is 4.17× greater than r at Earth's surface. The satellite clocks, therefore, run slower by only 0.54 seconds per century (2.25 / 4.17) relative to zero-gravity clocks — they run faster than clocks on Earth's surface. To match our clocks, GPS satellite clocks must be corrected by:

General relativistic correction: – 1.71 seconds per century

Another correction is required due to special relativity. GPS satellites orbit Earth twice each day, at a velocity of 13,280 km/h (8700 mph). According to special relativistic time dilation, their motion causes GPS satellite clocks to run slower than stationary clocks on Earth by the usual γ factor. The numbers are:

$$v = 1.29 \times 10^{-5} c$$

$$\gamma = 1 + 8.32 \times 10^{-11}$$

Special relativistic correction: + 0.26 seconds per century.

The two corrections are in opposite directions: special relativity makes the GPS clocks run slower, while general relativity makes them run faster, relative to clocks on Earth's surface.

While these corrections seem tiny, they actually have a dramatic effect.

Without correcting for general and special relativistic time dilation, GPS would have a cumulative error of 12 km (7 miles) per day. If GPS got you home on Monday night, you would be in the wrong town on Tuesday night, and by Friday, you might be out in the ocean.

Without Einstein's two theories of relativity, GPS would be worthless.

§17.3 Cosmic Time Dilation

The strongest gravity in our Solar System is at the surface of the Sun, where:

$$M / r = 0.000,002,11$$

$$r / M = 473{,}500$$

An imaginary clock on the Sun's surface would lose about 67 seconds per year. The Sun's effect on Earth's clocks is:

$$M / r = 1.4767 \text{ km} \ / \ 149.6 \text{ million km}$$

$$M / r = 1 \text{ in } 100 \text{ million}$$

$$M / r = 31.1 \text{ seconds per century}$$

This effect has no impact on the GPS system, since the GPS satellites and Earth are the same distance from the Sun, on average, and therefore are time dilated by the same amount.

The Andromeda Galaxy is 2.5 million light-years away and 1 trillion times more massive than our Sun. Andromeda slows our clocks by 200 seconds per century, *for now*. Since Andromeda is coming toward us, and is expected to collide with our Milky Way galaxy in about 4 billion years, its gravity will slow our clocks ever more severely. But, time dilation might not be our biggest concern.

Nothing in our universe is so remote that it is unaffected by gravitational time dilation due to other cosmic bodies.

We are all connected by gravity.

Chapter 18
Simpler Black Holes

When stars come to the end of their lives, they die spectacularly. Their remains are nature's most bizarre creations.

§18.1 Life & Death of Stars

Stars illuminate the heavens with the power of nuclear fusion burning within their cores.

Nuclear fusion is nature's ultimate fire, releasing 40 million times more energy per pound than fossil fuels. Fusion is the merging of small nuclei to create larger nuclei.

The dominant fusion reaction in stars is the creation of one helium nucleus from four hydrogen nuclei through several intermediate steps. Since all nuclei have positive electric charge, they repel one another through the electric force. This precludes nuclear fusion in normal conditions. But at the center of a star, conditions are far from normal.

The gravitational attraction of 10^{57} particles on each of the other 10^{57} particles drives stars' core temperatures to millions, or even billions, of degrees. Their extreme kinetic energy at those temperatures allows some nuclei to come extremely close together, close enough to be within the limited range of one another's strong nuclear force, whose attraction overwhelms the repulsive electric force. The weak force also plays a critical role in nuclear fusion by enabling protons in stellar cores to transform into neutrons plus other particles. With a balance of protons and neutrons, the newly-fused, larger nuclei are stable.

The complete hydrogen-to-helium fusion process converts about 0.7% of the original mass into heat and light. By comparison, burning fossil fuel typically converts only a few parts in ten billion of the original mass into heat.

Fusion gives stars long, stable lives, sustaining them for millions of years, or even millions of millions of years. The immense energy released by fusion maintains the pressure required to prevent the gravitational force from continuing to compress the star.

Eventually, however, each star will exhaust its nuclear fuel; its core temperature and pressure then wane. Ever relentless, gravity finally defeats pressure and compresses the stellar core. As the core shrinks, the gravitational attraction increases as the inverse-distance-squared. At some point, the core implodes catastrophically, releasing an unimaginable explosion of energy that blasts away the star's outer layers.

The collapsed cores of lower-mass stars become *white dwarfs*. The cores of more massive stars become *neutron stars*, and the cores of the most mammoth stars become *black holes*.

White dwarfs can be Earth-sized, 100 times smaller (in diameter) than their parent stars, with masses up to 1.4 Msun (Msun is the mass of our Sun). The density of a white dwarf can be 30 tons per teaspoon.

Neutron stars have radii of about 12 km, 10,000 times smaller than their parent stars, with masses of 1.4 to perhaps 3 Msun. The density of a neutron star can be billions of tons per teaspoon.

Black holes can be tens of billions of times more massive than our Sun. Stellar cores with masses of 6 Msun or more can, in 0.0001 seconds, collapse from 12 km across to "virtually nothing", as we will discuss shortly.

Black holes are the ultimate expression of the power of gravity.

Recall that gravitational potential energy is proportional to $-1/r$. Thus, if a collapsing core becomes 100 times smaller, its gravitational potential energy becomes 100 times more negative. Massive stars release more energy during their core collapse than they did in their entire prior lifetime. The core-collapse of the most massive stars create *supernovae*, some of the most violent events in our universe. A single supernova can, for brief periods, release more energy than an entire galaxy of 100 billion stars.

For more on stars, their lives of exquisite balance, and their spectacular deaths, see my ebooks ***We are Stardust*** and ***Black Holes, Supernovae & More*** or my print book ***Our Universe: The Known & The Unknown***.

§18.2 Black Hole Singularity

When a massive star's core collapses to "virtually nothing" exactly how small is "virtually nothing"?

All of a black hole's mass is compressed into an unimaginably small ball called a *singularity*. The singularity is as small as anything can be in our universe, probably a trillion, trillion times smaller than an atom. While no one knows for sure, most physicists believe there is a limit to how small anything can be — that limit is the *Planck length*, 1.6×10^{-35} m, 16 trillionths of a trillionth of a trillionth of a meter. Space and time may be quantized at that scale, similar to the quantization of electron orbits in atoms. If so, this quantization is all that prevents a black hole from collapsing to zero size — the collapse must stop at one quantum if nothing smaller exists.

Where does this tiny Planck length come from? See Section §18.7 for further discussion.

How can so much matter be compressed into such an infinitesimal volume? Quantum mechanics explains that particles of matter have nothing "solid" inside them. Particles are *wave packets*, whose wavelengths become smaller as their energy increases. The tremendous energy liberated by the gravitational collapse of a large star is more than enough to squeeze the wavelengths of all its particles down to virtually nothing. For more on particles, waves, and quantum mechanics see my ebook ***Quantum Mechanics 1: Small Steps with Particle & Waves*** or my print book ***Quantum Mechanics Strange Theory of the Micro-World***.

§18.3 Black Hole Event Horizon

Every black hole singularity is surrounded by an *event horizon*, sketched in Figure 18-1. The event horizon is not a physical object, but rather the limit of what we can see, and also the point of no return. Everything that happens inside the event horizon is beyond the view of everyone outside.

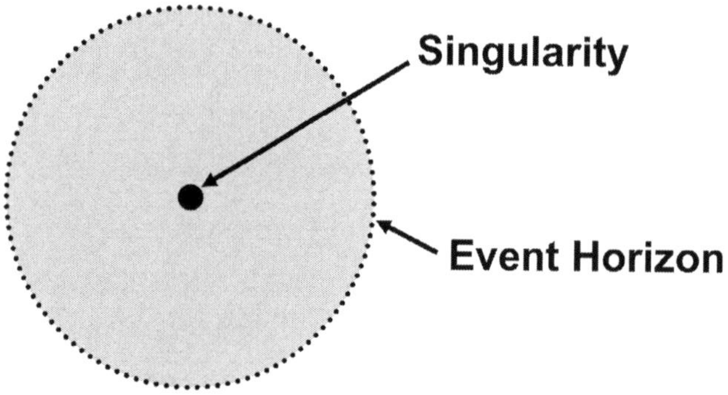

Figure 18-1 Sketch of Singularity and Event Horizon of a Black Hole

The event horizon of a black hole is the *locus* of points at which the escape velocity equals the speed of light, as we will discuss shortly.

On Earth, we are accustomed to a different sort of horizon — we gaze at the ocean and see a "line" separating water and sky. Our horizon is not a material object, but rather the limit of how far we can see.

Earth's horizon is due to the curvature of Earth's surface, and as observers move, the horizons they see move as well. (Heard the joke about a tourist who kept walking toward the horizon to get a closer shot?) A black hole's horizon is due to the gravitational curvature of spacetime; it remains in a fixed position surrounding the singularity. It is best not to get too close for that great photo.

The event horizon is the collection (locus) of all points in space where the escape velocity Vesc = 1. As we derived in Section §16.2, at a distance r from the center of a body of mass M, the minimum velocity an object must have to escape from that body's gravitational attraction, is:

$$\text{Vesc}^2 = 2\,M\,/\,r$$

For a non-rotating black hole, the radius of its event horizon, called the *Schwarzschild radius*, is given by:

Schwarzschild radius: $r = 2\,M$

With the c's and G's: $r = 2\,G\,M\,/\,c^2$

Let's consider an intrepid astronaut who elects to take a spacewalk and explore the inside of a black hole. We will watch from the mother ship, far from danger. Figure 18-2 shows what we observe at three instants in time, as our astronaut approaches the event horizon.

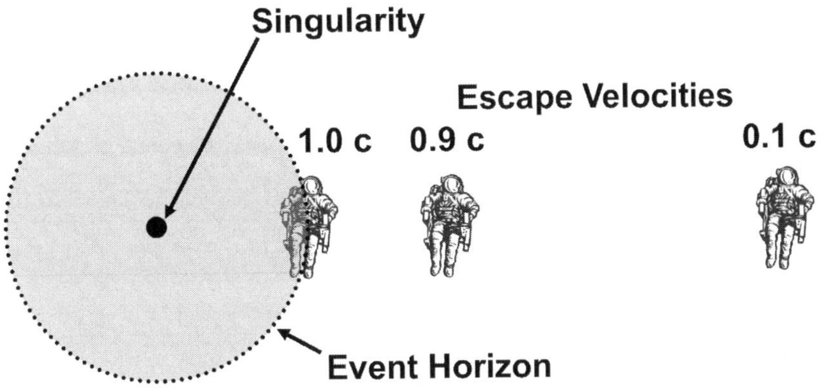

Figure 18-2 Astronaut Entering Event Horizon

Initially, the astronaut's escape velocity might be 0.1 c, 10% of the speed of light. Here, he can escape to safety if his NASA jet pack can propel him at 67 million mph (the manual says that is A-OK). But as the astronaut moves closer, r decreases and his escape velocity increases. At some point, the escape velocity equals 0.9 c, or 604 million mph (this is tempting fate). When the escape velocity is 1.0 c, the astronaut is at the event horizon, with no chance of escape.

Astronauts approaching from different directions would reach different points in space where their escape velocities equal the speed of light. The event horizon is the 3-D collection of all such points. If a black hole is not rotating, its event horizon is a spherical surface surrounding the singularity with a radius of r = 2 M. A black hole with a mass of 10 Msun (10 times the Sun's mass) has an event horizon radius of nearly 30 km (18 miles). The most massive black holes astronomers have found to date have masses up to 60 billion Msun and event horizon radii of 180 billion km, 1200 times the radius of Earth's orbit.

Inside the event horizon, $2\,M\,/\,r > 1$, and the escape velocity exceeds the speed of light. There, nothing can move fast enough to escape, not even light.

Just as everything outside the event horizon must move inexorably forward in time, everything inside the event horizon moves inexorably toward the singularity. Objects inside can no more resist falling into the singularity than we can resist getting older. We can understand this from the Schwarzschild metric: when $(1 - 2\,M\,/r)$ becomes negative, the coefficient of dt^2 switches from negative to positive, and the coefficient of dr^2 switches from positive to negative.

Inside the event horizon, the roles of time and radial distance are effectively swapped.

American physicist John Archibald Wheeler (1911-2008) coined the term *black hole*. He called it "black" because nothing comes out, not even light; hence, it appears absolutely black. He called it a "hole" because anything that enters the event horizon is gone forever, as if it fell out of our universe into an unobservable, bottomless abyss.

It turns out the astronaut in Figure 18-2 is a physicist (who else?) intent on discovering new science, and hoping for fame, glory, and a Nobel Prize.

To gather as much data as possible, the astronaut's spacesuit is equipped with environmental and biometric sensors that continually transmit data back to the mother ship.

Unfortunately, his plan has two key flaws. Firstly, the Nobel Prize is not awarded posthumously. Secondly, as Caltech's eminent black hole expert Professor Kip Thorne said, once our physicist crosses the event horizon, he suffers a fate even worse than death: the inability to publish.

As our astronaut approaches the black hole, he sees himself falling faster and faster into a black void. For him, the event horizon is nothing special, just another place in endless black space. He blows on by without even noticing. If our astronaut turns to look behind him, he might see bright blue light from countless stars that appear concentrated in the direction opposite the singularity. That light is also falling toward the singularity, becoming more energetic (bluer) and more focused as it nears oblivion.

For a black hole of mass 1 million Msun, our astronaut perceives it took him only about 15 seconds to travel from the event horizon to the singularity 2 million miles away.

§18.4 Gravitational Tidal Forces

At some point, our astronaut starts feeling the black hole's growing tidal forces. These gravitational forces are like those from our Moon that cause ocean tides on Earth, except a black hole's tidal forces are super-sized. For very massive black holes, the tidal forces become severe inside the event horizon, while for the least massive black holes, tidal forces become severe well outside the event horizon. This is because, as we discussed in Section§14.4, tidal forces are proportional to M / r^3, and as we see:

Event horizon radius: $r = 2M$

Tidal forces: $F \sim M / r^3$

Tidal forces at event horizon: $F \sim M / (8M^3) = 1 / (8M^2)$

While the tidal forces at the event horizons are greater for less massive black holes, all black holes have large values of M, and therefore enormous tidal forces as r becomes very small.

Since most of a person's mass is in their upper body, tidal forces pull our astronaut's head toward the singularity and pull his feet away from it, stretching him lengthwise. These forces also push in on his left, right, back, and front, squeezing him sideways. Soon he becomes what we would all prefer: tall and thin. Soon thereafter, he becomes *very tall* and *very thin*. Scientists call this is *spaghettification*. His body rapidly disintegrates, and later even his atoms are torn apart.

In the end, his remains reside forever within the incredibly small singularity — we think.

§18.5 End of Time

From our secure vantage point in the mother ship, we see the astronaut's journey evolving quite differently from what he sees.

At first, we see him falling faster and faster toward the black hole, just as he observes. But as he approaches the event horizon, our perceptions diverge from his. At some point, we see his velocity moderate, then begin decreasing, until eventually he is hardly moving at all.

Our perception is different from the astronaut's because time is relative. As gravity in his frame becomes stronger and stronger, we observe his time passing slower and slower. We observe his watch ticking slower, his heart beating slower, and his entire biochemistry becoming slower. As Einstein's theory of general relativity explains, in highly curved spacetime (in strong gravity where 2 M / r becomes a large fraction of 1), time runs slower.

We also see the image of our astronaut getting progressively redder and dimmer, as light reflecting from his body struggles to escape the black hole's immense gravity.

Eventually, we see our astronaut frozen motionless at the event horizon. There time completely stops, nothing ever changes, and even the concept of time is meaningless. We never see the astronaut die, because dying is a change, and change requires the passage of time. Yet, he does not seem alive either, because "life" seems meaningless without change.

His image quickly becomes too red and too dim for us to see. But in some sense, his image is preserved forever on the event horizon along with the image of everything else that ever "fell in".

§18.6 Gravity: Black Hole vs. Star

A common misconception about black holes, encouraged by Hollywood, is that their immense gravity can suck up planets, stars, entire galaxies, or even the entire universe.

While black holes do consume everything that enters their event horizons, it is important to realize just how small event horizons typically are. The spacetime curvature (the strength of gravity) due to any symmetric object is 2 M / r, as Schwarzschild proved. At the same distance r, the same mass M produces the same accelertion regardless of whether it is a black hole, a star, or the fruitcakes of Christmas past.

In Figure 18-3, the wide gray lines represent the acceleration due to the Sun's gravity at various distances from the Sun's center. The narrow black line represents the acceleration of gravity due to a black hole of the same mass at the same distance.

This is a log-log plot, which is very different from a conventional plot. Here each tick mark represents a change by a factor of ten. In a normal plot, the tick marks along each axis correspond to 0, 1, 2, 3 … ; here they correspond to 1, 10, 100, 1000 … . Log-log plots are used to represent phenomena that span an immense range of values.

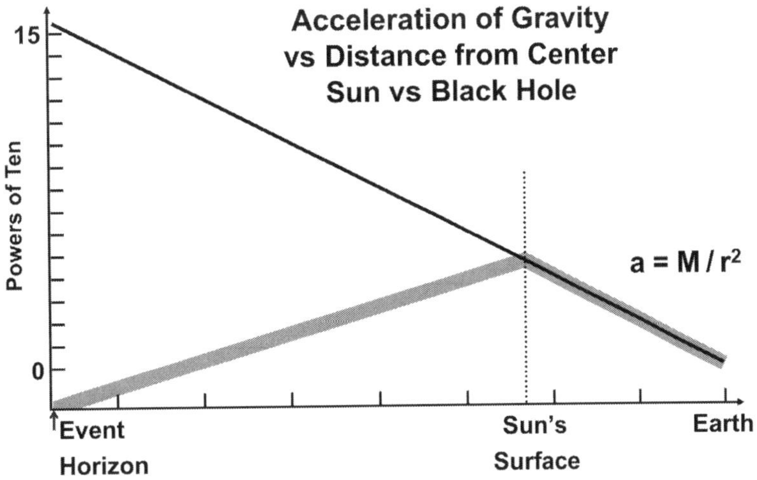

Figure 18-3 Gravity of Sun (Gray Stripe) & Black Hole (Black Line) of Equal Mass

Starting at Earth on the right side of Figure 18-3, the coincident gray and black lines show that the Sun and the black hole cause the same acceleration. Moving left, toward the center of gravity, both accelerations increase in lockstep. The situation changes, however, at the Sun's surface. As we penetrate inside the Sun, its gravitational attraction decreases.

The attraction of the Sun's gravity decreases at points closer and closer to its center because less of the Sun's mass is pulling inward. As discussed in Section §14.3, at a distance r from the center of a massive body, only the mass within r contributes to gravity. Once r is less than the body's radius, the mass inside r drops rapidly as r decreases. At the body's center, r = 0, the acceleration of gravity is zero, because there is the same amount of mass in every direction.

Conversely, the black hole's gravity continues increasing as r decreases, because it is impossible to get inside any of its mass. As Figure 18-3 shows, a black hole's gravity is greater than the Sun's gravity only *within* the Suns radius. At the black hole's event horizon, its gravity is 10 billion times greater than the Sun's surface gravity. But this is only because we can get much closer to the center of a black hole and still be outside all its mass.

It is safer to be 200,000 miles (320,000 km) from the center of a black hole of mass 1 Msun than to be the same distance from the center of our Sun, whose radius is 435,000 miles (700,000 km).

Just as there are stable orbits around stars, there are also stable orbits around black holes. The closest stable orbit around a non-rotating black hole is at r = 6 M, 3 times the radius of its event horizon.

If our Sun instantaneously became a black hole (we know of no way this could happen), Earth would continue to have a stable orbit exactly where it is now, but without sunlight. The closest stable orbit would be only 6 miles from the center of the collapsed Sun, but even there, no heat or light could be detected from our dead star.

For more about the astrophysical and quantum mechanical aspects of black holes see my ebook ***Black Holes, Supernovae & More*** or my print book ***Our Universe: The Known & The Unknown***.

§18.7 Planck Scale

German physicist Max Planck (1858-1947) observed some remarkable consequences of combining the primary constants of nature in various ways. These constants set the scales of the most important physical phenomena: G for gravity, c for relativity, and ħ for quantum mechanics. Their values are:

Gravitational constant:	$G = 6.674{,}1 \times 10^{-11}$ $m^3\, kg^{-1}\, s^{-2}$
Speed of light:	$c = 2.997{,}924{,}58 \times 10^{8}$ $m\, s^{-1}$
Reduced Planck constant:	$\hbar = 1.054{,}571{,}8 \times 10^{-34}$ $kg\, m^2\, s^{-1}$

These constants have awkward units; the last is mass multiplied by distance-squared divided by time. Additionally, all these units are anthropomorphic: seconds are determined by Earth's rotation; meters by Earth's circumference, and kilograms by the density of water. While reasonable for Earthlings, these units have no cosmic significance.

From these primary constants, Planck constructed three other quantities with much simpler units.

Planck length:	$\sqrt{(G\, \hbar / c^3)} = 1.616{,}23 \times 10^{-35}$ meters
Planck time:	$\sqrt{(G\, \hbar / c^5)} = 5.391{,}16 \times 10^{-44}$ seconds
Planck mass:	$\sqrt{(\hbar\, c / G)} = 2.176{,}47 \times 10^{-8}$ kilograms

Planck did not attribute any particular meaning to these bizarre numbers, but one feels they must mean something. G, c, and ħ are fundamental properties of nature, and the above combinations are the only ones that have simple units: distance, time, and mass. Physicists say these are *natural units*, the units of measure that do have cosmic significance.

There is one interpretation that is relevant to black holes.

In quantum mechanics, we discover that each particle is really a wave packet, with a characteristic wavelength λ that is inversely proportional to its momentum p according to:

De Broglie wavelength: $\lambda = h / p$

For a particle whose energy is far greater than its mass, $p \approx E / c$, and $E \approx h c / \lambda$.

Now, let's turn back to general relativity and black holes. From the Schwarzschild metric and Section §18.3, we know that when $r < 2 M$, Ω is negative, and time and radial distance switch roles. Inside the Schwarzschild radius, all objects must move toward the gravitational center, just as outside this radius, all objects must move forward in time. Inside this radius, the escape velocity exceeds c — it is as if everything were on an escalator moving inward at greater than light's speed.

Therefore, whenever a mass M is contained with a ball of radius $r < 2 M$, that mass must implode into a singularity, creating a black hole whose event horizon has radius 2 M. Additionally, since Einstein explained that all forms of energy curve spacetime, we can replace M in the prior statement with E / c^2. Hence, with all the constants, we can write:

$$r = 2 G M / c^2$$

Replacing M with E / c^2: $$r = 2 G E / c^4$$

Replacing E with $h c / \lambda$: $$r = 2 G h c / \lambda c^4$$

Replacing λ with 2 r: $$r = 2 G h / 2 r c^3$$

$$r^2 = G h / c^3$$

$$r = \sqrt{(G h / c^3)}$$

Above, we assumed E is contained within a ball of diameter λ, the wave packet wavelength. Our final result is close to the Planck length, differing by $\sqrt{(2\pi)}$, h instead of $\hbar$.

Our conclusions are:

- Any attempt to resolve a tiny region of radius x requires an amount of energy E inversely proportional to x: $E = h c / 2x$.
- If x is small enough, E is large enough to create a black hole of radius $r = G h / x c^3$.
- If r is greater than x, x is unobservable, shrouded within a black hole event horizon.
- The smallest observable size is, therefore, approximately the Planck length: $x = \sqrt{(G h / c^3)}$.

§18.8 Metric Failure at r = 2M

In Section 16.3, we discovered the Schwarzschild metric:

$$ds^2 = -\Omega \, dt^2 + dr^2 / \Omega + r^2 d\theta^2 + r^2 \sin^2\theta \, d\phi^2$$

Where: $$\Omega = 1 - 2 M / r$$

You may well be wondering: what happens when $r = 2 M$, and we divide dr^2 by zero?

It turns out, this is not a real physical problem, only a mathematical problem that we can easily fix.

This is a failure of this particular metric, but nothing physical goes to infinity at $r = 2 M$.

A clear example that is closer to home is the failure of Earth's latitude-longitude metric at each pole. At the North Pole, there is no east or west, or even north; everywhere one looks is south.

Nothing pathological happens to Earth's surface at the North Pole — it is not much different from another place 100 meters away (which must be 100 meters south). Our usual latitude-longitude metric has failed, but Earth's surface is still gently curved at the North Pole. The only problem is in our math — we need a new means of navigating because our normal coordinate system fails.

At the event horizon of a black hole, the Schwarzschild metric fails — it becomes singular (infinite). But on the event horizon, real spacetime is not singular. There, we just need another metric.

The Kruskal-Szekeres metric is one alternative:

$$ds^2 = -\Psi\ dv^2 + \Psi\ du^2 + r^2\, d\theta^2 + r^2 \sin^2\theta\, d\phi^2$$

Where:

$$\Psi = 32\ M^3\ e^{-r/2M} / r$$

For r <= 2 M:

$$u = e^{r/4M} \sinh(t / 4M) \sqrt{(1 - r/2M)}$$

$$v = e^{r/4M} \cosh(t / 4M) \sqrt{(1 - r/2M)}$$

$$(r/2M - 1)\ e^{r/2M} = u^2 - v^2$$

By mixing the coordinates of time and radial distance, this metric avoids the Schwarzschild metric's identity crisis of these coordinates swapping roles at r = 2 M.

It may not be pretty, but the Kruskal-Szekeres metric is mathematically well behaved on and within the event horizon.

Chapter 19
Our Expanding Universe

The Schwarzschild metric describes the curvature of spacetime around a single, massive, symmetric body. That metric enables our understanding of time dilation, stars, and collapsed stellar cores such as black holes. To understand the universe, the first step is, once again, to find the right metric.

§19.1 Robertson-Walker Metric

Two mathematicians, American Howard Percy Robertson and Englishman Arthur Geoffery Walker, developed a metric for a simple type of universe, one that is the same everywhere. Their philosophy was no location in the universe is special in any sense.

This philosophy, called the *cosmological principle*, is the antithesis of mankind's earliest beliefs. People once believed the geocentric model: Earth is the center of all existence. Later, in the heliocentric model, Earth was replaced by the Sun. Then, people believed the Milky Way galaxy was the entire universe. Thereafter, we discovered countless other galaxies, then clusters of galaxies, and ultimately superclusters of galaxy clusters.

The preeminence of our own location progressively diminished. The cosmological principle is the ultimate extension of that trend: there is no center; no location is intrinsically different from any other. All modern theories of cosmology assume the cosmological principle. Astronomers have confirmed that our universe, when viewed on a large scale, does appear consistent with the cosmological principle, at least to the precision of our current data. Key evidence of this is presented in Section §32.1.

Note that the cosmological principle does not claim the universe is the same at every moment in time, only that if the universe does change over time, the same changes must occur at all locations. Observations clearly demonstrate that the universe has indeed changed substantially throughout its history. Its expansion, decreasing rate of star formation, and increasing abundance of heavy elements are prime examples of these changes.

To be technically precise, the Robertson-Walker metric describes a *homogenous, isotropic* universe. Homogenous means every location has the same properties, such as density, temperature, pressure, and composition. Isotropic means, from each location, the universe looks the same in all directions.

A key requirement of the cosmological principle is that the curvature of the three dimensions of space must be the same at every location. Robertson and Walker achieved this uniformity by extending the perfect symmetry of a 2-D sphere (the surface of a 3-D ball) to a higher dimensional space.

Please try to imagine, because I cannot draw it, a 3-D *hypersphere*, the surface of a 4-D *hyperball*. At every point, a 2-D sphere looks the same in two orthogonal directions (latitude and longitude). Similarly, at every point, a 3-D hypersphere looks the same in three orthogonal directions.

A 3-D hypersphere has positive curvature, just like a 2-D sphere. One can also define a 3-D hypersurface with zero curvature (a hyperplane), or with negative curvature (a hyper-hyperboloid), each having the same curvature everywhere. One might say such 3-D hypersurfaces curve in a fourth spatial dimension that is not part of our universe. Recall the embedding diagram from Chapter 11, repeated below as Figure 19-1. The crosshatched, warped sheet is a 2-D representation of the 3-D space of our Solar System. The vertical dimension in this image is not within our universe; it serves only to help us better comprehend curvature.

We do not know, and do not need to know, whether or not our space actually curves into a fourth spatial dimension — the math is consistent with this, but that is not proof. All we know is that our space is non-Euclidean near massive objects.

The four dimensions of the Robertson-Walker metric are time and the 3 spatial dimensions of the hypersurface.

Figure 19-1 Earth Moving in Space Curved by the Sun

§19.2 Friedman Proves Universe is not Static

Alexander Friedman (1888-1925), a Russian cosmologist, mathematician, and meteorologist, published the second exact solution of Einstein's field equations of general relativity in 1922.

Friedman used the Robertson-Walker metric to derive an equation for the evolution of our universe. He showed that the universe must be either expanding or contracting. He proved a static universe, one whose size never changes, is impossible in Einstein's theory of gravity. (A static finite universe is impossible in Newton's theory of gravity as well.)

Einstein was not happy. He and most other physicists were deeply committed to the notion of a static universe — a universe that always was, and always will be, as it is now.

In 1927, Belgian Catholic priest and cosmologist Georges Lemaître (1894-1966) published the first theory of an ever-expanding universe, what we now call the Big Bang theory. Einstein was appalled, declaring: "Your calculations may be correct, but your physics is atrocious."

In 1929, American astronomer Edwin Hubble (1889-1953) published his observations that all distant galaxies are moving away from us, and therefore the universe really is expanding. That forced Einstein to concede. He declared that his commitment to a static universe was the "greatest blunder" of his career.

From Planck satellite measurements of the Cosmic Microwave Background radiation (CMB), we now also know that the global curvature of our universe is +0.001 ± 0.002, consistent with zero as precisely as we can measure. Spacetime is curved near massive celestial bodies, but in the vast emptiness that surrounds them, spacetime is flat, or very nearly so.

§19.3 FLRW Expansion Equation of Universe

Named after Friedman, Lemaître, Robertson, and Walker, the FLRW expansion equation is:

$$3\,H^2 = 8\,\pi\,\mu$$

Here, H is the Hubble expansion rate, the fractional growth rate of the universe: if the separation between two points increases from 1000 km to 1001 km over the course of 1 million years, H would be 0.1% per million years, or 1 per billion years. Also, μ is the energy density of the universe, its average energy per unit volume. The FLRW equation, derived in Appendix 8, is valid at every moment in time, even as H and μ change with cosmic expansion.

Our universe contains three distinct forms of energy that we will discuss next. But first, note that for a finite universe, the Hubble expansion rate H equals the expansion velocity v at the edge of the universe divided by the radius of the universe r. Additionally, the energy density of the universe equals its total equivalent mass M divided by its volume $4\pi r^3 / 3$.

Putting these factors into the FLRW equation yields:

$$3\,(v / r)^2 = 8\,\pi\,M\ /\ (4\,\pi r^3 / 3)$$

$$v^2 = 2\,M / r$$

Thus, we recover our old friend 2 M / r, and discover that our universe is expanding at its own escape velocity, to the precision of our current data, about ± 0.2%.

This really is the ideal expansion rate. Our universe has expanded rapidly enough to avoid re-collapsing under its own self-gravity, the attraction of each galaxy to every other galaxy. Yet it has expanded slowly enough to allow matter to condense, thereby forming stars, planets, and life.

§19.4 Three Forms of Energy

Our universe contains three distinct forms of energy:

- Radiation — comprising everything moving at light speed (photons), or nearly so (neutrinos).
- Matter (normal and dark) — comprising everything with velocity << light speed.
- Dark Energy — the energy of empty space.

Here is a very brief explanation of key terms.

Normal matter is the matter with which we are all familiar — material comprised of the elements on the Periodic Table, atoms made of protons, neutrons, and electrons, and the subatomic particles that make them. About 5% of the energy of our universe is now in the form of normal matter.

Dark matter is a substantial material entity that is not comprised of normal matter. We know very little about dark matter. It seems completely incapable of interacting through the electromagnetic force; hence, it neither emits nor absorbs light (thus the name "dark"). It does, however, exert and respond to gravity, which is how we know that it exists. About 26.5% of the energy of our universe is now in the form of dark matter.

Both normal and dark matter exert attractive gravitational forces.

Dark Energy is an invisible form of energy that appears to be massless, as is radiation. We say it is the energy of empty space (vacuum), but more precisely, dark energy is the energy that remains in any region of space after all forms of matter and radiation are removed. Quantum mechanics shows that even completely "empty" space is filled with *virtual particles* that appear spontaneously from "nothing" and disappear almost instantly. Dark energy interacts only gravitationally, as does dark matter. But because it is massless, dark energy also exerts pressure in a special way. The net effect is that dark energy exerts a *repulsive* gravitational force. This strange phenomenon is described in more detail in Section §19.7. About 68.5% of the energy of our universe is now in the form of dark energy.

For a thorough exploration of what is known and not known about dark matter and dark energy see my ebooks ***Our Universe 2: Redshift, Expansion & Dark Energy*** and ***Our Universe 3: CMB, Inflation & Dark Matter*** or my print book ***Our Universe: The Known & The Unknown***.

For our present purposes, the most important distinction between the three forms of energy is how their densities change with the size of the universe.

The density of matter, including normal matter and dark matter, varies as $1 / r^3$ — a constant number of particles divided by an increasing volume.

The density of radiation varies as $1 / r^4$. While the number of radiation particles does not change as the universe expands, their wavelengths stretch as space stretches. Hence radiation energy density varies as $1 / r^3$ (particle density) multiplied by $1 / r$ (the energy reduction factor for each particle of radiation). Recall that for light: $E = h / \lambda$, and as space stretches, λ is linearly proportional to r.

Finally, the density of dark energy never changes. Since every cubic meter of empty space is identical, dark energy's density is always a fixed amount.

§19.5 How Energy Drives Expansion

For each type of energy, I list below: its current energy density averaged throughout the universe, how that density varies with r, and how that energy contributes to the expansion of the universe.

Form of Energy	Density kg / meter³	Density change	Universe Growth rate
Dark Energy	7.5×10^{-27}	constant	$r \sim e^{at}$
Dark Matter	2.3×10^{-27}	$1/r^3$	$r \sim t^{2/3}$
Normal Matter	4.7×10^{-28}	$1/r^3$	$r \sim t^{2/3}$
Radiation	8.5×10^{-31}	$1/r^4$	$r \sim t^{1/2}$

For comparison, one hydrogen atom per cubic meter corresponds to a density of 1.67×10^{-27} kg / m³.

The rightmost column in this table shows how each form of energy impacts the expansion rate of our universe, as illustrated in Figure 19-2.

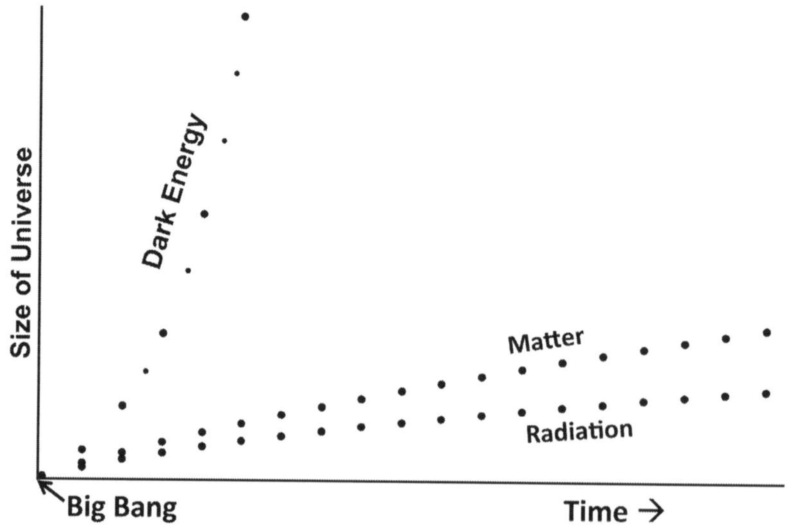

Figure 19-2 Expansion of Universes with One Form of Energy

Here, the three dotted curves show how the size of the universe would vary over time in the limiting cases of all energy being in each of the three forms. Notice how remarkably different is the impact of dark energy compared with matter and radiation.

Let's see how these calculations are done.

For example, if all the energy in the universe were dark energy, μ would be constant and the FLRW equation would be:

$$3\,v^2 / r^2 = 8\,\pi\,\mu$$

$$(dr/dt)^2 / r^2 = 8\,\pi\,\mu / 3$$

Let's define:

$$a^2 = 8\pi\,\mu / 3$$

$$dr / r = a\,dt$$

$$\ln(r) = a\,t$$

$$r = \exp\{\,a\,t\,\} = e^{at}$$

Would you like to try deriving the corresponding equations for $\mu \sim 1/r^3$ and $\mu \sim 1/r^4$?

Go for it. What have you got to lose?

§19.6 Changing Balance of Power

While our universe has always contained all three forms of energy, each form has dominated at different epochs.

In the beginning, when the universe was quite small, and with everything in much less space, dark energy played a minor role.

By about one second after the Big Bang, the annihilation of matter and antimatter was complete. Particles of radiation then became a billion times more numerous than particles of matter, and radiation energy dominated. As the universe expanded, the wavelengths of radiation particles stretched, and radiation energy density decreased with the fourth power of the radius of the universe.

Meanwhile, the energy density of matter decreased with only the third power of the radius of the universe. By about 60,000 years after the Big Bang, the energy density of radiation had fallen below that of matter. Matter then dominated for nearly 8 billion years.

As the universe continued expanding, the density of matter dropped ever lower, as shown in Figure 19-3. Eventually, about 6 billion years ago (6 Gya), the energy density of matter (solid curve) equaled the density of dark energy (dashed horizontal line). From then onward, dark energy has dominated the expansion of the universe. At this time, we know of nothing that might prevent dark energy from dominating the universe forevermore.

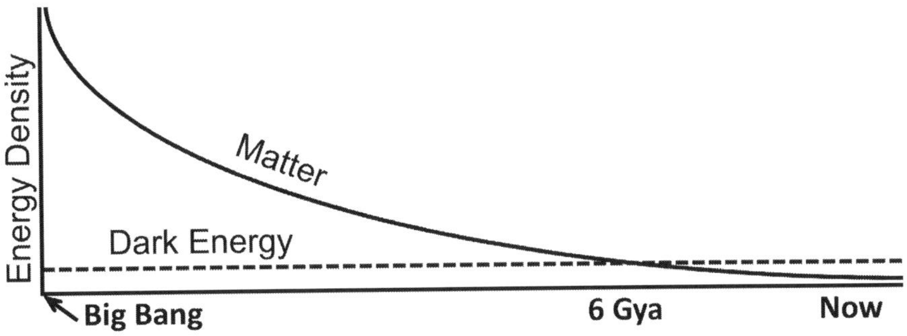

Figure 19-3 Changing Balance of Power

We have discovered a great deal with the simple expression 2M/r.

In following chapters, we will explore Einstein's field equations, delve deeper into the mathematics of general relativity, and see how the most important discoveries of general relativity are derived, including the Schwarzschild metric, the Kerr metric for frame-dragging, the Robertson-Walker metric, the FLRW expansion equation, and much more.

§19.7 Dark Energy & Negative Gravity

Let's now delve more deeply into the nature of dark energy and its impact on gravitation.

We described dark energy as the energy that remains in any region of space after all forms of matter and radiation are removed. Our intuition tells us that when every thing is removed nothing remains, but that is wrong — our intuition does not encompass quantum mechanics, which proves that "nothing" is not really nothing after all.

Quantum mechanics shows that even completely "empty" space is filled with *virtual particles* that appear spontaneously from "nothing" and disappear back into "nothing" almost instantly (in about a trillionth of a trillionth of a second). Virtual particles are a consequence of the Heisenberg uncertainty principle that states in part:

$$\Delta E \, \Delta t >= \hbar / 2$$

Here, ΔE is the uncertainty in the energy of any entity, and Δt is the time interval during which the uncertainty ΔE persists.

These uncertainties are inherent in nature, because every entity in nature is a *wave packet* that has both wave and particle properties. The wave packet of any real entity is a mixture of waves of different frequencies (hence different energy). This precludes an exact determination of its energy in any finite amount of time. In any finite time interval Δt, the energy of this entity does not have one single value, it has values that span a range $\Delta E = \hbar / (2 \, \Delta t)$. Regardless of how precise our instruments may be, we are doomed to measure a random energy value in a range ΔE.

This means energy need not be conserved exactly everywhere at every instant in time. If the total energy in any isolated region changes by an amount ΔE for a time interval Δt, this deviation is unobservable provided the product of ΔE and Δt is no more than $\hbar/2$. The numbers are:

$$\Delta E = \hbar / 2 \, \Delta t$$

$$\Delta E = 3.29 \times 10^{-16} \text{ eV-sec } / \, \Delta t$$

Macroscopically, it appears energy is conserved always and everywhere, for all practical purposes. The above equation allows energy conservation to be violated by less than 3.3×10^{-4} eV for no more than 1 trillionth of a second, which I allege is impossible to detect on a one-off basis. But virtual particles are not one-off, they occur with some probability everywhere and always.

Microscopically, an electron / antielectron pair, with total mass 1.022 MeV can spontaneously appear from "completely empty" space, persist for 3.22×10^{-22} seconds, and then disappear. Figure 19-4 is a Feynman diagram of such an electron / antielectron pair, with the antielectron interacting with a nearby proton during its extraordinarily brief existence. In this diagram, time increases vertically upward, and particle separation is schematically indicated horizontally.

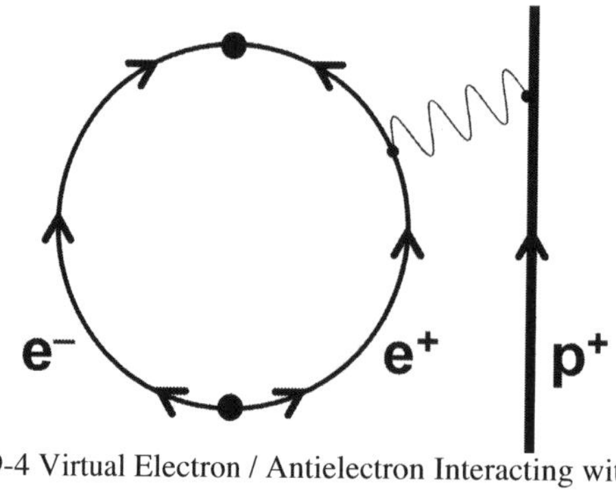

Figure 19-4 Virtual Electron / Antielectron Interacting with Proton

The roster of virtual particles includes not just electron / antielectron pairs. Every type of particle and its antiparticle can form virtual particle pairs in exactly the same way, with the duration of their existence set by their masses and the uncertainty principle.

By now, you may be thinking that I've been drinking too much Romulan Ale. Be that as it may, virtual particles are a real phenomenon of nature. The predictions of quantum theory, based on Feynman diagrams like Figure 19-4, have been experimentally confirmed to 12 decimal digits.

So what does all this have to do with dark energy?

I think virtual particles *are* dark energy. Others agree, but not all.

This is my reasoning. Since particles and their antiparticle both have positive mass and positive total energy, virtual particles, however brief their existence, constitute a gas that fills each cubic meter of space with an amount of energy that is greater than zero. Let's call that energy density ϱ.

As space expands, the additional space must also contain virtual particle energy. What is the source of this additional energy? Recall the thermodynamic equation:

$$P\ dV + dE = 0$$

This equation states: as an isolated gas expands and its volume increases by dV, its energy E divided by pressure P must decrease. For a normal gas, this means the gas cools as it expands. That, however, is not an option for the virtual particle gas; its energy is always ϱ V, its energy density ϱ multiplied by its volume V. As space expands, adding volume dV, the energy of the virtual particle gas increases by ϱ dV.

For virtual particles, our thermodynamic equation becomes:

$$P\ dV + \varrho\ dV = 0$$

$$P = -\varrho$$

Hence, the pressure of the virtual particle gas is negative, but equal in magnitude to its energy. We know of nothing else that has this property.

The next step is calculating the gravitational effect of a gas with positive energy and negative pressure. In Section §23.1, we discover that an ideal gas with energy density ϱ and pressure P exerts a gravitational force proportional to:

$$\varrho + 3P$$

Substituting P yields: $\varrho + 3(-\varrho) = -2\varrho$

Therefore, the virtual particle gas has negative gravity — it is repulsive, rather than attractive like all other gravitational forces.

It is the negative gravity of virtual particles that is pushing the universe apart, accelerating its expansion exponentially.

That is what I, and many other physicists, think. Those who are unconvinced want more evidence. Measurements of the CMB are consistent with virtual particles being the cause of dark energy, but clever theorists can always conceive alternative hypotheses that are consistent with any observations.

Perhaps the greatest impediment to broader acceptance of this notion is theorists' inability to calculate the right value of ϱ — they are off by 120 factors of 10 from the observed ϱ value. This may be the greatest embarrassment in the history of theoretical physics — and there have been many.

Job opportunities still exist in theoretical cosmology.

Hopefully, you will do better.

Chapter 20
Tensor Calculus

As elegant as the concepts and principles of general relativity are, its mathematics is often extremely difficult, perhaps the most complex in physics.

Tensor calculus empowers us to precisely and compactly manage that complexity.

Because many of our equations will be quite elaborate, and we will need to focus on subscripts and superscripts, I will set those equations in larger type.

§20.1 What is a Tensor?

Tensors are a generalization of vectors. Tensor calculus is a beautiful branch of mathematics that can elegantly and effectively describe many complex, multi-dimensional phenomena. Tensors are essential in general relativity, where 4-D spacetime curves, twists, and stretches differently at every point, at every instant, and in every direction. They are also employed in 3-D analyses of mechanics and wave propagation in anisotropic materials, those whose properties are different in different directions.

The key point is:

Tensor equations are equally valid in all coordinate systems

Any tensor equation that is valid in one coordinate system or reference frame is automatically valid without any modifications in all systems and frames, regardless of their rotation, distortion, or motion relative to the original coordinate system. That generality is one reason that the mathematics of general relativity is so challenging, but it is also one of the most powerful tools in solving problems. If we can identify a coordinate system in which we can solve a complex problem with a tensor equation, we have immediately solved the problem in all coordinate systems. General relativity is the primary branch of physics in which tensor equations are universally employed.

Tensors are arrays of components that transform properly between coordinate systems. In 3-D, they must transform according to Euclidian coordinate rotations. In 4-D spacetime, they must also transform according to the Lorentz transformation. Tensors can have one component or millions of components, with each component being a different function of all coordinates. Since their components vary, tensors have different values at different events in spacetime. They are therefore often called *tensor fields*.

Let's consider some quantities that are tensors, and others that are not. Temperature is a simple quantity that changes with time and location, making it a function of the four coordinates of spacetime. Its values are different in different coordinate systems, but the values do not change according to the rotation rules or the Lorentz transformation. Hence, temperature is not a tensor. Similarly, energy by itself is not a tensor. But, the proper combination of energy and momentum, shown below, is a tensor in 4-D spacetime that does transform properly.

$$(E / c,\ p_x,\ p_y,\ p_z)$$

Since coordinate transformations mix a tensor's components, the components must all have the same units of measure: they must all be in meters, or years, or m / s, or kg m^2 / s^3, ... The same requirement exists in 3-D, but it was automatically satisfied since each spatial dimension is inherently the same. But in 4-D, we must take care to ensure the extra component has the same units, as we did above with E/c.

Tensors are characterized by their *rank* and by the *dimensionality* of the space in which they are defined. In physics, the most common spaces are 3-D Euclidian, and 4-D spacetime. The most common tensors have rank 0, 1, 2, or 4. The largest tensor I know has rank 10 and has 1,048,576 components.

The simplest tensors are scalars; these are *rank 0* tensors. These include π, 7, 0, and your age — all numbers that have the same values in all coordinate systems.

We are also very familiar with *rank 1* tensors: vectors. Every proper 3-vector is a 3-D *rank 1* tensor, and every proper 4-vector is a 4-D rank 1 tensor.

An important rank 1 tensor is an object's 4-velocity: $u^\mu = dx^\mu/d\tau$, where τ is the object's proper time. The components of u are:

$$u^\mu = (\gamma\, c,\ \gamma\, dx/dt,\ \gamma\, dy/dt,\ \gamma\, dz/dt)$$

Here $\gamma = 1 / \sqrt{(1 - v^2 / c^2)}$ is the usual relativistic factor.

Multiplying u^μ by the object's rest mass m yields its energy-momentum 4-vector:

$$p^\mu = (E / c,\ \gamma\, m\, dx/dt,\ \gamma\, m\, dy/dt,\ \gamma\, m\, dz/dt)$$

§20.2 Tensor Ranks & Indices

Since tensors can have so many components, we use indices to avoid writing them all out individually. We adopt an indexing notation using superscripts and subscripts. Superscripted indices, such as x^α, are called *contravariant indices*, while subscripted indices, such as x_α, are called *covariant indices*.

Since superscripts look exactly the same as exponents, we try to avoid using exponents in tensor calculus whenever confusion might arise: x^2 always means the number 2 component of x, but dx^2 always means $(dx)^2$.

It would be very nice if the notation were completely unambiguous. Alas, a famous mathematician once said: "Life is short, but the alphabet is even shorter." Unfortunately, we long ago ran out of symbols and modifiers, so some must perform multiple duties. This seems tricky, but it will become easier than you might think.

A rank N tensor has N indices that must all range over the same set of allowed values. Tensors must have the same number of rows and columns, unlike matrices that may have a different number of rows and columns.

A rank N tensor has 4^N components in 4-D, and 3^N components in 3-D.

A common rank 1 tensor is the position 4-vector x^α, whose components are denoted:

$$x^0 = t$$

$$x^1 = x$$

$$x^2 = y$$

$$x^3 = z$$

We can also write this:

For *contravariant index* $\alpha = 0, 1, 2, 3$: $x^\alpha = (+c\,t,\ x,\ y,\ z)$

For *covariant index* $\alpha = 0, 1, 2, 3$: $x_\alpha = (-c\,t,\ x,\ y,\ z)$

Note that in general: x^α is not equal to x_α

In the absence of curvature, the only difference between superscripted and subscripted 4-vectors is the minus sign in component 0; hence for 3-D tensors, there is no difference between these two types of indices. Typically, 3-D tensors are written with subscripts only.

In curved 4-D spacetime, all components generally change when indices switch between subscripted and superscripted.

There is a critical difference between a *free index* σ that can have any value in the allowed range, and the *fixed indices*, such 0, 1, 2, 3 or t, x, y, z. The latter refer to specific components, whereas the former refer to any component corresponding to any possible value of σ.

Free indices are just labels. The specific letters we choose have no significance mathematically or physically: when they stand alone, x_β and x_μ mean exactly the same thing, as do $A_{\mu\sigma}$ and $A_{\beta\alpha}$.

The significance of free indices lies in the relationships they establish. For example, the equation below states: each component of B equals the component of A with the indices reversed — B equals the *transpose* of A. Just like a vector equation, this tensor equation would mean exactly the same thing for any letters we might substitute for μ and σ.

$$A_{\mu\sigma} = B_{\sigma\mu}$$

§20.3 Tensor Algebra

The sums and differences of tensors with the same rank and indices are also tensors. The product of two tensors is a tensor. But, there is no mathematical procedure for dividing one tensor by another, except in the trivial case of dividing by a scalar, a rank 0 tensor.

A simple, but common, tensor equation is:

$$L_\mu = 0$$

This establishes the same equation for all values of the free index μ. In this case, it means each component of L equals zero. Every tensor equation is valid for all values of every *unmatched* free index. By "unmatched", we mean a free index that appears only once in a single product term. In the following equation, μ is a matched free index, β is an unmatched free index, and the equation is valid for every value of β.

$$A_\beta{}^\mu \; B_\mu = 1$$

When a free index appears twice in one term, once as a superscripted index and once as a subcripted index, the *Einstein summation convention* directs us to sum over all values of that repeated index. This is similar to a vector dot product, and is called a *tensor contraction* (contraction in the sense that the result is a smaller tensor with two fewer indices). For example:

$$A_{\beta\mu} B^\mu = A_{\beta 0} B^0 + A_{\beta 1} B^1 + A_{\beta 2} B^2 + A_{\beta 3} B^3 = C_\beta$$

On the left side of this equation, μ appears twice, subscripted and superscripted, in the product term A B. This directs us to sum over all values of μ. The resulting sum, in this case, is the 4-vector C_β.

The tensor contraction of a rank N tensor with a rank M tensor has rank N + M – 2.

Let's take the tensor contraction, the tensor dot product, of the position 4-vector x^α with itself.

$$x^\alpha = (+c\,t,\; x,\; y,\; z)$$

$$x_\alpha = (-c\,t,\; x,\; y,\; z)$$

$$x^\alpha\; x_\alpha = -\,c^2\,t^2 + x^2 + y^2 + z^2$$

The minus sign in the subscripted 4-vector provides the essential relativistic sign, ensuring the constancy of the speed of light in all reference frames. If we replace position vector x_α with the displacement vector dx_α, its dot product yields the invariant interval, a vital quantity in relativistic analyses.

$$dx^\alpha\; dx_\alpha = -\,c^2\,dt^2 + dx^2 + dy^2 + dz^2$$

Like all other tensor equations, all tensor contractions are equally valid in every reference frame and coordinate system. Just as in 3-D, the dot product of any two 4-vectors (rank 1 tensors) is an invariant scalar, with the same value in all frames and coordinate systems.

In a proper tensor equation, each nonzero term must have the same set of unmatched free subscripted indices, and the same set of unmatched free superscripted indices.

Tensor equations are valid for all values of each unmatched index. Consider these examples:

$$A_\sigma{}^\mu x_\mu + C_\sigma = B_\sigma$$

$$A_\sigma{}^\mu x_\mu + C_\beta \neq B_\mu$$

The upper equation is valid for all values of σ, while the lower equation contains invalid mismatches in each of the three indices.

The *unit tensor* [1] is as important in tensor calculus as 1 is in arithmetic. The components of the unit tensor equal 1 when all indices are the same, and 0 otherwise. The rank 2, 4-D unit tensor is shown below; all diagonal components are 1 and all off diagonal components are 0.

$[1] = \delta_{\mu\sigma}$

	σ = t	x	y	z
μ = t	1	0	0	0
x	0	1	0	0
y	0	0	1	0
z	0	0	0	1

When a tensor is shown as an array of components, its components are typically enclosed in square brackets [ ], as above.

The symbol $\delta_{\sigma\mu}$, called the *Kronecker delta,* is often used in math and physics. The tensor contraction of the Kronecker delta is:

$$\delta_\mu{}^\mu = \delta_t{}^t + \delta_x{}^x + \delta_y{}^y + \delta_z{}^z = 4$$

Like matrices, all non-singular tensors have *inverses*: $A^{-1}{}_{\mu\sigma}$ is the inverse of $A_{\mu\sigma}$. The contraction of a tensor with its inverse always equals the unit tensor. This is as close to tensor division as tensor calculus gets.

$$A^{-1}{}_\sigma{}^\mu A_{\mu\beta} = A_\sigma{}^\mu A^{-1}{}_{\mu\beta} = [1] = \delta_{\sigma\beta}$$

We defined tensors as arrays of components that transform properly. In 3-D, that means they transform properly under Euclidean rotations.

Let's review 3-D rotations. We start with a point Q at (x, y, z) defined in a coordinate system S. We then rotate this *coordinate system* about its z-axis by angle θ, resulting in a new coordinate system S* with the same z-axis but new x- and y-axes. In S*, the coordinates of Q are (x*, y*, z*). The coordinate transformation equations are:

$$x^* = + x\cos\theta + y\sin\theta$$

$$y^* = - x\sin\theta + y\cos\theta$$

$$z^* = z$$

Note that we are rotating coordinate axes, keeping Q stationary. If the axes are stationary and Q rotates, the equations are the same as above except θ is replaced by –θ. We can write the coordinate rotation transformation as a tensor $Z_{jk}(\theta)$:

	k = x	y	z
j = x*	+cosθ	+sinθ	0
y*	– sinθ	+cosθ	0
z*	0	0	1

In this case, $Z^{-1}(\theta)$, the inverse of $Z(\theta)$, is simply $Z(-\theta)$, the *transpose* of Z: $(Z_{jk})^{-1} = Z_{kj}$.

In tensor notation, the above rotation equation becomes:

$$Q^*_j = Z_{jk}\ Q_k$$

Again the sum over k is implied. Without indices this is:

$$Q^* = Z\ Q$$

The requirement that a tensor transform properly under any rotation R can be written:

$$A^*_j = R_{jk}\ A_k$$

$$A^*_{jk} = R_{jn}\ R_{km}\ A_{nm}$$

$$A^*_{jk\mu} = R_{jn}\ R_{km}\ R_{\mu\sigma}\ A_{nm\sigma}$$

In each line, A* is the transformed tensor of the original tensor A. In the upper equation, for rank 1, we sum over all three values of k. In the middle equation, for rank 2, we sum over all 3×3 = 9 values of n and m. In the lower equation, for rank 3, we sum over all 3×3×3 = 27 combinations of n, m, and σ. For rank 2, the transform can also be written:

$$A^*_{jk} = R_{jn}\ R_{km}\ A_{nm} = R_{jn}\ R^{-1}_{mk}\ A_{nm}$$

$$A^* = R\ A\ R^{-1}$$

Note that since tensor equations are fully defined by paired free indices, the order of product terms is irrelevant, unlike matrix multiplication. However, the order of a tensor's indices is critical in most cases: in general $A^{\alpha\beta\gamma}$ is not equal to $A^{\beta\alpha\gamma}$. Keeping all the indices correctly subscripted or superscripted is also critical. Index positions are not interchangeable.

We can form tensors by combining any proper polar vectors (but not axial vectors). For example, if A_j and B_k are proper polar 3-vectors, then their product is a proper rank 2, 3-D tensor.

$$C_{jk} = A_j\ B_k$$

Let's show that C transforms properly for any rotation R by transforming each vector.

$$(A^*_n)\ (B^*_m) = (R_{nj}\ A_j)\ (R_{mk}\ B_k)$$

$$(A^*_n)\ (B^*_m) = R_{nj}\ R_{mk}\ A_j\ B_k$$

$$(A^*_n)\ (B^*_m) = R_{nj}\ R_{mk}\ C_{jk} = C^*_{nm}$$

Hence, C transforms properly and is therefore a tensor. The same logic applies to tensors of any rank. The following, for example, is a proper rank 6 tensor, although it has no physical meaning as far as I know.

$$F_{jk}\ \Lambda_{mn}\ R_{li}$$

Now, let's move up to 4-D, where all proper tensors must also transform according to the Lorentz transformation, the rank 2 tensor $\Lambda^\alpha{}_\mu$. To transform quantities in a stationary reference frame to quantities in a frame moving in the +x direction with velocity v = β c, the Lorentz tensor is:

$$\Lambda^\alpha{}_\mu = \begin{bmatrix} \gamma & -\beta\gamma & 0 & 0 \\ -\beta\gamma & \gamma & 0 & 0 \\ 0 & 0 & 1 & 0 \\ 0 & 0 & 0 & 1 \end{bmatrix}$$

Here, $\gamma = 1 / \sqrt{(1-\beta^2)}$ is the usual relativistic factor, α is the row index, and μ is the column index.

The Lorentz transformation of 4-vector x_μ is simply the tensor product of x_μ and $\Lambda^\mu{}_\alpha$.

$$X_\alpha = \Lambda^\mu{}_\alpha \; x_\mu$$

The Lorentz transformation of rank 2 tensor $a_{\mu\nu}$ is the tensor triple product:

$$A_{\sigma\phi} = \Lambda^\nu{}_\phi \; \Lambda^\mu{}_\sigma \; a_{\mu\nu}$$

Note that $\Lambda^\mu{}_\sigma$ replaces the μ index of $a_{\mu\nu}$ with σ, and $\Lambda^\nu{}_\phi$ replaces the ν index with ϕ.

Once you get used to tensor calculus, it all makes sense and becomes enjoyable.

§20.4 Tensor Symmetries

Most tensors have symmetry properties that reduce their complexity.

The Lorentz tensor, shown above, is symmetric in its two indices, meaning it is unchanged if the indices are exchanged: each $\Lambda^\mu{}_\alpha = \Lambda^\alpha{}_\mu$. This is because the Lorentz transformation itself is symmetric. The same equations transform x_μ in a stationary frame to X_μ in a moving frame, and vice versa — the only difference is the relative velocity β changes polarity.

Since the Lorentz tensor $\Lambda^\mu{}_\alpha$ is symmetric, it has only 10 independent components: the 4 diagonal components ($\Lambda^\mu{}_\mu$, $\mu = 0, 1, 2, 3$); and the 6 components above the diagonal (where $\alpha < \mu$). The 6 below-diagonal components (where $\alpha > \mu$) duplicate the corresponding above-diagonal components.

The rank 2 Faraday tensor is *antisymmetric* in its two indices, meaning $F_{\alpha\beta} = -F_{\beta\alpha}$. The Faraday tensor $F_{\mu\sigma}$, shown below, encapsulates relativistic electromagnetism.

$$F_{\mu\sigma} = \begin{array}{c|cccc} & \sigma = t & x & y & z \\ \hline \mu = t & 0 & -E_x/c & -E_y/c & -E_z/c \\ x & E_x/c & 0 & -B_z & +B_y \\ y & E_y/c & +B_z & 0 & -B_x \\ z & E_z/c & -B_y & +B_x & 0 \end{array}$$

With 4 components being zero, and 6 redundant components, the Faraday tensor has only 16 – 4 – 6 = 6 independent components. These are the 3 components of electric field ***E*** and the 3 components of magnetic field ***B***.

Some tensors of greater rank, which we will meet in later chapters, have more complex symmetry properties, being symmetric in some indices and antisymmetric in others.

Chapter 21
Metric Tensors

Metric tensors provide the interval between all pairs of nearby points, thereby completely defining the shape of any geometry. Before we can effectively analyze any physical situation in curved spacetime, we must identify the metric of that geometry.

The simplest metric in 4-D spacetime is the metric for flat (zero-curvature) spacetime, for which the interval ds^2 between any two events A and B is:

$$ds^2 = -c^2\,dt^2 + dx^2 + dy^2 + dz^2$$

Here, dx equals the x coordinate of event B minus the x coordinate of event A, and similarly for dt, dy, and dz. The 4-vector (c dt, dx, dy, dz) is the *displacement 4-vector*, the 4-vector from A to B.

Henceforth, we will return to using natural units in which c = G = 1.

§21.1 Most General Metric

In order to deal with all possible curved spacetime geometries, we must initially consider the most general form that a metric tensor might have. We will then employ mathematical logic to reduce it to its most compact form, without loss of generality.

The metric could in principle be the sum of any number of terms, each being the product of any combination of the coordinate differentials dt, dx, dy, and dz, raised to any combination of powers and multiplied by any coefficient. Examples of such terms are:

$$\begin{aligned} ds^2 &= a\,dt + b\,dt^2 + c\,dt^3 + e\,dt^4 + \ldots \\ &\quad + f\,dt\,dx + g\,dt^2\,dx + \ldots. \\ &\quad + h\,dx\,dy^4 + \ldots. \\ &\quad + j\,dz^2\,dy^2 + \ldots + q \end{aligned}$$

Here, a, b, c, e, f, g, h, j, and q are each functions of the 4 coordinates of spacetime.

Fortunately, we can discard most of these ugly terms. The key simplifying factor is the sum of the exponents of the coordinate differentials in any product term; call that sum S.

Firstly, because the interval from any event A to itself is zero, the metric cannot contain any terms with S = 0, such as the q term that has no coordinate differentials.

Secondly, the interval between events A and B must not depend on whether ds^2 is the AB interval or the BA interval. This means the metric cannot contain terms that change sign when A and B are swapped. Swapping A and B inverts the polarity of each coordinate differential, dx goes to – dx, hence S cannot be odd, as it is for terms such as dt and $dt^2\,dx$.

What remains are terms in which S is even and greater than zero.

We can next get rid of all terms where S is more than 2, because the metric equation focuses on incremental distances, and terms like dx^4 are negligible compared with dx^2 in the limit that dx goes to zero. In the usual calculus fashion, we will keep only the leading terms — the smallest powers of the differentials.

Logic has reduced an infinity of possibilties down to 16 terms. The result, shown below, is the most general metric capable of describing 4-D spacetimes of any geometry. It has 4×4 = 16 terms, each being the triple product of 2 coordinate differentials and a coefficient $g_{\alpha\beta}$.

$$\begin{aligned} ds^2 = & \; g_{tt}\, dt^2 + g_{tx}\, dt\, dx + g_{ty}\, dt\, dy + g_{tz}\, dt\, dz \\ & + g_{xt}\, dx\, dt + g_{xx}\, dx^2 + g_{xy}\, dx\, dy + g_{xz}\, dx\, dz \\ & + g_{yt}\, dy\, dt + g_{yx}\, dy\, dx + g_{yy}\, dy^2 + g_{yz}\, dy\, dz \\ & + g_{zt}\, dz\, dt + g_{zx}\, dz\, dx + g_{zy}\, dz\, dy + g_{zz}\, dz^2 \end{aligned}$$

Rather than continuing to write all these terms, let's employ the Einstein summation convention, reducing the sum of 16 terms to one tensor product that is shown below in three equivalent forms. This is why we use tensors.

$$ds^2 = dx^\mu\, dx_\mu = g_{\mu\sigma}\, dx^\mu\, dx^\sigma = g_{\mu\sigma}\, dx^\sigma\, dx^\mu$$

Here, the repeated indices μ and σ are each summed over their 4 possible values: t, x, y, and z.

We can also prove that the metric tensor is symmetric in any geometry. Consider two 4-vectors A and B, and their dot product.

$$A = (1, 0, 0, 0)$$

$$B = (0, 0, 0, 1)$$

$$A^\mu\, B_\mu = g_{\alpha\beta}\, A^\alpha\, B^\beta = g_{03}\, A^0\, B^3$$

Reversing the order of the dot product yields:

$$B^\mu\, A_\mu = g_{\alpha\beta}\, B^\alpha\, A^\beta = g_{30}\, A^0\, B^3$$

The dot product is a symmetric operation that must give the same result either way, hence:

$$g_{03} = g_{30}$$

The same logic applies to all other metric components.

In flat 4-D spacetime, the metric tensor is:

$$g_{\alpha\beta} = \begin{bmatrix} -1 & 0 & 0 & 0 \\ 0 & 1 & 0 & 0 \\ 0 & 0 & 1 & 0 \\ 0 & 0 & 0 & 1 \end{bmatrix}$$

You might wish to confirm that, in flat spacetime, the above tensor yields our familiar interval:

$$ds^2 = g_{\mu\sigma}\, dx^\mu\, dx^\sigma = -dt^2 + dx^2 + dy^2 + dz^2$$

It is important to remember that g_{tt} is negative, −1 in this case. This minus sign, and any others that appear in complex geometries, are part of the metric. These minus signs must be included in all calculations. Trying to add them later can easily yield erroneous results.

The metric can be employed to exchange superscripted and subscripted indices, called index gymnastics, as show here:

Lowering an index: $x_\mu = g_{\mu\alpha}\, x^\alpha$

Raising an index: $x^\mu = g^{\mu\alpha}\, x_\alpha$

§21.2 Diagonal Metric Tensors

The metric tensor is a 4×4 square matrix. Only in exceptional circumstances is the metric tensor *singular* (having a zero determinant). In all other cases, the metric tensor with subscripted indices has an *inverse tensor* with superscripted indices, and vice versa. The inverse tensor is defined by:

$$g^{\alpha\beta} \ g_{\alpha\mu} \ = \ \delta^{\beta}{}_{\mu}$$

Here $\delta^{\beta}{}_{\mu}$ is the Kronecker delta that equals 1 if $\beta = \mu$, and equals zero if β and μ are different.

One sometimes sees the prior equation written as:

$$g^{\alpha\beta} \ g_{\alpha\mu} \ = \ g^{\beta}{}_{\mu} \ = \ \delta^{\beta}{}_{\mu}$$

Here, the first operation, contracting on index α, can be thought of as raising the first index of $g_{\alpha\mu}$ and replacing α with β. The resulting metric tensor $g^{\beta}{}_{\mu}$ is equivalent to the Kronecker delta $\delta^{\beta}{}_{\mu}$.

Like other square matrices, any non-singular metric tensor can be *diagonalized*, meaning all non-diagonal components can be made zero with suitable transforms. That leaves only 4 nonzero terms that lie on the diagonal.

The invariant interval is then reduced to 4 terms, and the inverse metric is also diagonal according to:

$$g^{\alpha\alpha} \ = \ 1 \ / \ g_{\alpha\alpha}$$

Diagonalizing the metric tensor can mix the coordinates in surprising ways. For example, the time coordinate t and radial coordinate r might be replaced by the coordinates $u = t + r$ and $v = t - r$, leaving nothing that represents pure time.

Since the tensor calculus of general relativity works in any coordinate system, coordinate mixing is mathematically valid and can simplify our calculations even when it defies our intuition.

Chapter 22
Curvature Tensors

Einstein said gravity is not a force between objects, but rather the effect of objects moving through curved spacetime.

To analyze motion in geometries with curvature, we must employ tensor calculus. It is the only means of calculating how functions change along axes that twist and bend, more like overcooked spaghetti than rigid rulers

§22.1 Derivatives in Flat 3-D Space

Derivatives are the way we analyze change. Let's begin by reviewing the two types of derivatives most often used in the calculus of classical (pre-Einstein) physics.

The *common derivative* is the most basic. Consider the expression $\emptyset = x\,p - E\,t$, which plays a major role in both wave mechanics and quantum mechanics. The common derivative, written $d\emptyset/dt$, equals the ratio of the change in $\emptyset$, written $d\emptyset$, to the change in t, written dt. Derivatives are evaluated for infinitesimal changes, in the limit that $dt \rightarrow 0$, allowing us to neglect higher order terms like dt^2. If p and E are constant, and for the simple motion $x = v\,t$, the common derivative is:

$$\emptyset = x\,p - E\,t$$

For constant E and p: $$d\emptyset/dt = v\,p - E$$

The *partial derivative*, written $\partial\emptyset/\partial t$, is the same except that only one variable (t in this case) is allowed to change, while all other variables (x in this case) remain constant. Hence:

$$\partial\emptyset/\partial t = -E$$

§22.2 Derivatives in Curved Spacetime

Brevity is essential in managing the ponderous equations of tensor calculus. The following definitions and compact notations are employed for the four most common derivatives used in general relativity. Here, we are calculating the derivatives of tensor A at event x^μ in the limit that dx^μ or $\Delta \rightarrow 0$.

Derivative	Notation	Definition
Common	dA^α/dx^μ	$[\,A^\alpha(x^\mu + dx^\mu) - A^\alpha(x^\mu)\,]\,/\,dx^\mu$
Partial	$A^\alpha{}_{,\mu}$	$\partial A^\alpha/\partial x^\mu$
Directional	$\partial_v A^\alpha$	$[\,A^\alpha(x^\mu + \Delta\, v^\mu) - A^\alpha(x^\mu)\,]\,/\,\Delta$
Covariant	$A^\alpha{}_{\beta;\mu}$	$A^\alpha{}_{\beta,\mu} + A^\varepsilon{}_\beta\,\Gamma^\alpha{}_{\varepsilon\mu} - A^\alpha{}_\varepsilon\,\Gamma^\varepsilon{}_{\beta\mu}$

The common and partial derivatives are the same in curved or flat spacetime. However, neither of these derivatives is a tensor because their components do not transform in the proper Lorentz manner.

If a derivative is taken with respect to one value of the index μ, the derivative of A has the same rank as A. If a derivative is taken over all values of μ, the rank of the derivative is one more than the rank of A.

The *directional derivative* $\partial_v A^\alpha$ is the partial derivative of tensor A in the direction of 4-vector v^μ, which need not be a coordinate axis. For any event x^μ in 4-D spacetime and any 4-vector v^μ originating at x^μ, first compute the change in tensor A from event x^μ to event $x^\mu + v^\mu \Delta$, for some infinitesimal quantity Δ. This is written $A(x^\mu + \Delta v^\mu) - A(x^\mu)$. The directional derivative of A at x^μ along v^μ is that change divided by Δ, in the limit $\Delta \rightarrow 0$. The directional derivative has the same rank as A, but $\partial_v A^\alpha$ does not transform in the proper Lorentz manner, in general.

The essential feature of *the covariant derivative* is that it is a tensor, and does transform in the proper Lorentz manner. The covariant derivative fulfills the same role in general relativity that the partial derivative fulfills in classical physics. To achieve this, "correction" terms must be added to convert the normal partial derivative into a tensor.

One correction term is required for each index; examples are shown above for superscripted and subscripted indices. Note the signs of the correction terms: + for the superscripted index α, and – for the subscripted index β. These correction terms involve a new mathematical device $\Gamma^\alpha{}_{\beta\mu}$ called a *connection coefficient*, which is a 4×4×4 matrix, but is not a tensor. Γ is the capital Greek letter gamma. Connection coefficients encapsulate all the twisting and stretching of coordinate axes in curved spacetime. The equations for connection coefficients are derived in Section §A5.4. Two equivalent definitions are:

(1) $\Gamma^\alpha{}_{\beta\mu}$ is the α component of change in basis vector e_β when moving along e_μ

(2) $e_\alpha \Gamma^\alpha{}_{\beta\mu}$ equals the covariant derivative of e_β along e_μ

§22.3 Curvature Tensors

Here is the full ensemble of primary curvature matrices employed in general relativity. All except the connection coefficient Γ are tensors.

Connection coefficient:
$$\Gamma_{\alpha\beta\mu} = (g_{\alpha\beta,\mu} + g_{\alpha\mu,\beta} - g_{\beta\mu,\alpha}) / 2 + \{ c_{\alpha\beta\mu} + c_{\alpha\mu\beta} - c_{\beta\mu\alpha} \} / 2$$

Riemann tensor:
$$R^\alpha{}_{\beta\mu\varepsilon} = \Gamma^\alpha{}_{\beta\varepsilon,\mu} - \Gamma^\alpha{}_{\beta\mu,\varepsilon} + \Gamma^\alpha{}_{\mu\delta}\, \Gamma^\delta{}_{\beta\varepsilon} - \Gamma^\alpha{}_{\varepsilon\delta}\, \Gamma^\delta{}_{\beta\mu} - \{ \Gamma^\alpha{}_{\beta\delta}\, c_{\mu\varepsilon}{}^\delta \}$$

Ricci tensor:
$$R_{\alpha\beta} = R^\mu{}_{\alpha\mu\beta}$$

Curvature scalar:
$$R = R^\mu{}_\mu$$

Einstein tensor:
$$G_{\alpha\beta} = R_{\alpha\beta} - g_{\alpha\beta}\, R / 2$$

In a coordinate basis, the terms above in { }'s are zero. In a non-coordinate basis (see Section §22.5):

$$e_\alpha\, c_{\beta\mu}{}^\alpha = [\partial_\beta , \partial_\mu]$$

A rarely-used related scalar quantity is:

Curvature invariant:
$$I = R_{\alpha\beta\mu\varepsilon}\, R^{\alpha\beta\mu\varepsilon}$$

The distinction between coordinate and non-coordinate bases is discussed in Section §22.5. For now, be assured that the coordinate systems we are accustomed to are all coordinate bases.

In a coordinate basis, the connection coefficients $\Gamma^{\alpha}{}_{\beta\mu}$ are sums of partial derivatives of the metric $g_{\alpha\beta}$. While Γ does not transform according to the Lorentz transformation, and is therefore not a proper tensor, the derivatives, products, and sums of its components determine the rank 4 Riemann curvature tensor $R^{\alpha}{}_{\beta\mu\varepsilon}$. Contracting the Riemann tensor produces the rank 2 Ricci tensor $R_{\alpha\beta}$, and contracting that yields the curvature scalar R. Finally, the Ricci tensor, $R_{\alpha\beta}$ and the metric $g_{\alpha\beta}$ combine to yield the rank 2 Einstein tensor $G_{\alpha\beta}$.

Let's consider the number of independent components of these quantities in a coordinate basis in 4-D spacetime. $\Gamma^{\alpha}{}_{\beta\mu}$ is symmetric in β and μ; hence of its 4×4×4 = 64 components, 40 are independent and 24 are duplicates. The Riemann tensor has several symmetries that reduce its number of independent components from 256 to 20, but in practice, one must compute 21 Riemann components (calculating one more component is typically easier than employing all its symmetries). The Ricci tensor and the Einstein tensor are both symmetric in their two indices, hence each has 10 independent components and the other 6 are duplicates. Typically we only need to calculate the 4 diagonal components of Ricci and Einstein.

In general, in 4-D curved spacetime, we need to calculate a total of 82 potentially nonzero, independent components: 40 Γ, 21 Riemann, 10 Ricci, 1 R, and 10 Einstein.

If you think this is daunting, you are absolutely correct. Many problems quickly explode into a mind-numbing array of terms that ultimately, we hope, miraculously cancel in droves, leaving a manageable final result. Fortunately, most problems have further symmetries that reduce the burden somewhat. The challenge is not exotic math — we only add, multiply, and differentiate — instead, the real challenge is the immense number of calculations that must be done perfektly.

The most useful symmetries of the Riemann tensor are:

(1) Antisymmetric on exchange of first two indices: $R_{\alpha\beta\delta\varepsilon} = -R_{\beta\alpha\delta\varepsilon}$

(2) Antisymmetric on exchange of last two indices: $R_{\alpha\beta\delta\varepsilon} = -R_{\alpha\beta\varepsilon\delta}$

(3) Symmetric on exchange of first pair with second pair: $R_{\alpha\beta\delta\varepsilon} = +R_{\delta\varepsilon\beta\alpha}$

Two other Riemann tensor symmetries require significant additional computation, making them less useful: the tensor has antisymmetric properties in its last three indices, and in all four indices. The number of independent nonzero Riemann tensor components in n dimensions is:

$$n^2 (n^2 - 1) / 12$$

The antisymmetry of both the first pair and the last pair of indices means the Riemann tensor in 4-D is equivalent to a 6x6 rank 2 tensor $S_{\alpha\beta}$, where α and β each span the 6 pair combinations. For example:

In spherical coordinates: $\alpha\beta$ = $t\,r$, $t\,\theta$, $t\,\phi$, $\theta\,\phi$, $\phi\,r$, $r\,\theta$

Tensor $S_{\alpha\beta}$ is shown schematically below. Of the 36 components of S, we must compute 21, the 6 diagonal components (black dots) and 15 above-diagonal components (open circles). The below-diagonal components (gray dots) are duplicates due to Riemann's first pair / second pair exchange symmetry.

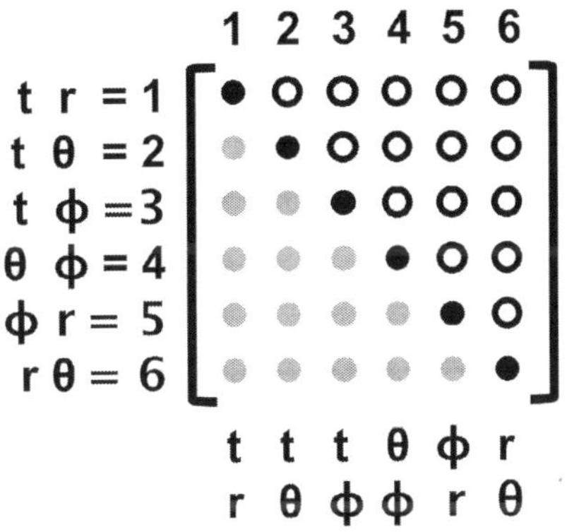

§22.4 Curvature of 2-D Sphere

Let's see how to apply the staggering array of mathematical tools presented above to the simplest curved geometry: a 2-dimensional sphere, the surface of a 3-D ball. This is like using a sledgehammer on a thumbtack, but it illustrates the process.

Let's use a coordinate basis, with polar angle θ, azimuthal angle β, and a fixed radius r. In what follows, θ and β are component labels, like t and x, to which the Einstein summation convention does not apply. Here, μ is the only repeated free index that we will sum over. The metric and its components are:

$$ds^2 = r^2\, d\theta^2 + r^2 \sin^2\theta\, d\beta^2$$

$$g_{\theta\theta} = r^2$$

$$g_{\beta\beta} = r^2 \sin^2\theta$$

$$g_{\theta\beta} = g_{\beta\theta} = 0$$

For our first curved spacetime calculation, I will show every intermediate step. In 2-D, the Γ tensor has $2^3 = 8$ components of which 6 are independent. Riemann has $2^4 = 16$ components of which only 1 is nonzero. The Ricci and Einstein tensors each have $2^2 = 4$ components of which 3 are independent.

First, let's compute the connection coefficients, many of which are zero.

$$2\,\Gamma_{\theta\theta\theta} = g_{\theta\theta,\theta} + g_{\theta\theta,\theta} - g_{\theta\theta,\theta} = g_{\theta\theta,\theta}$$
$$= \partial r^2/\partial\theta = 0$$

$$2\,\Gamma_{\theta\theta\beta} = g_{\theta\theta,\beta} + g_{\theta\beta,\theta} - g_{\theta\beta,\theta}$$
$$= \partial r^2/\partial\beta + 0 - 0 = 0$$

By symmetry of last 2 indices: $2\,\Gamma_{\theta\beta\theta} = 2\,\Gamma_{\theta\theta\beta} = 0$

$$2\,\Gamma_{\theta\beta\beta} = g_{\theta\beta,\beta} + g_{\theta\beta,\beta} - g_{\beta\beta,\theta}$$
$$= 0 + 0 - r^2\, \partial(\sin^2\theta)/\partial\theta = -2\, r^2 \sin\theta \cos\theta$$

$$2\,\Gamma_{\beta\theta\theta} = g_{\beta\theta,\theta} + g_{\beta\theta,\theta} - g_{\theta\theta,\beta}$$
$$= 0 + 0 - \partial r^2/\partial\theta = 0$$

$$2\,\Gamma_{\beta\theta\beta} = g_{\beta\theta,\beta} + g_{\beta\beta,\theta} - g_{\theta\beta,\beta}$$
$$= 0 + r^2\, \partial(\sin^2\theta)/\partial\theta - 0 = +2\, r^2 \sin\theta \cos\theta$$

By symmetry of last 2 indices: $2\,\Gamma_{\beta\beta\theta} = 2\,\Gamma_{\beta\theta\beta} = +2\, r^2 \sin\theta \cos\theta$

$$2\,\Gamma_{\beta\beta\beta} = g_{\beta\beta,\beta} + g_{\beta\beta,\beta} - g_{\beta\beta,\beta}$$
$$= 0 + 0 - 0 = 0$$

All this would have been gone a bit faster with some forethought. The only nonzero derivative of any metric component is:

$$g_{\beta\beta,\theta} = r^2\, \partial(\sin^2\theta)/\partial\theta = +2\, r^2 \sin\theta \cos\theta$$

Now let's raise the first index of each connection coefficient using the inverse metric; for a diagonal metric, this is very easy.

$$g^{\mu\sigma}\ g_{\mu\alpha} = \delta^{\sigma}_{\ \alpha}$$

In this equation, the only nonzero terms have $\sigma = \alpha$. Also, the only nonzero metric components have two equal indices. Hence, $\mu = \alpha$, and:

$$g^{\alpha\alpha} = 1 / g_{\alpha\alpha}$$

$$g^{\theta\theta} = 1 / g_{\theta\theta} = 1 / r^2$$

$$g^{\beta\beta} = 1 / g_{\beta\beta} = 1 / r^2 \sin^2\theta$$

We can now calculate the connection coefficients with the first index superscripted.

$g^{\theta\mu}$ is nonzero only if $\mu = \theta$: $\Gamma^{\theta}_{\ \beta\beta} = g^{\theta\mu}\ \Gamma_{\mu\beta\beta} = g^{\theta\theta}\ \Gamma_{\theta\beta\beta}$

$$\Gamma^{\theta}_{\ \beta\beta} = (-\,r^2 \sin\theta \cos\theta) / r^2 = -\sin\theta\cos\theta$$

$$\Gamma^{\beta}_{\ \theta\beta} = g^{\beta\mu}\ \Gamma_{\mu\theta\beta} = g^{\beta\beta}\ \Gamma_{\beta\theta\beta}$$

$$\Gamma^{\beta}_{\ \theta\beta} = \Gamma^{\beta}_{\ \beta\theta} = (+\,r^2 \sin\theta \cos\theta) / \ (r^2 \sin^2\theta) = +\cos\theta / \sin\theta$$

Next, we turn to the Riemann tensor. From the last equation in Section §22.3, we know this tensor has only 1 nonzero component in 2-D. We can also determine that because its first and last index pairs are antisymmetric. In 2-D, this excludes index pairs $\theta\theta$ and $\beta\beta$. Additionally, each $\theta\beta$ component equals -1 times the $\beta\theta$ component. Hence, the only independent component has indices $\theta\beta\theta\beta$. Let's compute this component.

$$R^{\theta}_{\ \beta\theta\beta} = \Gamma^{\theta}_{\ \beta\beta,\theta} - \Gamma^{\theta}_{\ \beta\theta,\beta} + \Gamma^{\theta}_{\ \mu\theta}\,\Gamma^{\mu}_{\ \beta\beta} - \Gamma^{\theta}_{\ \mu\beta}\,\Gamma^{\mu}_{\ \beta\theta}$$

Let's work on the third term, expanding the sum over μ.

$$\Gamma^{\theta}_{\ \mu\theta}\,\Gamma^{\mu}_{\ \beta\beta} = \Gamma^{\theta}_{\ \theta\theta}\,\Gamma^{\theta}_{\ \beta\beta} + \Gamma^{\theta}_{\ \beta\theta}\,\Gamma^{\beta}_{\ \beta\beta}$$

$$\Gamma^{\theta}_{\ \mu\theta}\,\Gamma^{\mu}_{\ \beta\beta} = (0)\,(-\sin\theta\cos\theta) + (0)\,(0) = 0$$

Now the fourth term, where we employ the symmetry $\Gamma^{\beta}_{\ \theta\beta} = \Gamma^{\beta}_{\ \beta\theta}$.

$$\Gamma^{\theta}_{\ \mu\beta}\,\Gamma^{\mu}_{\ \beta\theta} = \Gamma^{\theta}_{\ \theta\beta}\,\Gamma^{\theta}_{\ \beta\theta} + \Gamma^{\theta}_{\ \beta\beta}\,\Gamma^{\beta}_{\ \beta\theta}$$

$$\Gamma^{\theta}_{\ \mu\beta}\,\Gamma^{\mu}_{\ \beta\theta} = (0)\,(0) + (-\sin\theta\cos\theta)\,(\cos\theta / \sin\theta) = -\cos^2\theta$$

We can now continue with the Riemann component.

$$R^{\theta}_{\ \beta\theta\beta} = \Gamma^{\theta}_{\ \beta\beta,\theta} - \Gamma^{\theta}_{\ \beta\theta,\beta} + \Gamma^{\theta}_{\ \mu\theta}\,\Gamma^{\mu}_{\ \beta\beta} - \Gamma^{\theta}_{\ \mu\beta}\,\Gamma^{\mu}_{\ \beta\theta}$$

$$R^{\theta}_{\ \beta\theta\beta} = (-\sin\theta\cos\theta),\theta - (0),\beta + 0 - (-\cos^2\theta)$$

$$R^{\theta}_{\ \beta\theta\beta} = (-\cos^2\theta + \sin^2\theta) + \cos^2\theta$$

$$R^{\theta}_{\ \beta\theta\beta} = \sin^2\theta$$

The next step is the 3 independent Ricci components. Recall that $R^{\alpha\beta}_{\ \ \delta\varepsilon} = 0$ if either $\alpha = \beta$ or $\delta = \varepsilon$.

$$R^{\theta}_{\ \theta} = R^{\theta\mu}_{\ \ \theta\mu} = R^{\theta\theta}_{\ \ \theta\theta} + R^{\theta\beta}_{\ \ \theta\beta} = R^{\theta\beta}_{\ \ \theta\beta}$$

$$R^{\theta}_{\ \theta} = g^{\beta\mu}\ R^{\theta}_{\ \mu\theta\beta} = g^{\beta\theta}\ R^{\theta}_{\ \theta\theta\beta} + g^{\beta\beta}\ R^{\theta}_{\ \beta\theta\beta}$$

$$R^{\theta}_{\ \theta} = 0 + \sin^2\theta\ / (r^2 \sin^2\theta) = 1 / r^2$$

$$R^{\beta}{}_{\beta} = R^{\mu\beta}{}_{\mu\beta} = R^{\theta\beta}{}_{\theta\beta} + R^{\beta\beta}{}_{\beta\beta} = R^{\theta\beta}{}_{\theta\beta}$$
$$R^{\beta}{}_{\beta} = g^{\beta\mu}\, R^{\theta}{}_{\mu\theta\beta} = g^{\beta\theta}\, R^{\theta}{}_{\theta\theta\beta} + g^{\beta\beta}\, R^{\theta}{}_{\beta\theta\beta}$$
$$R^{\beta}{}_{\beta} = 0 + \sin^2\theta \,/\, (r^2 \sin^2\theta) = 1 / r^2$$

$$R^{\theta}{}_{\beta} = R^{\theta\mu}{}_{\beta\mu} = R^{\theta\theta}{}_{\beta\theta} + R^{\theta\beta}{}_{\beta\beta} = 0 + 0 = 0$$

Now the curvature scalar.

$$R = R^{\mu}{}_{\mu} = R^{\theta}{}_{\theta} + R^{\beta}{}_{\beta} = 2 / r^2$$

Finally, the Einstein tensor.

$$G_{\theta\theta} = R_{\theta\theta} - g_{\theta\theta}\, R / 2 = g_{\theta\mu}\, R^{\mu}{}_{\theta} - r^2 (2 / r^2) / 2$$
$$G_{\theta\theta} = g_{\theta\theta}\, R^{\theta}{}_{\theta} - 1 = r^2 (1 / r^2) - 1 = 0$$

$$G_{\beta\beta} = R_{\beta\beta} - g_{\beta\beta}\, R / 2 = g_{\beta\mu}\, R^{\mu}{}_{\beta} - (r^2 \sin^2\theta)(2 / r^2) / 2$$
$$G_{\beta\beta} = g_{\beta\beta}\, R^{\beta}{}_{\beta} - \sin^2\theta = r^2 \sin^2\theta\, (1 / r^2) - \sin^2\theta = 0$$

$$G_{\theta\beta} = R_{\theta\beta} - g_{\theta\beta}\, R / 2 = g_{\theta\mu}\, R^{\mu}{}_{\beta} - (0)(2 / r^2) / 2$$
$$G_{\theta\beta} = g_{\theta\theta}\, R^{\theta}{}_{\beta} = r^2 (0) = 0$$

In space without mass or energy, as we implicitly assumed here, the Einstein tensor G is always zero.

§22.5 Coordinate & Non-Coordinates Bases

Let's now discuss the meaning of *coordinate basis*. This is the most common type of coordinate system, in which the magnitudes of increments along each basis vector are independent of the values of the other coordinates. An example is normal Cartesian coordinates: a unit of length along the y axis is the same everywhere and does not depend on the values of x or z. This is not true in non-coordinate bases that we often employ in general relativity.

Compared below are two 4-D spherical coordinate systems.

Coordinate Basis

Interval: $ds^2 = -\, dt^2 + dr^2 + r^2\, d\theta^2 + r^2 \sin^2\theta\, d\phi^2$

Coordinates: t is time; r is radial distance from origin

θ is polar angle ($0 \rightarrow \pi$)

ϕ is azimuthal angle ($0 \rightarrow 2\pi$)

Non-Coordinate Basis

Interval: $ds^2 = -\, dt^2 + dr^2 + d\Theta^2 + d\phi^2$

Coordinates: t is time; r is radial distance from origin

$\Theta = r\, \theta$ — a scaled polar angle

$\Phi = r \sin\theta\, \phi$ — a scaled azimuthal angle

The non-coordinate basis above is *orthonormal*: all off-diagonal metric components are zero and the diagonal components are all ± 1. In this basis, all derivatives of the metric components, all $g_{\alpha\beta,\mu}$, are zero, simplifying the calculation of connection coefficients.

In a non-coordinate basis, we instead deal with *basis commutators*. Let's calculate the commutator of the two basis vectors r and Θ. We will employ the fact that all vectors are mathematically equivalent to directional derivatives. For clarity, I include a function A for the operators to act on, but even without any such functions, commutators are well-defined. With so much else to write, functions rarely appear in these calculations.

$$[\partial_r, \partial_\Theta]\, A = \partial_r\, \partial_\Theta\, A - \partial_\Theta\, \partial_r\, A$$

$$[\partial_r, \partial_\Theta]\, A = \partial/\partial r\, (\partial/r\theta)\, A - (\partial/r\theta)\, \partial/dr\, A$$

$$= -(1/r^2)\, \partial A/\partial\theta + (1/r)\, \partial^2 A/\partial r\partial\theta - (1/r)\, \partial^2 A/\partial r\partial\theta$$

$$= -(1/r^2)\, \partial A/\partial\theta$$

$$[\partial_r, \partial_\Theta]\, A = -(1/r)\, (\partial A/\partial r\theta)$$

Deleting the superfluous A: $[\partial_r, \partial_\Theta] = -(1/r)\, (\partial_\Theta)$

The resulting *commutation coefficients* are:

$$c_{r\Theta\Theta} = c_{r\Theta}{}^{\Theta} = -1/r$$

Figure 22-1, a copy of Figure 16-5, illustrates the distinction between the two types of bases, coordinate and orthonormal. The axes of a coordinate basis are what we are familiar with in classical Newtonian physics — the axes are unaffected by spacetime curvature, as if they exist in some external space that is uncurved. The coordinate axes are fixed references with which we can measure what happens in our curved spacetime. The coordinate r in this figure is an example; it measures un-stretched horizontal distances, such as the radius of the dotted circle linking A and E.

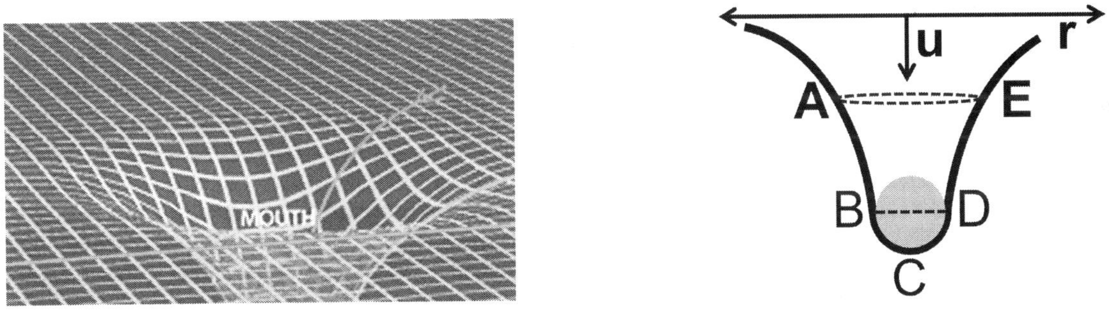

Figure 22-1 Curvature near Massive Gray Body

Conversely, the axes of an orthonormal basis lie within our curved spacetime; they twist and stretch as spacetime curves. We can use these axes to measure changes as they appear within our spacetime. The radial orthonormal coordinate in this figure measures distances along the dark black curve; the AE dotted circle's orthonormal radius is the path length from C to D to E.

In curved spacetime, a coordinate basis describes the curvature in the metric components $g_{\alpha\beta}$. Conversely, an orthonormal basis describes the curvature in the basis vectors, which stretch, contract or twist from one location to another.

We will gain a great deal of experience with both coordinate and non-coordinate bases, particularly in Appendices 5, 6, 7, and 8.

Appendix 6 provides the calculations of curvature tensors in coordinate and non-coordinate bases, for both general geometries, and for geometries with special symmetries.

Appendices 7 and 8 provide opportunities to practice your new skills.

Chapter 23
Einstein's Field Equations

Einstein's field equations equate spacetime curvature to the density of all forms of energy and stress, including mass, momentum, electromagnetic fields, and pressure. The field equations are:

$$G_{\alpha\beta} = 8\pi T_{\alpha\beta}$$

As American physicist John Archibald Wheeler described them, the field equations say: spacetime curvature tells mass and energy how to move, and mass and energy tell spacetime how to curve.

American physicist Brian Greene described the field equations as the choreography of the cosmic dance of spacetime with mass and energy.

Both G and T are rank 2 tensors that are defined at each event in 4-D spacetime. Because G and T both have 16 components, the tensor equation $G = 8\pi T$ is equivalent to 16 component equations — this is why we use the plural: field *equations*. Each tensor is symmetric in its two indices, thus each has 10 independent components and 6 duplicates.

The components of the Einstein tensor G are comprised of various combinations of Riemann tensor components. The Riemann components are sums of two types of terms: terms that are first derivatives of the connection coefficients Γ, and terms that are products of two Γ's. In turn, the Γ components are combinations of first derivatives of the metric tensor g. Thus, G is a combination of second derivatives of the metric g, and products of two first derivatives of g.

All this means, $G = 8\pi T$ constitutes 10 independent but interrelated, second-order, partial differential equations.

Unfortunately, no mathematical procedure exists that can analytically solve any second-order, partial differential equation.

Such equations can be integrated numerically with computers, but that is neither convenient, precise, nor easily affordable. These equations can be solved approximately, when gravity is weak, as in our Solar System. We explore that approach in Chapter 33.

In the 100 years since Einstein published his field equations, some of the best minds in physics have produced a handful of exact analytical solutions, in highly symmetric geometries. We have already explored the two most important solutions: the Schwarzschild metric in Section §16.3, and the FLRW equation in Section §19.3. The only other significant exact analytical solution is the Kerr-Newman metric that we explore in Chapter 29. That's a handful of solutions, if you are missing two fingers.

The Einstein tensor G does not represent the full curvature of spacetime; that requires the much larger, rank 4 Riemann tensor. Instead, G determines the changes in spacetime curvature at each specific event. Therefore, G is sometimes called the *moment of rotation* of spacetime curvature.

In the last chapter, we discovered how to calculate G, given a metric. Let's now explore T.

§23.1 Stress-Energy-Mass Tensor

The tensor T contains the density of all forms of energy, except gravitational energy — that would be double-counting.

Energy is conserved — it is neither created nor destroyed, but instead it moves from here to there, and changes form. This is reflected in T being colorfully described as a *river of 4-momentum* (energy and momentum) flowing through the universe.

More formally, T is called the stress-energy-mass-momentum tensor, or various subsets and combinations of those words.

In empty space, without mass, energy or electromagnetic fields, $T_{\alpha\beta} = 0$; simple enough.

The conservation of energy and momentum is encapsulated in a single tensor equation:

$$T^{\alpha\beta}{}_{;\beta} = 0$$

If we evaluate $T^{\alpha}{}_{\beta}$ in a reference frame moving with 4-velocity u^{β}, the result is:

$$T^{\alpha}{}_{\beta}\, u^{\beta} = -\, dp^{\alpha}/dV$$

The right side equals the density of 4-momentum in that reference frame.

Next, consider a collection of particles that comprise an ideal gas. In general relativity, *ideal gas* has the same meaning as in classical thermodynamics. The particles of an ideal gas interact only mechanically and gravitationally. An ideal gas is like a collection of idealized billiard balls. They may bounce off one another, but may not exchange internal energy that could be associated with internal vibration or chemical bonds. The pressure P within an ideal gas is isotropic, the same in every direction, in the gas rest frame. For an ideal gas:

$$T_{\alpha\beta} = (\varrho + P)\; u_{\alpha}\, u_{\beta} \;+\; P\, g_{\alpha\beta}$$

Here ϱ is the mass density of the gas, and u is the 4-vector velocity of the frame in which T is being evaluated.

In the rest frame of an ideal gas:

$$u_{\alpha} = (-1,\ 0,\ 0,\ 0)$$

$$T_{00} = \varrho$$

$$T_{11} = T_{22} = T_{33} = P$$

$$\text{all other } T_{\alpha\beta} = 0$$

In this case, the total gravitational attraction is proportional to the *trace* of T, which is:

$$T^{\mu}{}_{\mu} = \varrho + 3\,P$$

Another commonly used mass-energy tensor describes electromagnetic field energy.

$$4\pi\, T^{\mu\varepsilon} = F^{\mu\alpha}\, F^{\varepsilon}{}_{\alpha} \;-\; g^{\mu\varepsilon}\, F_{\alpha\beta}\, F^{\alpha\beta} / 4$$

Here, F is the Faraday tensor, representing the electromagnetic field, that is shown below.

$$F^{\alpha}{}_{\beta} = \begin{bmatrix} 0 & E_x & E_y & E_z \\ E_x & 0 & B_z & -B_y \\ E_y & -B_z & 0 & B_x \\ E_z & B_y & -B_x & 0 \end{bmatrix}$$

Chapter 24
Tensor Calculus in Action

General relativity relies extensively on Einstein's equivalence principle (see Chapter 13), which states: within a small enough region of spacetime, the acceleration of gravity is equivalent to a constant uniform acceleration. Einstein said this means free fall is the most natural state of motion, and physics is simplest in free-falling reference frames. "Free fall" means not subject to any external forces.

In general relativity, gravity is not considered a force, but the effect of curved spacetime.

§24.1 Geodesic Equation

The trajectories of free-falling objects are called *geodesics*.

Geodesics are the straightest possible paths through curved spacetime. On Earth's surface, geodesics are called great circle routes. Airplanes fly these routes to minimize travel time and fuel consumption, because they provide the shortest path between origin and destination.

In 4-D curved spacetime, the path of "least distance" is the path with the minimum interval ds^2. Equivalently, the shortest path is the one with the maximum proper time τ, since $d\tau^2 = -ds^2$. We can find that path by variational calculus — by varying the path to maximize the integral of τ:

$$\tau = \int du \sqrt{(-ds^2)}$$

$$\tau = \int du \sqrt{(-g_{\alpha\beta}\ dx^\alpha/du\ \ dx^\beta/du)}$$

Here, du is the increment along the path. The variational calculus, presented in Section §A5.5, yields the geodesic equation:

$$d^2x^\alpha/d\tau^2 + \Gamma^\alpha{}_{\beta\mu}\ dx^\beta/d\tau\ \ dx^\mu/d\tau = 0$$

Clearly, τ can be substituted with any other variable, such as t, by multiplying each term by $d\tau^2/dt^2$.

§24.2 Falling in Normal Gravity

Let's apply the geodesic equation to the simplest problem: the motion of an initially stationary small body of mass m at a distance r from the center of a large symmetric body of mass M. When stationary, the small body has 4-velocity:

$$u^\mu = (1,\ 0,\ 0,\ 0)$$

The Schwarzschild metric is:

$$ds^2 = -\Omega\, dt^2 + dr^2 / \Omega + r^2\, d\theta^2 + r^2 \sin^2\theta\, d\phi^2$$

$$\Omega = 1 - 2M / r$$

Since m falls directly toward M, $d\theta = d\phi = 0$. The key nonzero metric components and their nonzero derivatives are:

$$g_{tt} = -\Omega = -1 + 2M / r$$

$$g_{tt,r} = -2M / r^2$$

$$g_{rr} = 1 / \Omega$$

$$g_{rr,r} = (-1 / \Omega^2)(-2M)(-1 / r^2) = -2M / \Omega^2 r^2$$

Since the Schwarzschild metric is diagonal:

$$g^{tt} = 1 / g_{tt} = -1 / \Omega$$
$$g^{rr} = 1 / g_{rr} = \Omega$$
$$g^{\theta\theta} = 1 / g_{\theta\theta} = 1 / r^2$$
$$g^{\phi\phi} = 1 / g_{\phi\phi} = 1 / r^2 \sin^2\theta$$

Let's calculate the $\Gamma^r_{\alpha\beta}$ for all combinations of α and β equal to either t or r:

$$\Gamma^t_{tt} = g^{t\mu}\ \Gamma_{\mu tt} = g^{tt}\ \Gamma_{ttt}$$
$$\Gamma^t_{tt} = g^{tt}\ (g_{tt,t} + g_{tt,t} - g_{tt,t}) / 2 = 0$$

$$\Gamma^t_{tr} = \Gamma^t_{rt} = g^{t\mu}\ \Gamma_{\mu tr} = g^{tt}\ \Gamma_{ttr}$$
$$\Gamma^t_{tr} = g^{tt}\ (g_{tt,r} + g_{tr,t} - g_{tr,t}) / 2$$
$$\Gamma^t_{tr} = -\Omega^{-1}\ ([-2M / r^2] + 0 - 0) / 2 = +M / \Omega r^2$$

$$\Gamma^t_{rr} = g^{t\mu}\ \Gamma_{\mu rr} = g^{tt}\ \Gamma_{trr}$$
$$\Gamma^t_{rr} = g^{tt}\ (g_{tr,r} + g_{tr,r} - g_{rr,t}) / 2 = 0$$

$$\Gamma^r_{tt} = g^{r\mu}\ \Gamma_{\mu tt} = g^{rr}\ \Gamma_{rtt}$$
$$\Gamma^r_{tt} = g^{rr}\ (g_{rt,t} + g_{rt,t} - g_{tt,r}) / 2$$
$$\Gamma^r_{tt} = \Omega\ (0 + 0 - [-2M / r^2]) / 2 = +\Omega M / r^2$$

$$\Gamma^r_{tr} = \Gamma^r_{rt} = g^{r\mu}\ \Gamma_{\mu tr} = g^{rr}\ \Gamma_{rtr}$$
$$\Gamma^r_{tr} = g^{rr}\ (g_{rt,r} + g_{rr,t} - g_{tr,r}) / 2 = 0$$

$$\Gamma^r_{rr} = g^{r\mu}\ \Gamma_{\mu rr} = g^{rr}\ \Gamma_{rrr}$$
$$\Gamma^r_{rr} = g^{rr}\ (g_{rr,r} + g_{rr,r} - g_{rr,r}) / 2$$
$$\Gamma^r_{rr} = \Omega\ (-2M / \Omega^2 r^2) / 2 = -M / \Omega r^2$$

The r-component of the geodesic equation, with t instead of τ, is:

$$\partial^2 r / \partial t^2 = -\Gamma^r_{\alpha\beta}\ (\partial x^\alpha/\partial t)\ (\partial x^\beta/\partial t)$$

$$\partial^2 r / \partial t^2 = -\Gamma^r_{tt}\ (\partial t/\partial t)\ (\partial t/\partial t)$$
$$-2\,\Gamma^r_{tr}\ (\partial t/\partial t)\ (\partial r/\partial t)$$
$$-\Gamma^r_{rr}\ (\partial r/\partial t)\ (\partial r/\partial t)$$

Note that the Γ^r_{tr} contribution is doubled, once for $\alpha, \beta = t, r$ and once for $\alpha, \beta = r, t$.

For simplicity, let's assume the velocity of the falling body is always much less than the speed of light: $\partial r/\partial t << 1$. We can then retain only the $\alpha, \beta = t, t$ contribution.

$$\partial^2 r / \partial t^2 = -\Gamma^r_{tt}$$

$$\partial^2 r / \partial t^2 = -\Omega M / r^2$$

$$\partial^2 r / \partial t^2 = -(1 - 2M/r)\ (M/r^2)$$

The geodesic equation reduces to the Newtonian acceleration multiplied by the curvature term Ω.

The strongest gravity in our Solar System is at the Sun's surface, where $2M/r$ equals 4.2×10^{-6}. At Earth's surface, $2M/r$ equals 1.4×10^{-9}. Hence, distinguishing the effects of general relativity from those of Newtonian gravity requires high-precision observations.

But at the event horizon of a black hole, $2M/r = 1$, and the acceleration of gravity is zero, as seen by a stationary, remote observer. This is due to gravitational time dilation (see Section §18.5).

§24.3 Falling in Orthonormal Basis

Let's redo the falling body analysis of the prior section, but now in an orthonormal non-coordinate basis. In future chapters, we will employ such bases to solve some very challenging problems. Gaining experience by doing a simple problem will pay off later.

The Schwarzschild metric in a coordinate basis is:

$$ds^2 = -\Omega\, dt^2 + dr^2/\Omega + r^2\, d\theta^2 + r^2 \sin^2\theta\, d\phi^2$$

$$\Omega = 1 - 2M/r$$

The Schwarzschild metric in an orthonormal basis is:

$$ds^2 = -dT^2 + dR^2 + d\Theta^2 + d\Phi^2$$

Where the coordinate axes are "scaled" to absorb the metric components according to:

$$\omega = \sqrt{\Omega} = \sqrt{(1 - 2M/r)}$$

$$T = \omega t$$

$$R = r/\omega$$

$$\Theta = r\,\theta$$

$$\Phi = r\,(\sin\theta)\,\Phi$$

It is essential to note that these "scaling" factors are not constant; they are themselves functions of the 4-D spacetime coordinates.

The key advantage of an orthonormal basis is that the metric $g_{\alpha\beta}$ becomes the metric of flat spacetime:

$$g_{TT} = -1$$

$$g_{RR} = g_{\Theta\Theta} = g_{\Phi\Phi} = +1$$

The inverse metric is:

$$g^{TT} = -1$$

$$g^{RR} = g^{\Theta\Theta} = g^{\Phi\Phi} = +1$$

In a non-coordinate basis, our first step is calculating the *coordinate commutators* and *commutation coefficients* (see Sections §22.3 and §22.5). We do this by thinking of the basis vectors, unit vectors along each axis, as being directional derivatives, as mathematicians do. The basis vectors are:

$$e_T = \partial_T = \partial/\partial(\omega\, t) = (1/\omega)\, \partial/\partial t$$
$$e_R = \partial_R = \partial/\partial(r/\omega) = \omega\, \partial/\partial r$$
$$e_\Theta = \partial_\Theta = \partial/\partial(r\, \theta) = (1/r)\, \partial/\partial\theta$$
$$e_\Phi = \partial_\Phi = \partial/\partial(r \sin\theta\, \phi) = (1/r \sin\theta)\, \partial/\partial\phi$$

Let's start by calculating the commutator for basis vectors T and R.

$$[\partial_T, \partial_R] = (\partial/\omega\, \partial t)(\omega\, \partial/\partial r) - (\omega\, \partial/\partial r)(\partial/\omega\, \partial t)$$
$$[\partial_T, \partial_R] = 0 - \omega\, (\partial/\partial t)(\partial\, \omega^{-1}/\partial r)$$
$$[\partial_T, \partial_R] = -\omega\, (\partial/\partial t)(-\omega^{-2})(1/2\,\omega)(-2\,M)(-1/r^2)$$
$$[\partial_T, \partial_R] = (\partial/\partial t)(\omega^{-2})\, M/r^2$$
$$[\partial_T, \partial_R] = (\partial/\omega\partial t)\, M/\omega\, r^2 = (\partial_T)\, M/\omega\, r^2$$
$$[\partial_T, \partial_R] = e_\alpha\, c_{TR}{}^{\alpha}$$

Since this commutator contains only the ∂_T (e_T) term, only the $\alpha = T$ term contributes to the summation in the last equation. All other $c_{TR}{}^{\alpha}$, those where α is not T, are zero.

$$c_{TR}{}^{T} = M/\omega\, r^2$$
$$c_{TR}{}^{R} = c_{TR}{}^{\Theta} = c_{TR}{}^{\Phi} = 0$$

Now, let's calculate the commutators for the other 5 independent pairs of two different basis vectors.

$$[\partial_T, \partial_\Theta] = (\partial/\omega\, \partial t)(\partial/r\, \partial\theta) - (\partial/r\, \partial\theta)(\partial/\omega\, \partial t) = 0$$
$$c_{T\Theta}{}^{T} = c_{T\Theta}{}^{R} = c_{T\Theta}{}^{\Theta} = c_{T\Theta}{}^{\Phi} = 0$$

$$[\partial_T, \partial_\Phi] = (\partial/\omega\, \partial t)(\partial/r \sin\theta\, \partial\phi) - (\partial/r \sin\theta\, \partial\phi)(\partial/\omega\, \partial t)$$
$$[\partial_T, \partial_\Phi] = 0$$
$$c_{T\Phi}{}^{T} = c_{T\Phi}{}^{R} = c_{T\Phi}{}^{\Theta} = c_{T\Phi}{}^{\Phi} = 0$$

$$[\partial_R, \partial_\Theta] = (\omega\, \partial/\partial r)(\partial/r\, \partial\theta) - (\partial/r\, \partial\theta)(\omega\, \partial/\partial r)$$
$$[\partial_R, \partial_\Theta] = \omega\, (\partial/\partial\theta)(\partial\, r^{-1}/\partial r) - 0 = \omega\, (\partial/\partial\theta)(-r^{-2})$$
$$[\partial_R, \partial_\Theta] = (\partial/r\, \partial\theta)(-\omega/r) = (\partial_\Theta)(-\omega/r)$$
$$c_{R\Theta}{}^{\Theta} = -\omega/r$$
$$c_{R\Theta}{}^{T} = c_{R\Theta}{}^{R} = c_{R\Theta}{}^{\Phi} = 0$$

$$[\partial_R, \partial_\Phi] = (\omega\, \partial/\partial r)\,(\partial/ r \sin\theta\, \partial\phi) - (\partial/ r \sin\theta\, \partial\phi)\,(\omega\, \partial/\partial r)$$
$$[\partial_R, \partial_\Phi] = (\partial/\partial\phi)\,(\partial r^{-1}/\partial r)\, \omega / \sin\theta = (\partial/\partial\phi)\,(-r^{-2})\, \omega / \sin\theta$$
$$[\partial_R, \partial_\Phi] = (\partial/ r \sin\theta\, \partial\phi)\,(-\omega / r) = (\partial_\Phi)\,(-\omega / r)$$
$$c_{R\Phi}{}^{\Phi} = -\omega / r$$
$$c_{R\Phi}{}^{T} = c_{R\Phi}{}^{R} = c_{R\Phi}{}^{\Theta} = 0$$

$$[\partial_\Theta, \partial_\Phi] = (\partial/r\, \partial\theta)\,(\partial/ r \sin\theta\, \partial\phi) - (\partial/ r \sin\theta\, \partial\phi)\,(\partial/r\, \partial\theta)$$
$$[\partial_\Theta, \partial_\Phi] = (\partial/r^2\, \partial\phi)\,(\partial \sin\theta^{-1} /\partial\theta) - 0$$
$$[\partial_\Theta, \partial_\Phi] = (\partial/r^2\, \partial\phi)\,(-\sin\theta^{-2} \cos\theta)$$
$$[\partial_\Theta, \partial_\Phi] = (\partial/r \sin\theta\, \partial\phi)\,(-\cos\theta / r \sin\theta)$$
$$[\partial_\Theta, \partial_\Phi] = (\partial_\Phi)\,(-\cot\theta / r)$$
$$c_{\Theta\Phi}{}^{\Phi} = -\cot\theta / r$$
$$c_{\Theta\Phi}{}^{T} = c_{\Theta\Phi}{}^{R} = c_{\Theta\Phi}{}^{\Theta} = 0$$

The $c_{\beta\gamma}{}^{\alpha}$ commutation coefficients are antisymmetric in the first two indices $\beta\gamma$, because of the commutator $[\partial_\beta, \partial_\gamma]$. In 4-D, there are $4^3 = 64$ $c^\alpha{}_{\beta\gamma}$ coefficients, of which 16 are zero and 24 are duplicates with inverted polarity, due to $\beta\gamma$ antisymmetry. This leaves 24 independent coefficients, in general.

In this case, we found 4 coefficients that are nonzero and 20 that equal zero.

In an orthonormal basis, raising or lowering a T index, in any index position, multiplies $c^\alpha{}_{\beta\gamma}$ by -1, while raising or lowering indices R, Θ or Φ, in any position, leaves $c^\alpha{}_{\beta\gamma}$ unchanged.

The nonzero commutation coefficients are:

$$c_{TR}{}^{T} = M / \omega\, r^2 = -\, c_{TRT}$$
$$c_{R\Theta}{}^{\Theta} = -\omega / r = c_{R\Theta\Theta}$$
$$c_{R\Phi}{}^{\Phi} = -\omega / r = c_{R\Phi\Phi}$$
$$c_{\Theta\Phi}{}^{\Phi} = -\cot\theta / r = c_{\Theta\Phi\Phi}$$

The next step is calculating the connection coefficients $\Gamma^\alpha{}_{\beta\gamma}$, which are symmetric in the $\beta\gamma$ indices. For each α, there are 10 independent combinations of $\beta\gamma$, resulting in 40 total independent $\Gamma^\alpha{}_{\beta\gamma}$ components. Recall from Section §22.3, the equation for Γ, in an orthonormal basis where all derivatives of the metric $g_{\alpha\beta}$ are zero:

$$\Gamma_{\alpha\beta\mu} = (c_{\alpha\beta\mu} + c_{\alpha\mu\beta} - c_{\beta\mu\alpha}) / 2$$

Perhaps the best way to proceed is to identify which Γ's incorporate any of the 4 nonzero c's.

The only nonzero $c_{\alpha\beta\gamma}$ with a T index anywhere has indices TRT. There are 2 Γ's with those indices:

$$\Gamma^R{}_{TT} = (c_{RTT} + c_{RTT} - c_{TTR}) / 2 = c_{RTT}$$
$$\Gamma^R{}_{TT} = +M / \omega\, r^2$$

$$\Gamma^{T}{}_{RT} = (c_{TRT} + c_{TTR} - c_{RTT}) / 2 = c_{TTR}$$
$$\Gamma^{T}{}_{TR} = - M / \omega r^2$$

Two other nonzero c's have an R index; their indices are ΘRΘ and ΦRΦ. There are 4 Γ's with those indices:

$$\Gamma^{\Theta}{}_{R\Theta} = (c_{\Theta R\Theta} + c_{\Theta\Theta R} - c_{R\Theta\Theta}) / 2$$
$$\Gamma^{\Theta}{}_{R\Theta} = (+ \omega / r + 0 + \omega / r) / 2 = + \omega / r$$

$$\Gamma^{R}{}_{\Theta\Theta} = (c_{R\Theta\Theta} + c_{R\Theta\Theta} - c_{\Theta\Theta R}) / 2$$
$$\Gamma^{R}{}_{\Theta\Theta} = (- \omega / r - \omega / r - 0) / 2 = - \omega / r$$

$$\Gamma^{\Phi}{}_{R\Phi} = (c_{\Phi R\Phi} + c_{\Phi\Phi R} - c_{R\Phi\Phi}) / 2$$
$$\Gamma^{\Phi}{}_{R\Phi} = (+ \omega / r + 0 + \omega / r) / 2 = + \omega / r$$

$$\Gamma^{R}{}_{\Phi\Phi} = (c_{R\Phi\Phi} + c_{R\Phi\Phi} - c_{\Phi\Phi R}) / 2$$
$$\Gamma^{R}{}_{\Phi\Phi} = (- \omega / r - \omega / r - 0) / 2 = - \omega / r$$

The last nonzero c has indices ΦΘΦ. There are 2 Γ's with those indices:

$$\Gamma^{\Phi}{}_{\Theta\Phi} = (c_{\Phi\Theta\Phi} + c_{\Phi\Phi\Theta} - c_{\Theta\Phi\Phi}) / 2$$
$$\Gamma^{\Phi}{}_{\Theta\Phi} = (+ \cot\theta / r + 0 + \cot\theta / r) / 2 = + \cot\theta / r$$

$$\Gamma^{\Theta}{}_{\Phi\Phi} = (c_{\Theta\Phi\Phi} + c_{\Theta\Phi\Phi} - c_{\Phi\Phi\Theta}) / 2$$
$$\Gamma^{\Phi}{}_{\Theta\Phi} = (- \cot\theta / r - \cot\theta / r - 0) / 2 = - \cot\theta / r$$

The R-component of the geodesic equation is:

$$\partial^2 R / \partial T^2 = - \Gamma^{R}{}_{\alpha\beta} \, (\partial x^{\alpha} / \partial T) \, (\partial x^{\beta} / \partial T)$$

Assuming dΘ/dT = dΦ/dT = 0 and dR/dT << 1, the right side summation reduces to:

$$\partial^2 R / \partial T^2 = - \Gamma^{R}{}_{TT} \, (\partial T / \partial T) \, (\partial T / \partial T)$$
$$\partial^2 R / \partial T^2 = - M / \omega r^2$$
$$\partial^2 (r / \omega) / \omega^2 \partial t^2 = - M / \omega r^2$$
$$\partial^2 r / \partial t^2 = - \omega^2 M / r^2$$
$$\partial^2 r / \partial t^2 = - \Omega M / r^2$$

Which, thankfully, is exactly what we calculated in the coordinate basis.

§24.4 Gravity Bends Starlight

The first and most dramatic confirmation of general relativity was the observation of the bending of starlight during a total solar eclipse accomplished by Sir Arthur Eddington. That heroic story is described in Section §12.2. Now, with our tensor calculus expertise, we can compute general relativity's prediction of starlight bending.

Figure 24-1 shows light from a star far off to the left (the bold black line) bending slightly as it passes at a minimum distance b from the Sun's center — b is called the *impact parameter*. Let's define our coordinate system so that starlight enters parallel to the horizontal axis x, y is the vertical axis along which we will detect deflection, and the Sun is at x = y = 0.

Since the deflection angle of starlight by the Sun is very small, we can approximate this light's momentum 4-vector p^μ as:

With c = 1: $$p^\mu = (E,\ p,\ 0,\ 0)$$

The Sun is not perfectly spherically symmetric: it rotates once every 28 days at an average latitude, which slightly distends its equator. But this effect is very small, and for our present purposes the Schwarzschild metric is an excellent approximation.

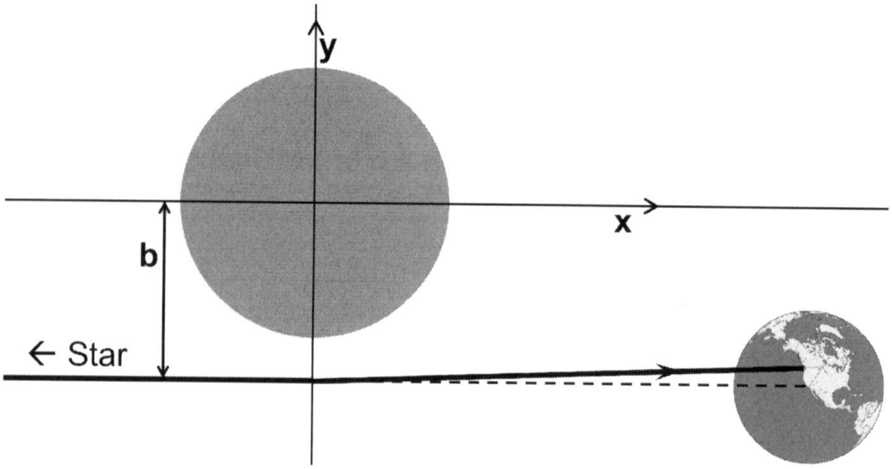

Figure 24-1 Star Light Bending in Sun's Gravity

In rectilinear coordinates, the Schwarzschild metric is:

$$ds^2 = -\Omega\, dt^2 + (dx^2 + dy^2 + dz^2) / \Omega$$
$$\Omega = 1 - 2\,M / r$$

Note that:

$$d\Omega/dy = (d\Omega/dr)\ (dr/dy)$$
$$d\Omega/dy = (-2\,M)(-1 / r^2)\ (1/2)\,(1/r)\,(2\,y)$$
$$d\Omega/dy = +2\,M\,y / r^3$$

We will need the following connection coefficients for this metric:

$$\Gamma^y{}_{tt} = g^{yy}\ \Gamma_{ytt} = g^{yy}\ (g_{yt,t} + g_{yt,t} - g_{tt,y}) / 2$$
$$\Gamma^y{}_{tt} = \Omega\ (0 + 0 + 2\,M\,y / r^3) / 2$$
$$\Gamma^y{}_{tt} = \Omega\,M\,y / r^3$$

$$\Gamma^y{}_{xx} = g^{yy}\ (g_{yx,x} + g_{yx,x} - g_{xx,y}) / 2$$

$$\Gamma^y{}_{xx} = \Omega\ (-1)(-1 / \Omega^2)\ (2\,M\,y / r^3) / 2$$

$$\Gamma^y{}_{xx} = (1 / \Omega)\ (M\,y / r^3)$$

$$\Gamma^y{}_{xt} = \Gamma^y{}_{tx} = g^{yy}\ (g_{yt,x} + g_{yx,t} - g_{tx,y}) / 2 = 0$$

In this case, M / r is only about 2 parts per million, so we neglect terms with the second or higher power of M / r. Replacing Ω with 1, the connection coefficients reduce to:

$$\Gamma^y{}_{tt} = \Gamma^y{}_{xx} = M\,y / r^3$$

Care must be taken with the geodesic equation for photons, because ds = dτ = 0 along their paths. Instead, let's take the geodesic equation for the y component of position of a particle of rest mass m, with m > 0 and 4-velocity v^α.

$$-\,d^2y/dt^2 = \Gamma^y{}_{\alpha\beta}\ (dx^\alpha/dt)\ (dx^\beta/dt)$$

Multiply both sides by m:

$$-\,m\,dv^y/dt = \Gamma^y{}_{\alpha\beta}\ m\ v^\alpha\ dx^\beta/dt$$

$$-\,d(p^y)/dt = \Gamma^y{}_{\alpha\beta}\ p^\alpha\ dx^\beta/dt$$

$$-\,dp^y = \Gamma^y{}_{\alpha\beta}\ p^\alpha\ dx^\beta$$

Mass m has been absorbed into momenta that remain valid quantities in the limit of m → 0. With the connection coefficients from above:

$$-\,dp^y = \Gamma^y{}_{tt}\ E\ dt + \Gamma^y{}_{xx}\ p\ dx$$

For photons, E = p and dt = dx.

$$-\,dp^y = (\Gamma^y{}_{tt} + \Gamma^y{}_{xx})\ p\ dx$$

$$-\,dp^y = 2\,M\ y\ p\ dx\ /\ r^3$$

Note there are two equal contributions to starlight deflection: one from the curvature of time and the other from the curvature of space. A slow-moving object, with velocity v << 1, is affected by time curvature, but is minimally affected by space curvature.

The y-value of the photon trajectory is very nearly –b everywhere, while the x-value effectively ranges from – ∞ to + ∞ (Earth's orbital radius is over 200 times the Sun's radius). The final p^y is given by the integral:

$$p^y = +2\,M\,b\,p \int_{-\infty}^{+\infty} dx / [\,b^2 + x^2\,]^{3/2}$$

$$p^y = 2\,M\,p \int_{-\infty}^{+\infty} dx / \{\,b^2\,[1 + (x\ / b)^2\,]^{3/2}\,\}$$

Let u = x / b, and du = dx / b.

$$p^y = 2\,M\,p \int_{-\infty}^{+\infty} du / \{\,b\,[1 + u^2\,]^{3/2}\,\}$$

$$p^y = (2\,M\,p / b)\ [u / (1 + u^2)^{1/2}]\,\Big|_{-\infty}^{+\infty}$$

$$p^y = (2\,M\,p / b)\ [1 - (-1)]$$

$$p^y / p = 4\,M / b$$

The ratio of final y-momentum to x-momentum is the deflection angle. Its value is maximum for light grazing the Sun's surface, where b = Sun's radius = 696,392 km. In natural units, the Sun's mass M equals 1.477 km.

The maximum deflection angle θ_{max} is therefore:

$$\theta_{max} = 4\ M / b = 4\ (1.477 / 696{,}392)$$

$$\theta_{max} = 8.5 \text{ micro-radians}$$

$$\theta_{max} = 1.74 \text{ arc-seconds}$$

$$\theta_{max} = 1 / 2068^{th} \text{ of one degree}$$

The deflection angle is the same for any photon energy and is inversely proportional to the impact parameter b, the photon's minimum distance to the center of mass M.

The most precise modern measurements of light bending by the Sun come from VLBI (Very Long Baseline Interferometer) studies of radio waves from distant quasars. The measured deflection confirms general relativity to an experimental precision of 450 parts per million. VLBI measured:

$$(4\ M / b) \times (0.999{,}83 \pm 0.000{,}45)$$

Some have tried to explain the bending of starlight without general relativity. They compute an equivalent photon mass, invoking $m = E / c^2$, and apply Newtonian gravity to calculate the force on m.

To paraphrase Wolfgang Pauli: "That idea isn't right; it isn't even Wrong!"

Pauli's message was that while we can learn from some mistakes, that mistake is so erroneous that it does not even provide a useful lesson.

This wrong approach is inconsistent with everything: precise measurements, Newton's theory, and Einstein's theory. It also gives the wrong answer — half the confirmed prediction of general relativity, because it misses half the physics.

This is like being afraid to jump across the Grand Canyon, and deciding instead to jump halfway.

Chapter 25
Orbits per Newton & Einstein

This chapter examines orbits of smaller bodies around much more massive objects. We begin with classical orbits and later compare these with orbits in the curved spacetime of general relativity.

In both cases, conservation of angular momentum ensures orbits lie within a plane defined by a line and a point: the line is the smaller body's angular momentum 3-vector, and the point is the center of the larger object of mass M, the source of the gravitational potential.

Let's assume throughout that the small body's mass m is nonzero but much less than M, and that the gravitational potential is spherically symmetric, which precludes a rotating source. We define spherical coordinates, with M at $r = 0$, and the orbital plane set by fixing $\theta = \pi / 2$.

The orbital variables are t, r, and ø. We will use natural units with $c = G = 1$.

§25.1 Newtonian Orbits

In the presence of mass M, the orbit of a much smaller body of mass m is determined by the conservation of its energy E and angular momentum L. In our coordinate system:

$$L = m\ r\ v_\text{ø}$$

$$L = m\ r^2\ (d\text{ø}/dt)$$

The energy of the orbiting body is the sum of its gravitational potential energy and its kinetic energy. We separate below the kinetic energy into radial motion (dr/dt) and azimuthal motion (dø/dt):

$$E = -\ M\,m / r\ +\ [\,(dr/dt)^2 + r^2\,(d\text{ø}/dt)^2\,]\ m / 2$$

$$2\,E / m = -\ 2\,M / r\ +\ (dr/dt)^2\ +\ L^2 / m^2\,r^2$$

Eqn. #1:
$$(dr/dt)^2 = 2\,E / m\ +\ 2\,M / r\ -\ (L / m\,r)^2$$

Define V:
$$(dr/dt)^2 = 2\,E / m\ -\ 2\,V$$

Eqn. #2:
$$V = -\,M / r + (L / m\,r)^2 / 2$$

V is called the *effective* potential. For a gravitationally-bound body, energy $E < 0$. The solutions to Eqn. #1 are then *ellipses*, as proven in §25.4. The polar coordinates r and ø of any point on the ellipse, and the ellipse's *eccentricity* ε and *semi-latus rectum* r_0 are given by:

Eqn. #3:
$$r = r_0\ /\ (1 + \varepsilon\cos\text{ø})$$

$$r_0 = L^2 / M\,m^2$$

Radius r reaches its minimum value, *periapsis*, at ø = 0, and reaches its maximum, *apapsis*, at ø = π. If the massive object M is the Sun, these limits are called *perihelion* and *aphelion*. If M is Earth, the limits are called *perigee* and *apogee*.

The effective potential V is plotted in Figure 25-1. The key feature is at small r, where V increases as r decreases, due to angular momentum conservation. As r decreases, the kinetic energy associated with rotation, $L^2 / 2\,m\,r^2$, increases faster than gravitational potential, $-\,M\,m / r$, decreases.

The rising effective potential at small r prevents bodies with large angular momenta from falling into the center. This is why planets have stable orbits.

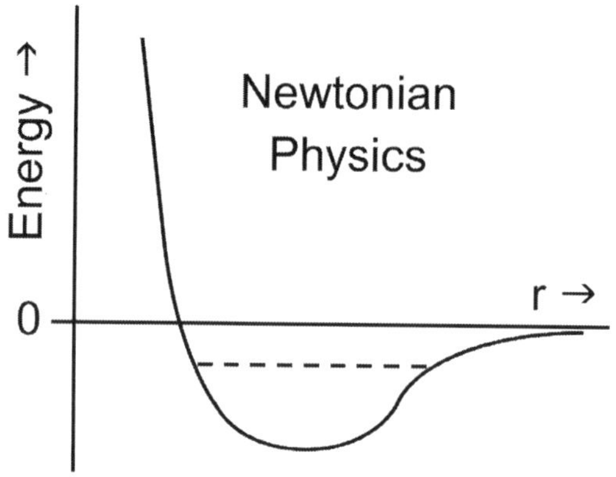

Figure 25-1 Effective Potential in Newtonian Gravity for Body with Angular Momentum

The dotted line in Figure 25-1 represents the total energy of an orbiting planet, such as Mercury. Mercury's kinetic energy is insufficient to escape the Sun's gravity, which is why its total energy is negative. But Mercury's angular momentum is large enough to prevent it from falling into the Sun.

Starting at a point along Mercury's orbit where its radius r is decreasing ($v_r < 0$), r eventually reaches its minimum at perihelion, where its radial kinetic energy is zero. This is where the left end of the dotted line meets the effective potential. Mercury's radial velocity v_r then passes through zero, and becomes positive. Mercury then moves away from the Sun. When its radius reaches aphelion, its radial velocity is again zero. This is where the right end of the dotted line meets the effective potential. Mercury then starts moving toward the Sun, again. Perihelion and aphelion are the two turning points of its orbit.

Earth's orbit is much less eccentric, being closer to a perfect circle. Its perihelion and aphelion are nearly equal, so its orbital radius and average surface temperature change only slightly throughout its orbit. If the dotted line of Earth's orbit were plotted in Figure 25-1, it would be much shorter and lie only slightly above the minimum of Earth's effective potential.

Far from gravitating sources, such as stars, most bodies have very modest kinetic energy. When a body enters a star's gravitational field, its kinetic energy increases but its angular momentum remains constant. The body will fall into the star if its effective potential is less than zero at the star's surface. If the star's radius is R, the star consumes everything with $(L / m)^2 < 2 M R$.

§25.2 Relativistic Orbits

In general relativity, for a planar orbit with $\theta = \pi / 2$ and $d\theta = 0$, the interval reduces to:

$$ds^2 = -\Omega\, dt^2 + dr^2 / \Omega + r^2\, d\phi^2$$

$$\Omega = 1 - 2M / r$$

Conservation laws in general relativity arise when all metric components are independent of a coordinate. We write this:

If for any μ and all α, β: $\quad \partial g_{\alpha\beta} / \partial x^\mu = 0$

Then: $\quad p_\mu$ is conserved $\rightarrow dp_\mu/d\tau = 0$

In the above metric: $\quad \partial g_{\alpha\beta} / \partial t = 0 \rightarrow p_t$ is conserved

$$p_t = E = g_{tt}\, p^t$$

$\partial g_{\alpha\beta} / \partial \phi = 0 \rightarrow p_\phi$ is conserved

$$p_\phi = L = g_{\phi\phi}\, p^\phi$$

To compute the orbit, let's start with the square of the body's 4-momentum, assuming $m > 0$:

$$-m^2 = g_{\alpha\beta}\ p^\alpha\ p^\beta$$

$$-m^2 = g_{tt}\ p^t\ p^t + g_{rr}\ p^r\ p^r + g_{\phi\phi}\ p^\phi\ p^\phi$$

$$-m^2 = g_{tt}\ (E / g_{tt})^2 + m^2\ (dr/d\tau)^2 / \Omega + g_{\phi\phi}\ (L / g_{\phi\phi})^2$$

$$0 = -E^2 / \Omega + m^2\ (dr/d\tau)^2 / \Omega + (L / r)^2 + m^2$$

$$(dr/d\tau)^2 = (E / m)^2 - \Omega\ (1 + L^2 / m^2 r^2)$$

Eqn. #4: $$(dr/d\tau)^2 = (E / m)^2 - V$$

Eqn. #5: $$V = (1 - 2M / r)\ (1 + L^2 / m^2 r^2)$$

$$V = 1 - 2M / r + L^2 / m^2 r^2 - 2 M L^2 / m^2 r^3$$

Eqn. #5, plotted in Figure 25-2, is very different from the classical effective potential of Eqn. #2. Note that energy E in relativity includes the rest mass m.

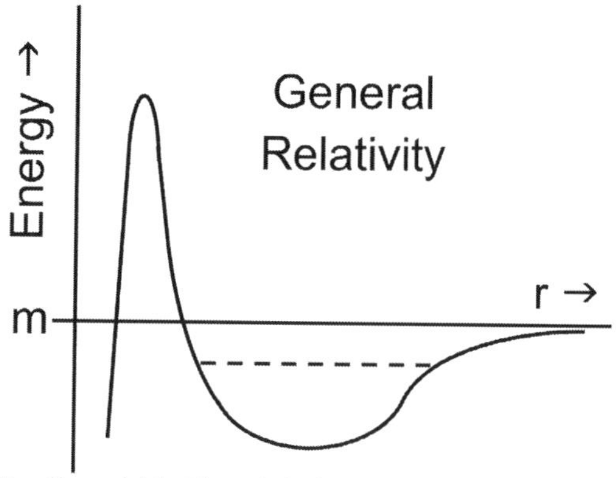

Figure 25-2 Effective Potential in Einstein's Gravity for Body with Angular Momentum

The key feature of the relativistic potential is its central pit. The effective potential drops through zero at $r = 2M$, and decreases rapidly for smaller r. This means there are no stable orbits very close to the central mass.

The central pit is not important for normal stars and planets, because their sizes are much larger than their masses (in natural units). Earth's radius is 1.4 billion times its mass M in natural units. Our Sun's radius is 470,000 times its mass M in natural units.

For white dwarfs, R is typically about 4000 M. For neutrons stars, R can approach the lower limit of 6 M. The event horizon of a non-spinning black hole is at $r = 2M$. But the size of a black hole's singularity, which contains all its mass, is a trillion, trillion times smaller than a hydrogen atom. For a thorough description of white dwarfs, neutron stars and black holes see my ebook ***Black Holes, Supernovae & More*** or my print book ***Our Universe: The Known & The Unknown***.

Let's next consider which orbits are possible and impossible around collapsed monsters such as black holes. We start by finding the maxima and minima of the relativistic potential. For brevity, define:

$$\mu = L / m M$$

Let's take the derivative of Eqn. #5, and find where it is zero.

$$V = 1 - 2M / r + \mu^2 M^2 / r^2 - 2 \mu^2 M^3 / r^3$$

$$dV/dr = +2M / r^2 - 2 \mu^2 M^2 / r^3 + 6 \mu^2 M^3 / r^4$$

$$dV/dr = 0 = +r^2 - \mu^2 M r + 3 \mu^2 M^2$$

The r-values of V's maximum and local minimum, near which orbits may exist, are:

$$r = \{ \mu^2 M \pm \sqrt{(\mu^4 M^2 - 12 \mu^2 M^2)} \} / 2$$

$$r = \mu^2 M \{ 1 \pm \sqrt{(1 - 12 / \mu^2)} \} / 2$$

For $\mu^2 < 12$, or equivalently $L < m M \sqrt{12}$, the prior equation has no solutions for real values of r. Hence, there are no stable orbits at any r. Any body with angular momentum below this limit eventually plummets into the central pit and never returns.

For $\mu^2 = 12$, a stable circular orbit exists at r = 6 M, three times the Schwarzschild radius.

For $\mu^2 > 12$, V has a local minimum at r > 6 M, and stable elliptical orbits exist in that vicinity.

A body approaching a black hole from far away is consumed if its $(L / m)^2 < 12 M^2$. Compare that to the corresponding equation we found above for falling into a normal star: $(L / m)^2 < 2 M R$. For our Sun:

$$R = 470{,}000 \ M$$

$$2 M R = 940{,}000 \ M^2 = 12 \ M^2 \ (280)^2$$

This means L / m, the angular momentum per unit mass, must be 280× higher to avoid being consumed by our Sun than to avoid being consumed by a black hole of the same mass.

Black holes do not fulfill the Hollywood image of insatiable cosmic devourers.

To derive the relativistic equation for stable orbits, rearrange the left side of Eqn. #4 using the definition of L:

$$m r^2 \, d\phi/d\tau = L$$

$$(dr/d\tau) \, (d\phi/dr) = L / m r^2$$

$$(dr/d\tau)^2 = L^2 \, (dr/d\phi)^2 / \{ m^2 r^4 \}$$

Putting this into Eqn. #4 yields:

$$(dr/d\tau)^2 = (E / m)^2 - 1 + 2 M / r - L^2 / m^2 r^2 + 2 M L^2 / m^2 r^3$$

$$(dr/d\phi)^2 = E^2 r^4 / L^2 - m^2 r^4 / L^2 + 2 M m^2 r^3 / L^2 - r^2 + 2 M r$$

Eqn. #6: $$(dr/d\phi)^2 = (E^2 - m^2) r^4 / L^2 + 2 M m^2 r^3 / L^2 - r^2 + 2 M r$$

Eqn. #7: $$\phi = \int dr / \sqrt{\{(E^2 - m^2) r^4 / L^2 + 2 M m^2 r^3 / L^2 - r^2 + 2 M r\}}$$

§25.3 Precession of Mercury

Better mathematicians than I can integrate Eqn. #7, at least approximately, to compute $\Delta\phi$ the change in ϕ between two consecutive perihelia. If $\Delta\phi = 2\pi$, the orbit's ellipse remains fixed, as in Newtonian physics. But due to spacetime curvature, we find (to first order):

$$\Delta\phi = 2\pi \ (1 + 3 M / r_0)$$

Here, r_0 is semi-latus rectum from Eqn. #3; r_0 is the orbit's radius if the orbit is circular. For Mercury's orbit around our Sun, the prediction of general relativity is:

$$\Delta\phi = 2\pi + 43.0 \text{ arc-seconds per century}$$

This compares well to the measured value of:

$$\Delta\phi = 2\pi + 42.5 \pm 0.9 \text{ arc-seconds per century}$$

The slight excess, $\Delta\phi - 2\pi$, causes the ellipse of Mercury's orbit to *precess* — the ellipse itself rotates slowly around the Sun. The precession due to general relativity amounts to one full 360° turn every 12 million orbits of Mercury, which takes about 3 million Earth years.

The orbit of Mercury also precesses due to gravitational tugs from the other planets, which are expected in both Newton's and Einstein's theories of gravity. Astronomers estimate that precession is 531.6 ± 0.7 arc-seconds per century.

The oblateness of the Sun also causes a precession of Mercury's orbit, but that is estimated to be only 0.025 arc-seconds per century.

The observed precession of Mercury, the sum of all effects, is 574.10 ± 0.65 arc-seconds per century.

All orbiting bodies precess. Being closest to the Sun, Mercury precesses more than other bodies in our Solar System. Venus's precession due to general relativity is 8.62 arc-seconds per century, and Earth's is 3.84 arc-seconds per century.

§25.4 Proof Newtonian Orbits are Ellipses

We prove here that the orbits of planets around their stars are elliptical, in Newtonian gravity. This proof also applies to any small body in the gravitational field of a much more massive body.

Let the small body have mass m, and the more massive body have mass M. It is important to note that the orbit lies entirely within a plane. At any instant in time, the velocity vector of mass m and the center of mass M define a unique plane. The gravitational force always lies entirely within that plane; hence, mass m never experiences any force component that would drive it out of this plane.

We can therefore analyze the orbit in 2-D polar coordinates, with mass M at the center, r being the distance from the center of M to m, and ø being the azimuthal angle defining the position of m around its orbit.

In Section §25.1, we derived the equation for the radial velocity of mass m:

$$(dr/dt)^2 = 2E/m + 2M/r - (L/mr)^2$$

Here, L is the angular momentum of the small body, and E is its energy, kinetic plus potential.

As this equation shows, celestial mechanics is messy, even in Newtonian gravity. One small trick is to move variables from denominators to numerators. In this case, we express the right side above in terms of the reciprocal of r. We define ϱ:

$$\varrho = 1/r$$

The prior equation becomes:

$$(dr/dt)^2 = 2E/m + 2M\varrho - (L\varrho/m)^2$$

$$m^2 (dr/dt)^2 = 2Em + 2m^2 M\varrho - L^2\varrho^2$$

Let's assume the angular momentum L of mass m is nonzero (else it would fall directly into the star). From Section §25.1, L is given by:

$$L = m\, r^2\, (d\phi/dt)$$

Angle ø is the integral of dø/dt.

$$\phi = \int (L / m\, r^2)\, dt$$

$$\phi = \int (L / m\, r^2)\, (dt/dr)\, dr$$

$dr = -d\varrho / \varrho^2$:

$$\phi = \int (L / m\, r^2)\, (dt/dr)\, (-r^2\, d\varrho)$$

$$\phi = -\int L\, d\varrho \,/\, m\, (dr/dt)$$

Substituting m (dr/dt) from above:

$$\phi = -\int L\, d\varrho \;/\; \surd\{ 2\,E\,m \;+\; 2\,m^2\,M\,\varrho \;-\; L^2\,\varrho^2 \}$$

Let's rewrite this slightly by dividing numerator and denominator by L.

$$\phi = -\int d\varrho \,/\, \surd\{ 2\,E\,m \,/\, L^2 \;+\; 2\,m^2\,M\,\varrho \,/\, L^2 \;-\; \varrho^2 \}$$

Trust me — we are making progress.

In math tables, we find:

$$d\,[\text{arc-cos}(x \,/\, a)] \,/\, dx \;=\; 1 \,/\, \surd(a^2 - x^2)$$

$$\text{arc-cos}(x \,/\, a) \;=\; \int dx \,/\, \surd(a^2 - x^2)$$

To employ this integral, we must make the square root in the ϕ integral have the form: $\surd(a^2 - x^2)$. Another small trick is recognizing that the expression in our square root is a quadratic in ϱ, and that everything but ϱ is constant.

To preserve some degree of sanity, let's define:

$$B \;=\; m^2\;M \;/\; L^2$$

$$C \;=\; 2\;E\;m \;/\; L^2$$

The expression within the square root becomes:

$$C \;+\; 2\,B\,\varrho \;-\; \varrho^2$$

The roots of this quadratic are:

$$\varrho \;=\; B \;\pm\; \surd[\,B^2 \;+\; C\,]$$

One more definition:

$$D^2 \;=\; B^2 \;+\; C \;=\; m^4\;M^2 \;/\; L^4 \;+\; 2\;E\;m \;/\; L^2$$

The quadratic in ϱ equals the product of (ϱ – root #1) and (ϱ – root #2).

$$C \;+\; 2\,B\,\varrho \;-\; \varrho^2 \;=\; -\,\{\varrho \;-\; B \;-\; D\}\;\{\varrho \;-\; B \;+\; D\}$$

$$C \;+\; 2\,B\,\varrho \;-\; \varrho^2 \;=\; D^2 \;-\; (\varrho \;-\; B)^2$$

Voilà! We have the difference of two squares and can employ the math table's integral. Note that since B is constant: $d\varrho = d(\varrho - B)$.

$$\phi \;=\; -\int d(\varrho - B) \;/\; \surd\{\,D^2 \;-\; (\varrho - B)^2\,\}$$

$$\phi \;=\; \text{arc-cos}\{\,(\varrho - B) \;/\; D\,\}$$

$$\cos\phi \;=\; (\varrho - B) \;/\; D$$

$$D\;\cos\phi \;=\; \varrho \;-\; B$$

$$\varrho \;=\; B \;+\; D\;\cos\phi$$

Replacing ϱ with 1 / r yields:

$$r \;=\; 1 \,/\, \{\,B \;+\; D\;\cos\phi\,\}$$

$$r \;=\; (1 \,/\, B) \;/\; \{\,1 \;+\; (D \,/\, B\,)\cos\phi\,\}$$

For your convenience, I will restore Newton's constant G (one for each M) in the final equations.

Now define ε, the *eccentricity* of the ellipse, and r_0, the *semi-latus rectum* of the ellipse.

$$\varepsilon^2 = D^2 / B^2 = (m^4 M^2 / L^4 + 2 E m / L^2) / (m^4 M^2 / L^4)$$

$$\varepsilon^2 = 1 + 2 E L^2 / m^3 G^2 M^2$$

$$r_0 = (1 / B) = L^2 / m^2 G M$$

$$r = r_0 / (1 + \varepsilon \cos\phi)$$

Our final equation shows the orbit is an ellipse, illustrated in Figure 25-3, just as Kepler claimed.

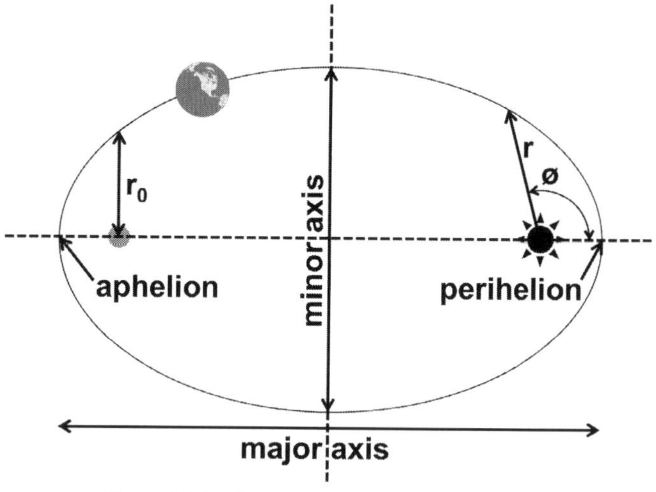

Figure 25-3 Elliptical Obit of Planet around Sun

The planet's radial distance ranges from:

Aphelion, at $\phi = \pi$: $r = r_0 / (1 - \varepsilon)$

Perihelion, at $\phi = 0$: $r = r_0 / (1 + \varepsilon)$

From measured values of the aphelion A, and perihelion P, one can calculate r_0 and ε:

$$r_0 = 2 A P / (A + P)$$

$$\varepsilon = (r_0 A P / 2) / (A - P)$$

Chapter 26
Riemann Normal Coordinates & Symmetries

Einstein's equivalence principle states gravity is equivalent to a uniform acceleration, within a small enough region of spacetime.

Einstein went further, saying that in a free-falling reference frame, the effects of gravity disappear, within a small enough region.

§26.1 Riemann Normal Coordinates

Let's expand on Einstein's insight expressed above.

At any specific event P^β, and for any chosen 4-velocity u^μ, there is a geodesic curve along the path of a free-falling body that passes through P with velocity u. Let's select four mutually orthogonal geodesics as our coordinate axes.

These axes form a free-falling reference frame called *Riemann normal coordinates* (RNC).

Within a small enough region, Einstein said there is no effect of gravity in such a reference frame — in this frame, all accelerations are zero. Recall the geodesic equation from Section §24.1:

$$d^2x^\alpha/d\tau^2 \;+\; \Gamma^\alpha{}_{\beta\mu}\; dx^\beta/d\tau\; dx^\mu/d\tau \;=\; 0$$

For all accelerations to be zero at P, for any set of velocities $dx^\beta/d\tau$, all $\Gamma^\alpha{}_{\beta\mu}$ must be zero:

In RNC, for all α, β, and μ: $\quad \Gamma_{\alpha\beta\mu} \;=\; 0$

Let's next discover what effect this has on the first derivatives of the metric g. For any α, β, μ:

$$\Gamma_{\alpha\beta\mu} \;=\; (\, g_{\alpha\beta,\mu} \;+\; g_{\alpha\mu,\beta} \;-\; g_{\beta\mu,\alpha}\,)\,/\,2$$

$$\Gamma_{\beta\mu\alpha} \;=\; (\, g_{\beta\mu,\alpha} \;+\; g_{\beta\alpha,\mu} \;-\; g_{\mu\alpha,\beta}\,)\,/\,2$$

$$\Gamma_{\alpha\beta\mu} \;+\; \Gamma_{\beta\mu\alpha} \;=\; (\, g_{\alpha\beta,\mu} \;+\; g_{\beta\alpha,\mu}\,)\,/\,2 \;=\; g_{\alpha\beta,\mu}$$

Since all $\Gamma_{\alpha\beta\mu}$ are zero, we have:

In RNC, for all α, β, and μ: $\quad g_{\alpha\beta,\mu} \;=\; 0$

These conditions imply:

$$g_{\alpha\beta} \;=\; \eta_{\alpha\beta} \;+\; A_{\alpha\beta\mu\sigma}\; dx^\mu\; dx^\sigma \;+\; O(\,[dx]^3\,)$$

$$\Gamma_{\alpha\beta\gamma} \;=\; 0 \;+\; B_{\alpha\beta\gamma\mu}\; dx^\mu \;+\; O(\,[dx]^2\,)$$

The first equations says metric g equals the flat spacetime metric η plus terms proportional to coordinate displacements to the second power, with A being some array of coefficients. The notation $O([dx]^3)$ means third and higher powers of the coordinate displacements. The second equation is similar; all $\Gamma_{\alpha\beta\gamma}$ equal zero, their flat spacetime values, plus first and higher power terms in the displacements.

The key point is: at any event P, there exists a reference frame in which spacetime curvature is zero. Furthermore, the first derivative of curvature is zero in every direction, and as we move away from P in any direction, curvature changes only slowly, being proportional to second and higher powers of the coordinate displacements. Such a reference frame is called Riemann normal coordinates.

In RNC, covariant derivatives and partial derivatives become identical, because all $\Gamma_{\alpha\beta\mu}$ are zero.

Since all the partial derivatives of the metric tensor $g_{\alpha\beta}$ are zero at event P, so are all the covariant derivatives of $g_{\alpha\beta}$.

Even in Riemann normal coordinates, second and higher order derivatives of g and first and higher order derivatives of Γ are not zero, in general.

$$g_{\alpha\beta,\mu\sigma} \text{ may not be } 0$$

$$\Gamma_{\alpha\beta\mu,\sigma} \text{ may not be } 0$$

Any tensor equation that is valid at an event P in Riemann normal coordinates is valid at all other events, since the choice of P is arbitrary. But the set of coordinates that provide zero curvature at P may not be the same set of coordinates that provide zero curvature at other events. If one coordinate system ensured zero spacetime curvature everywhere, there would be no gravity — clearly that is not true.

Equations valid in Riemann normal coordinates cannot in general be integrated or differentiated, because, in a fixed coordinate system, they may be valid at only a single event.

§26.2 Symmetries of Riemann Tensor

Employing Riemann normal coordinates simplifies the proof of the symmetry properties of the Riemann tensor. At any point P, this tensor reduces to:

$$R_{\alpha\beta\mu\sigma}(P) = \Gamma_{\alpha\beta\sigma,\mu} - \Gamma_{\alpha\beta\mu,\sigma}$$

$$R_{\alpha\beta\mu\sigma}(P) = (g_{\alpha\beta,\sigma\mu} + g_{\alpha\sigma,\beta\mu} - g_{\beta\sigma,\alpha\mu} - g_{\alpha\beta,\mu\sigma} - g_{\alpha\mu,\beta\sigma} + g_{\beta\mu,\alpha\sigma}) / 2$$

Eqn. #1:

$$R_{\alpha\beta\mu\sigma}(P) = (g_{\alpha\sigma,\beta\mu} - g_{\beta\sigma,\alpha\mu} - g_{\alpha\mu,\beta\sigma} + g_{\beta\mu,\alpha\sigma}) / 2$$

Above, "(P)" emphasizes that these equations are valid only at a specific event P in specific coordinates, but not in general.

The Riemann tensor is antisymmetric in its first and last pair of indices, as we can now easily prove. In Eqn. #1, exchanging α and β swaps the first term with the second term, and swaps the third with the fourth. Those swaps invert the polarity of each term on the right side of the equation, and thus invert the polarity of $R_{\alpha\beta\mu\sigma}$. Additionally, exchanging μ and σ swaps the first and third terms, and swaps the second and fourth terms, again inverting all polarities. Hence:

$$R_{\alpha\beta\mu\sigma} = -R_{\beta\alpha\mu\sigma}$$

$$R_{\alpha\beta\mu\sigma} = -R_{\alpha\beta\sigma\mu}$$

Let's next prove that the Riemann tensor is symmetric on the exchange of the first index pair with the second index pair. In Eqn. #1, exchanging α and μ and exchanging β and σ swaps the first and fourth terms, and swaps the second and third terms, with no change of sign. Hence:

$$R_{\alpha\beta\mu\sigma} = +R_{\mu\sigma\beta\alpha}$$

The Riemann tensor also has an antisymmetric property in its last 3 indices. Antisymmetry is simple when dealing with two indices: $F_{\alpha\beta} = -F_{\beta\alpha}$. But there are 3-factorial = 6 permutations of three indices.

The antisymmetric property of the last 3 indices is: the sum of the 3 even permutations minus the sum of the 3 odd permutations equals zero. A permutation is deemed even or odd depending on the number of exchanges of adjacent indices. For example, if abc is an even permutation, acb would be an odd permutation since it results from 1 exchange, and bca would be an even permutation because it requires 2 exchanges: abc → bac → bca. Since the sum of all properly-signed permutations is zero, it makes no difference which permutation is originally declared "even".

The sum of all permutations with appropriate signs is denoted by putting [ ] around the indices in question; this notation can be employed with any number of indices.

$$R_{\alpha[\beta\gamma\delta]} = R_{\alpha\beta\gamma\delta} + R_{\alpha\gamma\delta\beta} + R_{\alpha\delta\beta\gamma} - R_{\alpha\beta\delta\gamma} - R_{\alpha\gamma\beta\delta} - R_{\alpha\delta\gamma\beta}$$

$$R_{\alpha[\beta\gamma\delta]} = 2\,(R_{\alpha\beta\gamma\delta} + R_{\alpha\gamma\delta\beta} + R_{\alpha\delta\beta\gamma})$$

The first equation reduces to the second because of the antisymmetry of the last index pair. Below, we expand this equation in Riemann normal coordinates, and discover that each term cancels another of the opposite sign. Note that:

Since g is symmetric: $g_{\beta\gamma,\alpha\delta} = g_{\gamma\beta,\alpha\delta}$

Partial derivatives commute: $g_{\alpha\beta,\gamma\delta} = g_{\alpha\beta,\delta\gamma}$

Expanding the prior equation:

$$\begin{aligned} R_{\alpha[\beta\gamma\delta]} = &(g_{\alpha\delta,\beta\gamma} - g_{\beta\delta,\alpha\gamma} - g_{\alpha\gamma,\beta\delta} + g_{\beta\gamma,\alpha\delta}) \\ &+ (g_{\alpha\beta,\gamma\delta} - g_{\gamma\beta,\alpha\delta} - g_{\alpha\delta,\gamma\beta} + g_{\gamma\delta,\alpha\beta}) \\ &+ (g_{\alpha\gamma,\delta\beta} - g_{\delta\gamma,\alpha\beta} - g_{\alpha\beta,\delta\gamma} + g_{\delta\beta,\alpha\gamma}) \end{aligned}$$

$$\begin{aligned} R_{\alpha[\beta\gamma\delta]} = &(+\, g_{\alpha\delta,\beta\gamma} - g_{\alpha\delta,\gamma\beta}) + (-\, g_{\beta\delta,\alpha\gamma} + g_{\delta\beta,\alpha\gamma}) \\ &+ (-\, g_{\alpha\gamma,\beta\delta} + g_{\alpha\gamma,\delta\beta}) + (+\, g_{\beta\gamma,\alpha\delta} - g_{\gamma\beta,\alpha\delta}) \\ &+ (+\, g_{\alpha\beta,\gamma\delta} - g_{\alpha\beta,\delta\gamma}) - (+\, g_{\gamma\delta,\alpha\beta} - g_{\delta\gamma,\alpha\beta}) \end{aligned}$$

Hence:

$$R_{\alpha[\beta\gamma\delta]} = 0$$

The Riemann tensor has a similar antisymmetric property in all 4 indices:

$$R_{[\alpha\beta\gamma\delta]} = 0$$

This means that the sum of all 12 even index permutations minus the sum of all 12 odd permutations equals zero. The summation is accomplished by adding the 6 permutations in which α is the first index, subtracting the 6 with β first, adding the 6 with γ first, and subtracting the 6 with δ first:

$$R_{[\alpha\beta\gamma\delta]} = R_{\alpha[\beta\gamma\delta]} - R_{\beta[\alpha\gamma\delta]} + R_{\gamma[\alpha\beta\delta]} - R_{\delta[\alpha\beta\gamma]}$$

The signs above arise from the number of exchanges needed to bring each index to first position from the initial sequence $\alpha\beta\gamma\delta$: α needs no exchanges, β requires 1 exchange, γ requires 2, and δ requires 3. For n exchanges, the sign of the permutation is $(-1)^n$.

Since we just proved the Reimann tensor is antisymmetric in the last 3 indices, each of the 4 terms on the right side of the above equation equals zero.

Therefore, the Riemann tensor is antisymmetric in all four indices.

$$R_{[\alpha\beta\gamma\delta]} = 0$$

Given the Riemann tensor's other symmetries, its antisymmetry in its last three indices, and its antisymmetry in all four indices, are not two independent properties — they are mathematically equivalent.

§26.3 Bianchi Identity

The last symmetry property of the Riemann tensor that we discovered in the prior section, the asymmetry of all four indices, plays a critical role in general relativity.

The Bianchi identity states that the covariant derivative of the Riemann tensor has an antisymmetric property in its last 3 indices:

$$R_{\sigma\beta[\gamma\delta;\varepsilon]} = 0$$

The antisymmetry of the Riemann tensor's last index pair reduces this summation from 6 to 3 terms:

$$R_{\sigma\beta[\gamma\delta;\varepsilon]} = 2\,(R_{\sigma\beta\gamma\delta;\varepsilon} + R_{\sigma\beta\delta\varepsilon;\gamma} + R_{\sigma\beta\varepsilon\gamma;\delta})$$

In Riemann normal coordinates, all $\Gamma_{\alpha\beta\gamma} = 0$, and we may replace covariant derivatives with partial derivatives.

$$\begin{aligned} R_{\sigma\beta[\gamma\delta;\varepsilon]} = 2\,(& \Gamma_{\sigma\beta\delta,\gamma\varepsilon} - \Gamma_{\sigma\beta\gamma,\delta\varepsilon} \\ & + \Gamma_{\sigma\beta\varepsilon,\delta\gamma} - \Gamma_{\sigma\beta\delta,\varepsilon\gamma} \\ & + \Gamma_{\sigma\beta\gamma,\varepsilon\delta} - \Gamma_{\sigma\beta\varepsilon,\gamma\delta}) \end{aligned}$$

$$\begin{aligned} R_{\sigma\beta\gamma[\delta;\varepsilon]} = \; & (g_{\sigma\beta,\delta\gamma\varepsilon} + g_{\sigma\delta,\beta\gamma\varepsilon} - g_{\beta\delta,\sigma\gamma\varepsilon}) \\ & - (g_{\sigma\beta,\gamma\delta\varepsilon} + g_{\sigma\gamma,\beta\delta\varepsilon} - g_{\beta\gamma,\sigma\delta\varepsilon}) \\ & + (g_{\sigma\beta,\varepsilon\delta\gamma} + g_{\sigma\varepsilon,\beta\delta\gamma} - g_{\beta\varepsilon,\sigma\delta\gamma}) \\ & - (g_{\sigma\beta,\delta\varepsilon\gamma} + g_{\sigma\delta,\beta\varepsilon\gamma} - g_{\beta\delta,\sigma\varepsilon\gamma}) \\ & + (g_{\sigma\beta,\gamma\varepsilon\delta} + g_{\sigma\gamma,\beta\varepsilon\delta} - g_{\beta\gamma,\sigma\varepsilon\delta}) \\ & - (g_{\sigma\beta,\varepsilon\gamma\delta} + g_{\sigma\varepsilon,\beta\gamma\delta} - g_{\beta\varepsilon,\sigma\gamma\delta}) \end{aligned}$$

Regrouping terms:

$$\begin{aligned} R_{\sigma\beta\gamma[\delta;\varepsilon]} = \; & (g_{\sigma\beta,\delta\gamma\varepsilon} + g_{\sigma\beta,\varepsilon\delta\gamma} + g_{\sigma\beta,\gamma\varepsilon\delta} \\ & - g_{\sigma\beta,\gamma\delta\varepsilon} - g_{\sigma\beta,\delta\varepsilon\gamma} - g_{\sigma\beta,\varepsilon\gamma\delta}) \\ & + (g_{\sigma\gamma,\beta\varepsilon\delta} - g_{\sigma\gamma,\beta\delta\varepsilon}) \\ & + (g_{\sigma\delta,\beta\gamma\varepsilon} - g_{\sigma\delta,\beta\varepsilon\gamma}) \\ & + (g_{\sigma\varepsilon,\beta\delta\gamma} - g_{\sigma\varepsilon,\beta\gamma\delta}) \\ & + (g_{\beta\gamma,\sigma\delta\varepsilon} - g_{\beta\gamma,\sigma\varepsilon\delta}) \\ & + (g_{\beta\delta,\sigma\varepsilon\gamma} - g_{\beta\delta,\sigma\gamma\varepsilon}) \\ & + (g_{\beta\varepsilon,\sigma\gamma\delta} - g_{\beta\varepsilon,\sigma\delta\gamma}) \end{aligned}$$

In the first two lines above, $g_{\sigma\beta,\delta\gamma\varepsilon}$ occurs 6 times, once with each permutation of $\delta\gamma\varepsilon$. Since partial derivatives commute, each term has equal magnitude, and the three positive and three negatives terms cancel.

Each of the last six lines contains two terms that similarly cancel. Hence, we have proven the Bianchi identity:

$$R_{\sigma\beta[\gamma\delta;\varepsilon]} = 0$$

Chapter 27
Conservation Laws from Geometry

One of the most important results of general relativity is its insight into the conservation of energy and momentum, an essential principle of physics.

Let's see how these results are due to geometry, beginning with properties of the metric and the Einstein tensor G.

§27.1 Covariant Derivative of Metric

An important property of the metric is that its covariant derivative equals zero. We show here that this is consistent with the equations for the connection coefficients given in Section §22.3, and derived in Appendix §A5.4.

$$g^{\alpha\beta}{}_{;\nu} = g^{\alpha\beta}{}_{,\nu} + g^{\mu\beta}\,\Gamma^{\alpha}{}_{\mu\nu} + g^{\alpha\varepsilon}\,\Gamma^{\beta}{}_{\varepsilon\nu}$$

$$g^{\alpha\beta}{}_{;\nu} = g^{\alpha\beta}{}_{,\nu} + g^{\mu\beta}\,g^{\alpha\varepsilon}\,\Gamma_{\varepsilon\mu\nu} + g^{\alpha\varepsilon}\,g^{\beta\mu}\,\Gamma_{\mu\varepsilon\nu}$$

$$\begin{aligned} g^{\alpha\beta}{}_{;\nu} = {} & g^{\alpha\beta}{}_{,\nu} \\ & + g^{\mu\beta}\,g^{\alpha\varepsilon}\,(g_{\varepsilon\mu,\nu} + g_{\varepsilon\nu,\mu} - g_{\mu\nu,\varepsilon})\,/\,2 \\ & + g^{\alpha\varepsilon}\,g^{\beta\mu}\,(g_{\mu\varepsilon,\nu} + g_{\mu\nu,\varepsilon} - g_{\varepsilon\nu,\mu})\,/\,2 \end{aligned}$$

Recall the metric's symmetry: $g_{\mu\varepsilon} = g_{\varepsilon\mu}$. In the prior equation, the last two terms in the middle line cancel the last two terms in the lower line, thus yielding:

$$g^{\alpha\beta}{}_{;\nu} = g^{\alpha\beta}{}_{,\nu} + g^{\mu\beta}\,g^{\alpha\varepsilon}\,g_{\varepsilon\mu,\nu}$$

Next, recall the product rule of differentiation:

$$(A\,B\,C),d = (A,d)\,B\,C + A\,(B,d)\,C + A\,B\,(C,d)$$

or:

$$A\,B\,(C,d) = (A\,B\,C),d - (A,d)\,B\,C - A\,(B,d)\,C$$

Applying the last equation to the rightmost term in the prior equation for $g^{\alpha\beta}{}_{;\nu}$, we obtain:

$$\begin{aligned} g^{\alpha\beta}{}_{;\nu} = {} & g^{\alpha\beta}{}_{,\nu} + (g^{\mu\beta}\,g^{\alpha\varepsilon}\,g_{\varepsilon\mu})_{,\nu} \\ & - g^{\mu\beta}{}_{,\nu}\,g^{\alpha\varepsilon}\,g_{\varepsilon\mu} - g^{\mu\beta}\,g^{\alpha\varepsilon}{}_{,\nu}\,g_{\varepsilon\mu} \end{aligned}$$

Now use $g^{\alpha\varepsilon}\,g_{\varepsilon\mu} = \delta^{\alpha}{}_{\mu}$.

$$\begin{aligned} g^{\alpha\beta}{}_{;\nu} = {} & g^{\alpha\beta}{}_{,\nu} + (g^{\mu\beta}\,\delta^{\alpha}{}_{\mu})_{,\nu} \\ & - g^{\mu\beta}{}_{,\nu}\,\delta^{\alpha}{}_{\mu} - \delta^{\beta}{}_{\varepsilon}\,g^{\alpha\varepsilon}{}_{,\nu} \end{aligned}$$

$$g^{\alpha\beta}{}_{;\nu} = g^{\alpha\beta}{}_{,\nu} + g^{\alpha\beta}{}_{,\nu} - g^{\alpha\beta}{}_{,\nu} - g^{\alpha\beta}{}_{,\nu}$$

$$g^{\alpha\beta}{}_{;\nu} = 0$$

This confirms the important fact that the covariant derivative of the metric tensor is zero.

§27.2 Divergence of G is Zero

Using the symmetry properties of the Riemann tensor established in Section §26.2, let's evaluate the *divergence* of the Einstein tensor G.

The divergence of any tensor is the covariant derivative of that tensor, with a tensor contraction of the derivative index with one of the tensor's original indices. An example is:

Divergence of tensor A: $A^{\alpha}{}_{\beta;\alpha}$

The Einstein tensor G before contracting the Riemann tensor is:

$$G^{\alpha}{}_{\beta} = R^{\alpha\mu}{}_{\beta\mu} - \delta^{\alpha}{}_{\beta} \, R^{\mu\varepsilon}{}_{\mu\varepsilon} \, / \, 2$$

Let's take the covariant derivative of G, and then contract on its first index.

$$G^{\alpha}{}_{\beta;\alpha} = R^{\alpha\mu}{}_{\beta\mu;\alpha} - \delta^{\alpha}{}_{\beta} \, R^{\mu\varepsilon}{}_{\mu\varepsilon;\alpha} \, / \, 2$$

Next, contract $\delta^{\alpha}{}_{\beta}$ in the last term, thereby replacing α with β.

$$G^{\alpha}{}_{\beta;\alpha} = R^{\alpha\mu}{}_{\beta\mu;\alpha} - R^{\mu\varepsilon}{}_{\mu\varepsilon;\beta} \, / \, 2$$

Using the Bianchi identity, we replace the last term with two new terms, each with a different even permutation of $\mu\varepsilon\beta$.

$$G^{\alpha}{}_{\beta;\alpha} = R^{\alpha\mu}{}_{\beta\mu;\alpha} - (- R^{\mu\varepsilon}{}_{\varepsilon\beta;\mu} - R^{\mu\varepsilon}{}_{\beta\mu;\varepsilon}) \, / \, 2$$

Now re-label two summation indices: change μ to α in the middle term, and change ε to α in the last term.

$$G^{\alpha}{}_{\beta;\alpha} = R^{\alpha\mu}{}_{\beta\mu;\alpha} + (R^{\alpha\varepsilon}{}_{\varepsilon\beta;\alpha} + R^{\mu\alpha}{}_{\beta\mu;\alpha}) \, / \, 2$$

Next, swap the lower index pair $\varepsilon\beta$ in the middle term, and swap the upper index pair $\mu\alpha$ in the last term. Both swaps introduce a minus sign since the Riemann tensor is antisymmetric in its first and last index pair.

$$G^{\alpha}{}_{\beta;\alpha} = R^{\alpha\mu}{}_{\beta\mu;\alpha} + (- R^{\alpha\varepsilon}{}_{\beta\varepsilon;\alpha} - R^{\alpha\mu}{}_{\beta\mu;\alpha}) \, / \, 2$$

We now have three identical terms on the right, which net to zero with the 1 / 2 and minus signs. Recall that it makes no difference how summation indices are labeled; summing over all values of ε is the same as summing over all values of μ.

Hence, the divergence of the Einstein tensor is always zero.

$$G^{\alpha}{}_{\beta;\alpha} = 0$$

§27.3 Geometry Conserves Momentum & Energy

The Bianchi identity ensures that the divergence of the Einstein tensor G equals zero, always and everywhere. Through Einstein's field equations this ensures that the divergence of the mass-energy tensor T is zero, also always and everywhere.

$$8\pi \, T^{\alpha}{}_{\beta;\alpha} = G^{\alpha}{}_{\beta;\alpha} = 0$$

The physical significance of T having zero divergence is that energy and momentum are conserved — two of the most important principles of physics.

Let's discover why the origin of these hallowed conservation principles are due to the geometry of spacetime.

Our 4-D spacetime is a *compact manifold*, as properly described in the mathematics of topology.

Roughly speaking, compact manifolds are spaces of any number of dimensions that have no voids, and have smooth, concave boundaries.

More technically, for any pair of points A and B within a compact manifold, every point along the line connecting A and B is also within the manifold.

Within a compact manifold of any dimensionality, Gauss's theorem relates the divergence of a tensor within a volume V to properties of that tensor on the surface S that encloses V.

In 4-D, any infinitesimal 3-D surface d^3S has an associated 4-vector perpendicular to it that we denote d^3S_ε. For any rank 2 tensor $X^{\alpha\varepsilon}$, the tensor product of X with the perpendicular to d^3S is:

$$X^{\alpha\varepsilon}\ d^3S_\varepsilon$$

In this notation, the general form of Gauss's theorem in 4-D is:

$$\int_V X^{\alpha\varepsilon}{}_{;\varepsilon}\ d^4V = \int_S X^{\alpha\varepsilon}\ d^3S_\varepsilon$$

Here, the left integral is over the 4-D volume V, and the right integral is over the 3-D surface S.

Perhaps the most comprehensive book on general relativity is ***Gravitation*** by Charles Misner, Kip Thorne, and John Archibald Wheeler. They describe the Einstein tensor G as the moment of rotation of spacetime curvature, and they relate G to a rank 3 tensor; call it tensor $M^{\alpha\beta\varepsilon}$. Tensor M at event x^μ is the contraction of the cross product of x^μ with the *double dual* of the Riemann tensor. Dual tensors are discussed in Section §30.4. I will not reproduce their derivation here, but these authors prove the following equation:

$$G^{\alpha\beta} = \int_V M^{\alpha\beta\varepsilon}{}_{;\varepsilon}\ d^4V = \int_S M^{\alpha\beta\varepsilon}\ d^3S_\varepsilon$$

Above, we dealt with 4-D volume V bounded by 3-D surface S. Next, let's apply Gauss's theorem to the divergence of G, treating S as a 3-D volume bounded by a 2-D surface Σ.

$$\int_S G^{\alpha\beta}{}_{;\beta}\ d^3S_\alpha = \int_\Sigma G^{\alpha\beta}\ d^2\Sigma_\beta$$

Here is the key point: a boundary has no boundary — the boundary of a boundary is zero.

Within a compact manifold, the boundary S of a 3-D volume V is a continuous surface that has no edges and no holes. If one starts at any point P on S, and proceeds in any direction across surface S, one forever loops around S in some manner, but never reaches the end of S — S has no boundary.

Here are two familiar examples. The 2-D surface of our 3-D Earth has no edges or voids — it is continuous, complete, and boundary-less. The boundary of a 2-D disk is a 1-D circle that has neither a beginning nor an end.

This argument applies in any number of dimensions.

The above defined surface Σ is identically zero, because it is the boundary of S, which in turn is the boundary of V.

That is the fundamental message of the Bianchi identity.

Since the area of every such 2-D surface Σ equals zero, the above equation shows that the divergence of G must be zero throughout every volume, by Gauss's theorem. This also means the divergence of T must be zero everywhere, and therefore energy and momentum are always conserved.

The "river" of 4-momentum T flows from here to there, from past to future, but never disappears into a drain or emerges from a spout.

Energy and momentum conservation result from 4-D spacetime being a compact manifold, in which the boundary of a boundary is zero.

Chapter 28
Photon Orbits Near Black Holes

In Chapter 25, we examined the orbits of a small body of mass m around a much more massive body of mass M. We defined the following:

- Orbital plane defined by a fixed $\theta = \pi / 2$
- t, coordinate time (not proper time)
- r, radial distance from the center of M
- ø, azimuthal angle within the orbital plane
- L, small body's angular momentum

We derived the following equations for relativistic orbits (Eqn.'s #1 through #3 in Chapter 25 pertain to Newtonian orbits):

$$L = m\, r\, v_ø = m\, r^2\, (dø/dt)$$

Eqn. #4: $$(dr/d\tau)^2 = (E / m)^2 - V$$

Eqn. #5: $$V = (1 - 2\, M / r)\, (1 + L^2 / m^2 r^2)$$

$$V = 1 - 2\, M / r + L^2 / m^2 r^2 - 2\, M\, L^2 / m^2 r^3$$

Eqn. #6: $$(dr/dø)^2 = (E^2 - m^2)\, r^4 / L^2 + 2\, M\, m^2 r^3 / L^2 - r^2 + 2\, M\, r$$

Eqn. #7: $$ø = \int dr \,/ \sqrt{\{(E^2 - m^2)\, r^4 / L^2 + 2\, M\, m^2 r^3 / L^2 - r^2 + 2\, M\, r\}}$$

The key feature of effective gravitational potential V, plotted below, is the central pit, where V approaches $-\infty$ as r approaches zero, as shown in this copy of Fig 25-2.

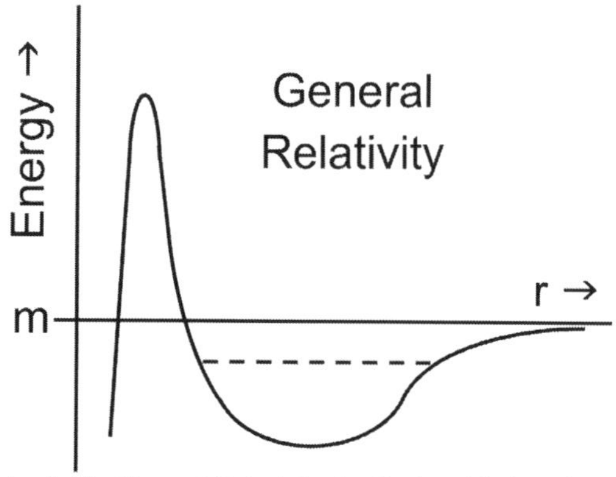

Effective Potential in General Relativity for Body with Angular Momentum

This central pit distinguishes relativistic and Newtonian physics. The pit in effective potential V becomes significant as r approaches 3 M, hence this critical feature of relativistic physics is primarily important for black holes and neutron stars. For less compact objects, nothing can come close enough to the object's center to pass the maximum positive V, created by the angular momentum term. For example, the surface of our Sun is at r = 470,000 M, far outside the central pit.

Let's therefore focus here on the most interesting orbits, those near black holes.

In Chapter 25, we also found the maximum and local minimum of potential V, where the derivative of V is zero:

$$V = 1 - 2M/r + L^2/m^2 r^2 - 2ML^2/m^2 r^3$$

We define $\mu M = L/m$: $$V = 1 - 2M/r + \mu^2 M^2/r^2 - 2\mu^2 M^3/r^3$$

$$0 = dV/dr = +2M/r^2 - 2\mu^2 M^2/r^3 + 6\mu^2 M^3/r^4$$

Multiply by $r^4/2M$: $$0 = +r^2 - \mu^2 M r + 3\mu^2 M^2$$

The solutions of this quadratic equation are:

Eqn. #8: $$r = \mu^2 M \{1 \pm \sqrt{(1 - 12/\mu^2)}\}/2$$

The plus sign above corresponds to the local minimum V, and the minus sign to the maximum V.

For $\mu^2 < 12$, there are no stable orbits at any real values of r.

For $\mu^2 = 12\,(1+\varepsilon)$ and $\varepsilon \ll 1$, we can approximate the square root as: $\sqrt{\{1 - (1 - \varepsilon)\}} = \sqrt{\varepsilon}$. The equation then becomes:

$$r = \mu^2 M \{1 \pm \sqrt{\varepsilon}\}/2$$

The maxiumum of V is at r slightly less than 6 M, and the local minimum of V is at r slightly more than 6 M, where stable orbits exist, the closest possible stable orbits near a non-rotating black hole.

For $\mu^2 = 12$, the maxiumum and local minimum of V are both at r = 6 M; an orbit there is metastable, like a knife balanced on its tip.

§28.1 Photon Orbits

Now, let's consider orbits of massless particles: photons.

As shown in Figure 28-1, we define the *impact parameter* b = L / E, the photon's angular momentum divided by its energy. For photons, E equals its linear momentum p (with c = 1), and we have L = b p. Impact parameter b is the distance of closest approach to the gravitational center along the photon's original trajectory, the path it would follow in flat spacetime. Also shown are the photon's radial distance r, and azimuthal angle ø.

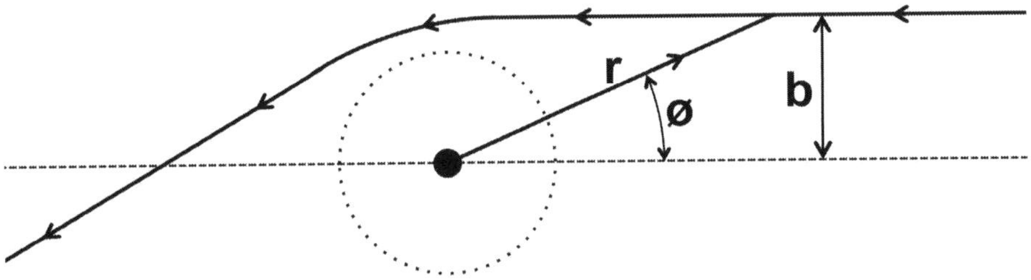

Figure 28-1 Photon Approaching Black Hole (Black Dot inside dotted Event Horizon)

Recall Eqn. #6:

$$(dr/dø)^2 = (E^2 - m^2)\, r^4/L^2 + 2M m^2 r^3/L^2 - r^2 + 2Mr$$

Now take the limit as m goes to zero:

$$(dr/dø)^2 = r^4/b^2 - r^2(1 - 2M/r)$$

$$(dr/dø)^2/r^4 = 1/b^2 - (1 - 2M/r)/r^2$$

Eqn. #9: $$(dr/dø)^2/r^4 = 1/b^2 - 1/B^2$$

Where: $$1/B^2 = (1 - 2M/r)/r^2$$

Figure 28-2 is a sketch of $1 / B^2$ versus radial distance r.

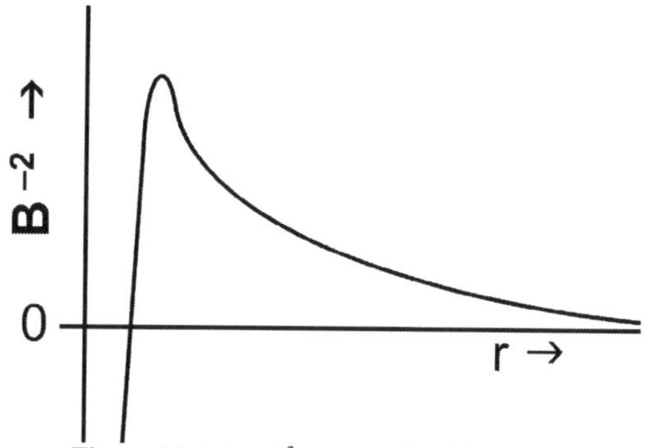

Figure 28-2 ($1 / B^2$) vs. Radial Distance r

Let's explore the consequences of Eqn. #9 by finding the maximum of $1 / B^2$.

$$0 = dB^{-2}/dr = d/dr \{ (1 - 2M/r) / r^2 \}$$

$$0 = d/dr \{ 1/r^2 - 2M/r^3 \}$$

$$0 = -2/r^3 - (2M)(-3/r^4)$$

Multiply both sides by $(r^4 / 2)$:

$$0 = -r + 3M$$

$$r = 3M$$

At its maximum, B^{-2} equals:

$$(B^{-2})_{max} = (1 - 2M/3M) \,/\, 9M^2$$

$$(B^{-2})_{max} = (1/3) \,/\, 9M^2$$

$$(B^{-2})_{max} = 1 / 27M^2$$

$$(B^{-1})_{max} = 1 / 3M\sqrt{3}$$

The maximum of B^{-2} corresponds to a minimum B, given by:

$$B_{min} = 3M\sqrt{3} \approx 5.196\,M$$

Any photon approaching the black hole from far away must initially have a negative $(dr/d\phi)$ — as ϕ increases, the photon moves closer to the black hole.

A photon approaching with impact parameter $b < B_{min}$ has $b^{-2} > (B^{-2})_{max}$. The right side of Eqn. #9 is then always greater than zero. This means $(dr/d\phi)$ cannot pass through zero, and dr must always remain negative. Such photons spiral into the black hole.

A photon approaching with impact parameter $b > B_{min}$ has $b^{-2} < (B^{-2})_{max}$. These photons eventually reach a radius r at which $(dr/d\phi) = 0$, their *periapsis*. Here r stops decreasing, and starts increasing. These photons, in effect, bounce off, deflected by the effective potential V of the gravitating source M.

Let's next consider the fate of a photon emitted just outside a black hole's event horizon.

If the photon's initial trajectory points directly toward the singularity, the photon plummets into oblivion. If it points in exactly the opposite direction, the photon escapes from the black hole's gravitational field.

Where is the dividing line separating oblivion from escape?

In Figure 28-3, the dotted circle represents the black hole's event horizon, and the black dot represents its singularity.

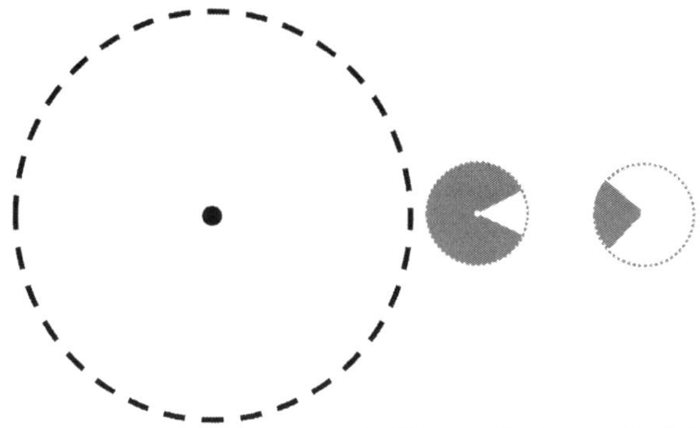

Figure 28-3 Survival Zones in White for Photons Emitted at Disk Centers

Consider a photon starting at the center of the Pacman-like partial disk closest to the event horizon. Let r be its distance from the singularity, with $2\ M < r < 3\ M$ — outside the event horizon but inside the maximum of B^{-2}. This photon will escape from the black hole only if its initial trajectory lies within the white portion of the disk, Pacman's mouth. If it is emitted within the gray portion, it will enter the event horizon and be consumed.

A photon starting farther away, $r > 3\ M$, has more white than gray in its survival disk, as shown by the rightmost partial disk in Figure 28-3.

For photons starting at $r < 3\ M$, the survival criterion is $\sin(\alpha) < 3\ M\sqrt{3}\ /\ B(r)$, where α is the angle between the photon's initial trajectory and the radius vector $\boldsymbol{r}$; α is the half-angle of the white survival zone. Photons starting in the white zone eventually escape the black hole, possibly after spiraling around it several times. For photons starting farther away with $r > 3\ M$, the criterion is reversed: α becomes the half-angle of the much smaller gray zone of oblivion on the side of the disk nearest the black hole.

If an enormous blast of radiation is released just outside a black hole's event horizon, photons will be emitted with a wide range of directions and b values. Some may come directly toward our telescopes, some may orbit once before coming toward us, and some may orbit multiple times before escaping. In such circumstances, we might detect a periodic series of flashes of decreasing intensity.

Photons with initial $r < 3\ M$ that start in the survival disk's gray zone are doomed. Even if they are initially moving outward, with $dr > 0$, the right side of Eqn. #9 eventually becomes zero, dr passes through zero and becomes negative. Those photons reach apapsis, and then spiral into the black hole, possibly after several orbits.

Perhaps the most bizarre scenario happens to photons approaching from afar with values of b only slightly greater than $3\ M\sqrt{3}$. These photons are effectively "reflected" by the black hole's gravity, but not before spiraling around the black hole, perhaps many times.

One can even whimsically imagine a black hole being a bizarre type of mirror. When exposed to an intense flash of light from afar, a black hole redirects some of that light back to the source. Most of the redirected photons return after looping around the backside of the black hole; these photons rotate through 180 degrees, one-half of a full turn. Fewer return after 1.5 full turns, and even fewer after 2.5 turns, etc. The fraction f of the original light that is redirected back to the source and captured by a telescope of aperture d near the source is:

$$f = \pi\,(d\,M^2 / R^3) \times (1.75 + 0.0029 + 0.0000055 + \ldots)$$

Here R is the distance between the source and a black hole of mass M. The decreasing numerical series corresponds to photons making 0.5, 1.5, 2.5, ... full turns.

If this were practical, and it certainly is not, one could dream of seeing images of Earth's past reflected by a remote black hole. The light received today from a "mirror" 40 million light-years away might reveal dinosaurs swimming in the seas of Montana.

Chapter 29
Rotating Black Holes

So far we have discussed black holes that do not rotate and have zero net electric charge, although we were not explicit about that last requirement.

Particles inside the event horizon of a black hole cannot interact with anything outside the event horizon through either the strong or weak forces, each of which has a subatomic range. That leaves only gravity and electromagnetism.

Theoretical calculations show that, after stabilizing following their initial collapse, black holes have only three properties that can affect the exterior world:

- Mass M
- Angular momentum J
- Electric charge Q

These three parameters completely define black holes, as perceived outside their event horizons. This limited diversity has been described as "black holes have no hair" — no ornate or colorful traits.

§29.1 Kerr-Newman Metric

The most complex black holes have spin and charge, in addition to mass. The Schwarzschild metric that we have relied upon up to now pertains only to the simplest black holes — those with $J = Q = 0$. The Kerr-Newman metric, developed nearly 50 years after the Schwarzschild metric, describes the full complexity of black holes of all types. As before, we employ spherical coordinates, with coordinate time t, radial distance r, polar angle θ, and azimuthal angle ϕ. This metric is:

$$\begin{aligned} ds^2 = &- [dt - S \sin^2\theta \, d\phi]^2 \, \Delta / \Sigma \\ &+ [(r^2 + S^2) \, d\phi - S \, dt]^2 \sin^2\theta / \Sigma \\ &+ dr^2 \, \Sigma / \Delta + \Sigma \, d\theta^2 \end{aligned}$$

Where:

$$\Delta = r^2 - 2 M r + S^2 + Q^2$$

$$\Sigma = r^2 + S^2 \cos^2 \theta$$

$$S = J / M$$

Above, both charge Q and spin S are always less than or equal to M, in natural units.

Large bodies, such as stars, have a strong tendency to be electrically neutral. The electric force is so much stronger than gravity that unbalanced charges are rapidly neutralized. We expect the same for black holes. Let's therefore simplify somewhat by setting $Q = 0$. With $Q = 0$, this is called the Kerr metric.

Grouping terms with the same coordinate differentials (see Section §29.6 for this math), we obtain:

$$\begin{aligned} ds^2 = &- dt^2 [1 - 2 M r / \Sigma] + dr^2 \, \Sigma / \Delta + \Sigma \, d\theta^2 \\ &+ d\phi^2 \, [(r^2 + S^2)^2 - S^2 \Delta \sin^2\theta] \sin^2\theta / \Sigma \\ &+ 2 \, dt \, d\phi \, [-2 M r S \sin^2\theta] / \Sigma \end{aligned}$$

The Kerr metric has the form:

$$ds^2 = dt^2\ g_{tt} + d\phi^2\ g_{\phi\phi} + 2\, dt\, d\phi\ g_{t\phi}$$
$$+ dr^2\ g_{rr} + d\theta^2\ g_{\theta\theta}$$

This employs the metric symmetry:

$$g_{t\phi} + g_{\phi t} = 2\, g_{t\phi}$$

§29.2 Orbits in Kerr Metric

Previously, in analyzing Newtonian orbits and relativistic orbits near non-rotating black holes, we were free to choose coordinates in which the orbital plane was $\theta = \pi / 2$. Here, for a rapidly rotating black hole, it behooves us to align our coordinate axes so that $\theta = 0$ points in the direction of the black hole's angular momentum. This means the black hole rotates about the ϕ-axis.

Azimuthal orbits, those where ϕ cycles throughout its total range of 0 to 2π radians, are only meaningful when the polar angle θ is not at either end of its range: 0 or π. Let's therefore address orbits for which $\sin\theta > 0$.

We define the angular velocity $\Omega = d\phi/dt$. (This Ω is not the Ω of the Schwarzschild metric.) The interval along the path of a body orbiting a black hole is:

$$ds^2 = X\ dt^2 + dr^2\ g_{rr} + d\theta^2\ g_{\theta\theta}$$

Where:

$$X = g_{tt} + \Omega^2\ g_{\phi\phi} + 2\,\Omega\ g_{t\phi}$$

Since g_{rr} and $g_{\theta\theta}$ are both > 0, and since the interval must be negative for any particle with mass, X must be negative.

Since $g_{\phi\phi} > 0$, the equation for X is quadratic in Ω, and we can factor it as:

$$X = g_{\phi\phi}\,(\Omega - \Omega_{min})\ (\Omega - \Omega_{max})$$

X is zero at two values of Ω, and X is negative only if Ω is in between those values. Hence for massive particles, Ω must be between these limits. The two roots of the quadratic equation are:

$$X = \{ - g_{t\phi} \pm \sqrt{(\, g_{t\phi}^{\ 2} - g_{tt}\ g_{\phi\phi}\,)}\,\} / g_{\phi\phi}$$

Let's define:

$$\omega = - g_{t\phi} / g_{\phi\phi}$$
$$\Omega_{max} = \omega + \sqrt{(\omega^2 - g_{tt} / g_{\phi\phi})}$$
$$\Omega_{min} = \omega - \sqrt{(\omega^2 - g_{tt} / g_{\phi\phi})}$$

The allowed range of Ω is:

$$\Omega_{min} < \Omega < \Omega_{max}$$

In this range, X is negative.

Let's examine the polarity of ω. Recall the expression for $g_{t\phi}$:

$$g_{t\phi} = -[\, 2\, M\, r\, S \sin^2\theta] / \Sigma$$
$$g_{t\phi} = -[\, 2\, M\, r\, S \sin^2\theta] / (\, r^2 + S^2 \cos^2\theta\,)$$

The denominator of Σ is positive because all its terms are squared. The numerator is negative because every term within the [ ]'s is positive. Hence, $g_{t\phi}$ is negative.

Now consider $g_{\phi\phi}$:

$$g_{\phi\phi} = [\,(r^2 + S^2)^2 - S^2\,\Delta\,\sin^2\theta]\,\sin^2\theta\,/\,\Sigma$$

$$\Delta = r^2 - 2\,M\,r + S^2 + Q^2$$

We are assuming $Q = 0$, and have shown above that Σ is positive. Evidently, the quantity in [ ]'s in the upper equation is minimum when $\sin^2\theta = 1$. Expanding that quantity, assuming $\sin^2\theta = 1$, yields:

$$[\,] = r^4 + 2\,r^2\,S^2 + S^4 - r^2\,S^2 + 2\,M\,r\,S^2 - S^4$$

$$[\,] = r^4 + r^2\,S^2 + 2\,M\,r\,S^2$$

Clearly, this is always positive. Therefore, for any $\theta > 0$:

$$\omega = -\,g_{t\phi}\,/\,g_{\phi\phi} > 0$$

Far from the black hole, where $r >> M$ and $r >> S$:

$$\Delta = r^2 - 2\,M\,r + S^2 \rightarrow r^2$$

$$\Sigma = r^2 + S^2\cos^2\theta \rightarrow r^2$$

$$g_{tt} = -1 + 2\,M\,r\,/\,\Sigma \rightarrow -1$$

$$g_{\phi\phi} = [\,(r^2 + S^2)^2 - S^2\,\Delta\,\sin^2\theta]\,\sin^2\theta\,/\,\Sigma \rightarrow r^2\sin^2\theta$$

$$g_{t\phi} = [\,-2\,M\,r\,S\,\sin^2\theta]\,/\,\Sigma \rightarrow -2\,M\,S\,\sin^2\theta\,/\,r$$

$$\omega = -\,g_{t\phi}\,/\,g_{\phi\phi} \rightarrow 2\,M\,S\,/\,r^3$$

$$\omega^2 - g_{tt}\,/\,g_{\phi\phi} \rightarrow 4\,M^2\,S^2\,/\,r^6 + 1\,/\,r^2\sin^2\theta \rightarrow 1\,/\,r^2\sin^2\theta$$

$$\omega \pm \sqrt{(\omega^2 - g_{tt}\,/\,g_{\phi\phi})} \rightarrow 2\,M\,S\,/\,r^3 \pm 1\,/\,r\sin\theta$$

$$r\,\sin\theta\;\Omega_{max} \rightarrow +1$$

$$r\,\sin\theta\;\Omega_{min} \rightarrow -1$$

The last two equations mean that, far from a rotating black hole, orbital speeds in either azimuthal direction are restricted only by the speed of light.

But closer in, the situation is dramatically different.

As r decreases, Ω_{min} becomes less negative and Ω_{max} becomes less positive. Both changes reduce the allowed range of rotational frequency. At some r, Ω_{min} becomes 0. This occurs when:

$$g_{tt} = -1 + 2\,M\,r\,/\,\Sigma = 0$$

$$\Omega_{min} = \omega - \sqrt{(\omega^2 - g_{tt}\,/\,g_{\phi\phi})} = \omega - \sqrt{(\omega^2)} = 0$$

$$\Omega_{max} = \omega + \sqrt{(\omega^2 - g_{tt}\,/\,g_{\phi\phi})} = \omega + \sqrt{(\omega^2)} = 2\,\omega$$

When $g_{tt} = 0$:

$$2\,M\,r = \Sigma$$

$$2\,M\,r = r^2 + S^2\cos^2\theta$$

$$r = M + \sqrt{(M^2 - S^2\cos^2\theta)}$$

This is called the *static limit*. Within the static limit, all free-falling objects orbit the black hole in the same direction as its spin S. This is because space swirls around the black hole, twisted by its gravity.

This phenomenon is called *frame dragging*.

As r continues decreasing, frame dragging intensifies. At a certain r, the allowed range of orbital motion reduces to a single rotational frequency given by:

$$\omega^2 = g_{tt} / g_{\phi\phi}$$

When $\omega^2 = g_{tt} / g_{\phi\phi}$:

$$\Omega_{min} = \Omega = \Omega_{max}$$

At this r, only one rotational frequency becomes possible. Section §29.7 presents a full calculation of r, and shows that its value is:

$$r = M + \sqrt{(M^2 - S^2)}$$

What happens at this radius?

Since $\Omega_{min} = \Omega = \Omega_{max}$, the quadratic in Ω equals zero. Hence:

$$X = g_{tt} + \Omega^2 g_{\phi\phi} + 2\,\Omega\, g_{t\phi} = 0$$

This eliminates three terms from the interval. For a photon, the remaining interval is:

$$ds^2 = 0 = dr^2\, g_{rr} + d\theta^2\, g_{\theta\theta}$$

This equation requires $dr = d\theta = 0$. We discover that r is the radius of the event horizon — the place where time stops, as seen by remote observers. This precludes any change in either r or θ.

§29.3 Ergosphere & Static Limit

To recap, a rotating black hole is surrounded by two critical surfaces. The locations of these surfaces are:

Static limit: $$r = M + \sqrt{(M^2 - S^2 \cos^2\theta)}$$

Event horizon: $$r = M + \sqrt{(M^2 - S^2)}$$

The event horizon of every black hole is spherical, but its size is affected by its rotation rate. If the black hole does not rotate at all, if $S = 0$, its event horizon radius r equals 2 M, the Schwarzschild radius. If its rotation rate is modest, $S << M$, the radius of the event horizon is only slightly less than 2 M. But if S is maximal, $S = M$, the radius of the event horizon shrinks to $r = M$, half the Schwarzschild radius.

As shown in cross section in Figure 29-1, the static limit is oblate rather than spherical, due to the cosine term. The static limit coincides with the event horizon where $\theta = 0$ and $\theta = \pi$, at the top and bottom of the white circle, at both poles of the vertical axis of rotation. The static limit bulges outward at the equator, as does the surface of a rotating star.

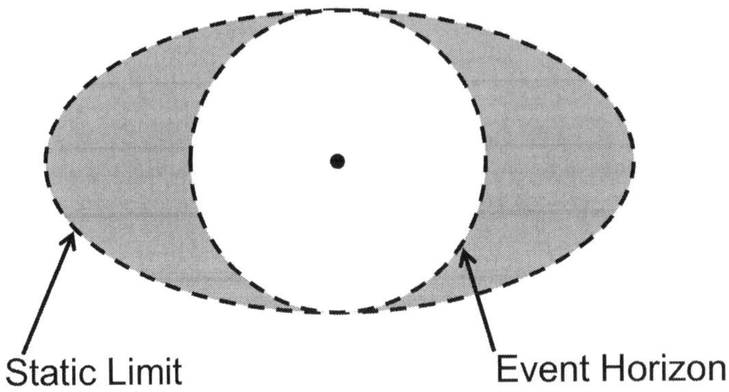

Figure 29-1 Event Horizon & Static Limit of Rotating Black Hole

The gray area in Figure 29-1, everything between the static limit and the event horizon, is called the *ergosphere*.

Within the ergosphere, all free-falling objects orbit the black hole in the same direction as its spin S. A spaceship entering the ergosphere is accelerated, and forced to orbit faster and faster as it approaches the event horizon.

If, due to its enormous angular momentum, a spaceship within the ergosphere manages to avoid entering the event horizon, it can escape from the black hole, at least in principle.

However, as with a non-rotating black hole, anything that enters the event horizon can never escape.

§29.4 Photon Orbits

Let's imagine a hollow toroidal tube within a black hole's ergosphere that has θ always = $\pi/ / 2$ and a constant r, as seen in cross section in Figure 29-2. Let the tube's interior be perfectly reflecting, making it a *light pipe*, a conduit carrying light around in either direction.

If we flash a light at event ($t = 0$, r, θ, $\phi = 0$), two light beams will move through the light pipe, one toward $+\phi$ and one toward $-\phi$.

Where will the two light beams meet, as they race around the light pipe in opposite directions?

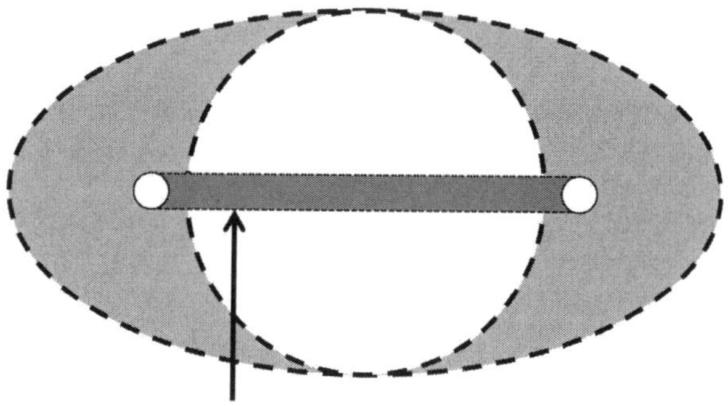

Figure 29-2 Photons in Fiber Optic Ring in Ergosphere

Let's examine the invariant interval for such photons. Since they are constrained to move within the light pipe, $dr = d\theta = 0$. We again define their angular velocity as $\Omega = d\phi/dt$. The invariant interval is reduced to:

$$0 = ds^2 = dt^2\, g_{tt} + d\phi^2\, g_{\phi\phi} + 2\, d\phi\, dt\, g_{t\phi}$$

Dividing by dt^2:

$$0 = g_{tt} + \Omega^2\, g_{\phi\phi} + 2\,\Omega\, g_{t\phi}$$

This is the equation for X that we had in Section §29.2. The solutions are:

$$\omega = -\,g_{t\phi} / g_{\phi\phi}$$

$$\Omega = \omega \pm \sqrt{\{\omega^2 - g_{tt} / g_{\phi\phi}\}}$$

This means photons moving in the same direction as the black hole's spin have a higher apparent speed, $\omega + \sqrt{\{\ldots\}}$, than the photons moving in the opposite direction, $\omega - \sqrt{\{\ldots\}}$.

This is not because the constancy of the speed of light is violated.

Instead, this is because space is rotating with angular velocity ω.

Photons in both directions are moving *through space* at v = 1, but space is carrying the photons, and everything else, toward + ϕ. An observer stationary in local space sees photons going at the same speed in both directions.

The two flashes return to their "starting point" simultaneously *if* what we call the "starting point" also rotates at ω.

§29.5 Confirmation of Frame-Dragging

Gravity Probe B ("GPB") may be history's longest-lasting physics experiment; it extended over 47 years and cost $750 million.

In 2011, GPB confirmed the frame dragging of Earth's gravity. The prediction of general relativity and the measured value are:

	Earth's Frame Dragging (in milliarcseconds per year)
General relativity prediction:	39.2
GPB measured value:	37.2 ± 7.2

GPB also confirmed the orbital precession of the GPB satellite due to Earth's gravity, which they call the *geodetic effect*, the predicted and measured values are:

	Earth's Precession Effect (in milliarcseconds per year)
General relativity prediction:	6606
GPB measured value:	6602 ± 18

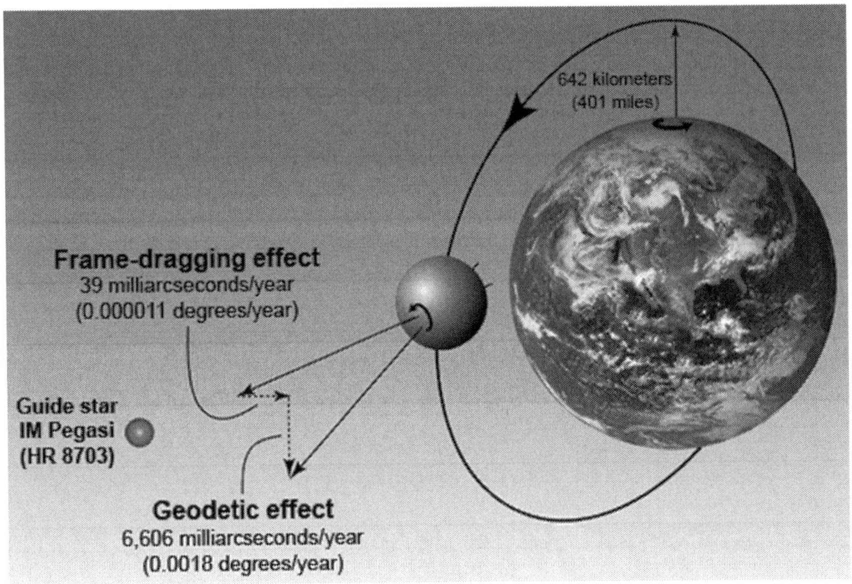

Figure 29-3 Schematic of Gravity Probe B Experiment

GPB thus confirmed both Einstein's theory of general relativity and the Kerr metric for spacetime curvature due to a rotating massive body.

§29.6 Rearranging Kerr Metric

This section presents the mathematics of rearranging the Kerr metric so that terms with the same combination of coordinate differentials are placed together. If this were not tensor calculus, this rearrangement would be trivial. We assume, as above, that the black hole is electrically neutral (Q = 0).

As above:

$$\Delta = r^2 - 2\,M\,r + S^2$$

$$\Sigma = r^2 + S^2 \cos^2\theta$$

$$S = J / M$$

Start with the original Kerr metric:

$$\begin{aligned} ds^2 = &- [dt - S \sin^2\theta \, d\phi]^2 \, \Delta / \Sigma \\ &+ [\,(r^2 + S^2)\, d\phi - S\, dt]^2 \sin^2\theta / \Sigma \\ &+ dr^2\, \Sigma / \Delta + \Sigma\, d\theta^2 \end{aligned}$$

First, expand the [ ]'s.

$$\begin{aligned} ds^2 = &- [dt^2 - 2\,S \sin^2\theta \; dt \; d\phi + S^2 \sin^4\theta \; d\phi^2] \; \Delta / \Sigma \\ &+ (r^2 + S^2)^2 \; d\phi^2 \; \sin^2\theta / \Sigma \\ &- 2\,S\,(r^2 + S^2)\; dt \; d\phi \; \sin^2\theta / \Sigma \\ &+ S^2\, dt^2 \; \sin^2\theta / \Sigma \\ &+ dr^2 \; \Sigma / \Delta + \Sigma \; d\theta^2 \end{aligned}$$

Group like differentials:

$$\begin{aligned} ds^2 = &- dt^2 \; [\Delta / \Sigma - S^2 \sin^2\theta / \Sigma\,] \\ &+ dr^2 \; \Sigma / \Delta + \Sigma\, d\theta^2 \\ &+ d\phi^2 \; [- S^2 \; \sin^4\theta \; \Delta / \Sigma + (r^2 + S^2)^2 \; \sin^2\theta / \Sigma\,] \\ &+ dt \; d\phi \; [2\,S \sin^2\theta \; \Delta / \Sigma - 2\,S \; (r^2 + S^2) \; \sin^2\theta / \Sigma\,] \end{aligned}$$

Make some substitutions:

$$\begin{aligned} ds^2 = &- dt^2 \; [r^2 - 2\,M\,r + S^2 - S^2 \sin^2\theta\,] / \Sigma \\ &+ dr^2 \; \Sigma / \Delta + \Sigma \; d\theta^2 \\ &+ d\phi^2 \; [\,(r^2 + S^2)^2 - S^2 \sin^2\theta \; \Delta\,] \sin^2\theta / \Sigma \\ &+ dt \; d\phi \; [2\,S \sin^2\theta \; (r^2 - 2\,M\,r + S^2)\,] / \Sigma \\ &- dt \; d\phi \; [2\,S \sin^2\theta \; (r^2 + S^2)\,] / \Sigma \end{aligned}$$

Simplify lines 1, 4 and 5:

$$\begin{aligned} ds^2 = &- dt^2 \; [r^2 - 2\,M\,r + S^2 \cos^2\theta\,] / \Sigma \\ &+ dr^2 \; \Sigma / \Delta + \Sigma \; d\theta^2 \\ &+ d\phi^2 \; [\,(r^2 + S^2)^2 - S^2 \; \sin^2\theta \; \Delta\,] \sin^2\theta / \Sigma \\ &+ dt \; d\phi \; [2\,S \; \sin^2\theta\,(-2\,M\,r)\,] / \Sigma \end{aligned}$$

Replace $r^2 + S^2 \cos^2\theta$ with Σ:

$$ds^2 = -\,dt^2\,[1 - 2\,M\,r\,/\,\Sigma\,]$$
$$+\,dr^2\,\Sigma\,/\,\Delta + \Sigma\,d\theta^2$$
$$+\,d\phi^2\,[\,(r^2 + S^2)^2 - S^2\sin^2\theta\,\Delta\,]\sin^2\theta\,/\,\Sigma$$
$$+\,2\,dt\,d\phi\,[-2\,M\,r\,S\,\sin^2\theta]\,/\,\Sigma$$

Return to Section §29.2 for further discussion of the Kerr metric in this form.

§29.7 Where Is $\Omega_{min} = \Omega_{max}$?

Here, we calculate the value of r at which:

$$\omega^2 = g_{tt}\,/\,g_{\phi\phi}$$
$$(-g_{t\phi}\,/\,g_{\phi\phi})^2 = g_{tt}\,/\,g_{\phi\phi}$$
$$(\,g_{t\phi})^2 = g_{tt}\,g_{\phi\phi}$$

Substituting the equation for each metric component yields:

$$[-2\,M\,r\,S\,\sin^2\theta\,]^2\,/\,\Sigma^2 = -[\,1 - 2\,M\,r\,/\,\Sigma\,]$$
$$\times\,[\,(r^2 + S^2)^2 - S^2\,\Delta\,\sin^2\theta]\,\sin^2\theta\,/\,\Sigma$$

Multiply both sides by $\Sigma^2 / \sin^2\theta$:

$$4\,M^2\,r^2\,S^2\,\sin^2\theta = [\,2\,M\,r - \Sigma\,]$$
$$\times\,[r^4 + 2\,r^2\,S^2 + S^4 - S^2\,(r^2 - 2\,M\,r + S^2)\,\sin^2\theta]$$

$$4\,M^2\,r^2\,S^2\,\sin^2\theta = [2\,M\,r - r^2 - S^2\,\cos^2\theta]$$
$$\times\,[r^4 + (2 - \sin^2\theta)\,r^2\,S^2 + S^4\,\cos^2\theta + 2\,M\,r\,S^2\,\sin^2\theta]$$

Expand the product of two [ ]'s:

$$4\,M^2\,r^2\,S^2\,\sin^2\theta = 2\,M\,r^5 + 2\,M\,r^3\,S^2\,(2 - \sin^2\theta)$$
$$+\,2\,M\,r\,S^4\,\cos^2\theta + 4\,M^2\,r^2\,S^2\,\sin^2\theta - r^6$$
$$-\,(2 - \sin^2\theta)\,r^4\,S^2 - r^2\,S^4\,\cos^2\theta$$
$$-\,2\,M\,r^3\,S^2\,\sin^2\theta - S^2\,r^4\,\cos^2\theta$$
$$-\,r^2\,S^4\,\cos^2\theta\,(2 - \sin^2\theta)$$
$$-\,S^6\,\cos^4\theta - 2\,S^4\,M\,r\,\cos^2\theta\,\sin^2\theta$$

Cancel / combine like terms:

$$0 = +\,2\,M\,r^5 + 4\,M\,r^3\,S^2\,\cos^2\theta$$
$$+\,2\,M\,r\,S^4\,\cos^4\theta - r^6 - (1 + 2\cos^2\theta)\,r^4\,S^2$$
$$-\,r^2\,S^4\,\cos^2\theta\,(2 + \cos^2\theta) - S^6\,\cos^4\theta$$

$$0 = + 2Mr\{ r^4 + 2r^2 S^2 \cos^2\theta + S^4 \cos^4\theta \}$$
$$- r^6 - r^4 S^2 - 2r^4 S^2 \cos^2\theta - 2r^2 S^4 \cos^2\theta$$
$$- r^2 S^4 \cos^4\theta - S^6 \cos^4\theta$$

$$0 = + 2Mr\{ r^2 + S^2 \cos^2\theta \}^2 - r^4 (r^2 + S^2)$$
$$- 2r^2 S^2 \cos^2\theta\ (r^2 + S^2) - S^4 \cos^4\theta\ (r^2 + S^2)$$

$$0 = + 2Mr\{ r^2 + S^2 \cos^2\theta \}^2$$
$$- \{ r^4 + 2r^2 S^2 \cos^2\theta + S^4 \cos^4\theta \} (r^2 + S^2)$$

$$0 = 2Mr\{ r^2 + S^2 \cos^2\theta\}^2 - \{ r^2 + S^2 \cos^2\theta\}^2 (r^2 + S^2)$$
$$0 = \{ 2Mr - r^2 - S^2\} \{ r^2 + S^2 \cos^2\theta \}^2$$

Since the term in the right { } is always > 0, the left { } term must be zero.

$$0 = 2Mr - r^2 - S^2$$

When $\omega^2 = g_{tt} / g_{\phi\phi}$, the equation for Ω:

$$\Omega = \omega \pm \sqrt{\{\omega^2 - g_{tt} / g_{\phi\phi}\}}$$

Becomes:

$$\Omega = \omega$$

And:

$$\Omega_{min} = \Omega = \Omega_{max}$$

This means rotation can only occur at one specific frequency: $\Omega = \omega$.

Return to the end of Section §29.2, to find out why this r defines the event horizon.

Chapter 30
Electromagnetism in General Relativity

We now explore the tensors used in general relativity for phenomena involving the 3-vector electric field $\hat{\boldsymbol{E}}$ and the 3-vector magnetic field $\boldsymbol{B}$.

Since physicists use E for both energy and electric fields, I will denote the electric field as $\hat{\boldsymbol{E}}$, slightly differently to minimize confusion.

To avoid clutter, I will use natural units (c = 1). To restore the c's, multiply charge density ϱ by c, and divide all components of electric field $\hat{\boldsymbol{E}}$ by c.

§30.1 Principal Tensors

We begin with the 4-vector electric current J^α and the 4-vector potential A^α.

For electric current 3-vector: $\boldsymbol{J} = (J_x, J_y, J_z)$

At event x^μ:

$$J^\alpha(x^\mu) = (\varrho, J_x, J_y, J_z)$$

$$A^\alpha(x^\mu) = \int_V J^\alpha(x^\mu)\, dV / r$$

$$A^{\mu;\alpha}{}_\mu - A^{\alpha;\mu}{}_\mu = 4\pi J^\alpha$$

Here, ϱ is the charge density, the integral is over all 3-space, and r is the 3-D distance from integration volume dV to x^μ. The vector potential A defines the antisymmetric Faraday tensor:

$$F_{\alpha\beta} = A_{\beta;\alpha} - A_{\alpha;\beta}$$

Whose components are:

$$F^{\alpha\beta} = \begin{bmatrix} 0 & E_x & E_y & E_z \\ -E_x & 0 & B_z & -B_y \\ -E_y & -B_z & 0 & B_x \\ -E_z & B_y & -B_x & 0 \end{bmatrix}$$

Faraday Tensor

§30.2 Principal Equations

In 4-D spacetime, we write Maxwell's equations as:

$$F_{[\alpha\beta;\mu]} = 0$$

$$F^{\alpha\beta}{}_{;\beta} = 4\pi J^\alpha$$

The conservation of electric charge is ensured because the divergence of the current 4-vector is always zero:

$$J^\alpha{}_{;\alpha} = 0$$

The Lorentz force, the combined effect of electric and magnetic fields, accelerates a body with electric charge q, 4-velocity u, and mass m according to:

$$m\ du^{\beta}/d\tau\ =\ q\ F^{\beta\alpha}\ u_{\alpha}$$

Here, τ is proper time, the time measured in charge q's rest frame.

The Lorentz gauge, the standard convention for setting the divergence of vector potential A^{α}, is:

$$A^{\alpha}{}_{;\alpha}\ =\ 0$$

Below are some examples of Maxwell's equations for selected index combinations:

Divergence of ***B***: $0\ =\ F_{[12;3]}\ =\ F_{12,3}\ +\ F_{23,1}\ +\ F_{31,2}\ =\ B_{z,z}\ +\ B_{x,x}\ +\ B_{y,y}$

Divergence of $\hat{\boldsymbol{E}}$: $4\pi\,\varrho\ =\ J^0\ =\ F^{01}{}_{,1}\ +\ F^{02}{}_{,2}\ +\ F^{03}{}_{,3}\ =\ \hat{E}_{x,x}\ +\ \hat{E}_{y,y}\ +\ \hat{E}_{z,z}$

Z comp. of curl $\hat{\boldsymbol{E}} = -\,\partial B/\partial t$: $0 = F_{[01;2]} = F_{01,2} + F_{12,0} + F_{20,1} = -\ \hat{E}_{x,y} + B_{z,t} + \hat{E}_{y,x}$

X comp. of curl $\boldsymbol{B} = J + \partial\hat{E}/\partial t$: $J^1\ =\ F^{10}{}_{,0}\ +\ F^{12}{}_{,2}\ +\ F^{13}{}_{,3}\ =\ -\ \hat{E}_{x,t}\ +\ B_{z,y}\ -\ B_{y,z}$

§30.3 Electromagnetic Mass-Energy Tensor

The mass-energy-stress tensor for electromagnetic fields is:

$$4\,\pi\ T^{\mu\varepsilon}\ =\ F^{\mu\alpha}\,F^{\varepsilon}{}_{\alpha}\ -\ g^{\mu\varepsilon}\ F_{\alpha\beta}\ F^{\alpha\beta}\ /\ 4$$

§30.4 Dual Tensors & Exterior Derivative

An even more elegant statement of Maxwell's equations is possible with *dual tensors* and the *exterior derivative*. Two dual tensors are:

$$*F_{\alpha\beta} = F^{\mu\nu}\ \varepsilon_{\alpha\beta\mu\nu}$$

$$*J_{\alpha\beta\mu} = J^{\nu}\ \varepsilon_{\alpha\beta\mu\nu}$$

Here, the asterisk is the *duality operator* and $\varepsilon_{\alpha\beta\mu\nu}$ is the completely antisymmetric *Levi-Civita* tensor, defined as:

$$\varepsilon_{\alpha\beta\mu\nu}\ =\ +1 \text{ for } \alpha\beta\mu\nu \text{ an even permutation of } 0123$$

$$\varepsilon_{\alpha\beta\mu\nu}\ =\ -1 \text{ for } \alpha\beta\mu\nu \text{ an odd permutation of } 0123$$

$$\varepsilon_{\alpha\beta\mu\nu}\ =\ 0 \text{ otherwise}$$

The dual of a rank N tensor is a tensor of rank 4 – N.

The *exterior derivative* ***d*** is an extension of the covariant derivative. When applied to a tensor of rank N, the exterior derivative produces a completely antisymmetric tensor of rank N+1. For a rank N tensor of the form $O_{\alpha\ldots\beta}$, where $\alpha\ldots\beta$ are the N indices, the $\alpha\ldots\beta\mu$ component of the resulting rank N+1 tensor ***d*** O is defined by:

$$(\boldsymbol{d}\,O)_{\alpha\ldots\beta\mu}\ =\ O_{[\alpha\ldots\beta;\mu]}$$

The components of the dual Faraday tensor are shown below; duality exchanges E and B. The signs of individual components of F and *F depend on whether the indices are superscripted, subscripted, or mixed.

$$^*F_{\alpha\beta} = \begin{bmatrix} 0 & B_x & B_y & B_z \\ -B_x & 0 & -E_z & E_y \\ -B_y & E_z & 0 & -E_x \\ -B_z & -E_y & E_x & 0 \end{bmatrix}$$

Dual Tensor of the Faraday Tensor

With these definitions, the Faraday tensor and Maxwell's equations become:

$$F = -\,\boldsymbol{d}A$$

$$\boldsymbol{d}F = 0$$

$$\boldsymbol{d}\,{}^*F = 4\,\pi\ {}^*J$$

§30.5 Electromagnetic Scalars

Two important electromagnetic invariant scalars are:

$$\hat{E}^2 - B^2 = -\ F^{\alpha\beta}\ F_{\alpha\beta}\ /\ 2$$

$$\hat{\boldsymbol{E}} \times \boldsymbol{B} = F^{\alpha\beta}\ {}^*F_{\alpha\beta}\ /\ 4$$

Chapter 31
Inside a Star

This chapter explores the internal structure of stars.

We may reasonably assume a star is comprised of a spherically symmetric, ideal gas at rest in the star's reference frame. This assumption is valid when rotation, stress, and matter flow are negligible, and when energy flows slowly within the star.

These four functions of radius r define the structure of a star:

Mass-energy density: $\mu(r)$

Isotropic pressure: $P(r)$

Baryon number density: $n(r)$

Gas 4-velocity: $u^\varepsilon(r)$

The first three are defined in the gas rest frame, where $u^\varepsilon = (1, 0, 0, 0)$. Baryon number is the number of protons plus the number of neutrons. Protons and neutrons have nearly equal masses, both about 1800 times the electron's mass. We need not explicitly track electrons — they take care of themselves by following protons, thereby maintaining charge neutrality.

§31.1 Ideal Gas Equations

Section §23.1 explores ideal gases, collections of discreet parts that interact mechanically and gravitationally but do not exchange internal energy. While the first "ideal gases" that physicists studied were indeed gaseous, the same principles apply to fluids, to electrons and ions in a plasma, to molecules in thermodynamics, and even to galaxies in cosmology. All these can be called ideal gases.

Let's now highlight important properties of 4-velocities in general, and of ideal gases in particular. These properties have an essential role in the structure of stars.

We already know that the dot product of any 4-velocity u^ε with itself equals –1. Since the length of an object's 4-vector never changes, its acceleration $u^\varepsilon{}_{;\alpha}$ must always be perpendicular to u^ε, as shown by:

$$2\, u^\varepsilon{}_{;\alpha}\, u_\varepsilon = (u^\varepsilon\, u_\varepsilon)_{;\alpha}$$

$$2\, u^\varepsilon{}_{;\alpha}\, u_\varepsilon = (g_{\varepsilon\beta}\, u^\varepsilon\, u^\beta)_{;\alpha} = (-1)_{;\alpha} = 0$$

When the volume V of a gas expands or contracts, the 4-velocities of neighboring gas particles diverge or converge correspondingly. In terms of proper time τ, this is written:

$$u^\varepsilon{}_{;\varepsilon} = (dV/d\tau) / V$$

The conservation of baryon number is written:

$$0 = d(n\, V)/d\tau$$

$$0 = V\ dn/d\tau + n\ dV/d\tau$$

$$(dn/d\tau) / n = -\ (dV/d\tau) / V = -\ u^\varepsilon{}_{;\varepsilon}$$

Now recall from Section §23.1, the stress-energy-mass tensor of an ideal gas:

$$T^{\varepsilon\nu} = (\mu + P)\ u^\varepsilon\ u^\nu \ +\ g^{\varepsilon\nu}\ P$$

For this gas, the conservation equation for 4-momentum, fully expanded, is:

$$0 = T^{\varepsilon\nu}{}_{;\nu}$$

$$0 = (\mu + P)_{,\nu}\ u^\varepsilon\ u^\nu \ +\ (\mu + P)\ u^\varepsilon{}_{;\nu}\ u^\nu$$
$$+\ (\mu + P)\ u^\varepsilon\ u^\nu{}_{;\nu} \ +\ g^{\varepsilon\nu}\ P_{,\nu}$$

For scalar quantities, like μ and P, there is no distinction between partial derivatives and covariant derivatives.

Now, let's project the above equation onto u (take the dot product with u). We do this by multiplying everything by u_ε and contracting on ε:

$$0 = (\mu + P)_{,\nu}\ u^\varepsilon\ u^\nu\ u_\varepsilon \ +\ (\mu + P)\ u^\varepsilon{}_{;\nu}\ u^\nu\ u_\varepsilon$$
$$+\ (\mu + P)\ u^\varepsilon\ u^\nu{}_{;\nu}\ u_\varepsilon \ +\ g^{\varepsilon\nu}\ P_{,\nu}\ u_\varepsilon$$

Since $u^\varepsilon{}_{;\alpha}\ u_\varepsilon = 0$, and $u^\varepsilon\ u_\varepsilon = -1$:

$$0 = -\ (\mu + P)_{,\nu}\ u^\nu \ +\ 0$$
$$-\ (\mu + P)\ u^\nu{}_{;\nu} \ +\ u^\nu\ P_{,\nu}$$

Cancel $\pm\ u^\nu\ P_{,\nu}$ terms:

$$\mu_{,\nu}\ u^\nu = -(\mu + P)\ u^\nu{}_{;\nu}$$

Since $u^\varepsilon{}_{;\varepsilon} = -\ (dn/d\tau)\ /\ n$:

$$\mu_{,\nu}\ u^\nu = +(\mu + P)\ (dn/d\tau)\ /\ n$$

Therefore, in the rest frame of an ideal gas, where $u^\varepsilon = (1, 0, 0, 0)$, the local version of energy conservation is written:

$$d\mu/d\tau = (\mu + P)\ (dn/d\tau)\ /\ n$$

Next, let's take the conservation equation $T^{\varepsilon\nu}{}_{;\nu} = 0$ and project it onto a surface perpendicular to the 4-velocity of the gas. This projection is accomplished by multiplying everything by:

$$g_{\alpha\varepsilon} + u_\alpha\ u_\varepsilon$$

Let's check that first:

$$u^\varepsilon\ (g_{\alpha\varepsilon} + u_\alpha\ u_\varepsilon) = u_\alpha - u_\alpha = 0$$

Thus u^ε and $(g_{\alpha\varepsilon} + u_\alpha\ u_\varepsilon)$ are indeed orthogonal, just as advertised.

The projection of $T^{\varepsilon\nu}{}_{;\nu} = 0$ onto $(g_{\alpha\varepsilon} + u_\alpha\ u_\varepsilon)$ is:

$$0 = (g_{\alpha\varepsilon} + u_\alpha\ u_\varepsilon)\ T^{\varepsilon\nu}{}_{;\nu}$$

$$0 = (g_{\alpha\varepsilon} + u_\alpha\ u_\varepsilon)\ (\mu + P)_{,\nu}\ u^\varepsilon\ u^\nu$$
$$+\ (g_{\alpha\varepsilon} + u_\alpha\ u_\varepsilon)\ (\mu + P)\ u^\varepsilon{}_{;\nu}\ u^\nu$$
$$+\ (g_{\alpha\varepsilon} + u_\alpha\ u_\varepsilon)\ (\mu + P)\ u^\varepsilon\ u^\nu{}_{;\nu}$$
$$+\ (g_{\alpha\varepsilon} + u_\alpha\ u_\varepsilon)\ g^{\varepsilon\nu}\ P_{,\nu}$$

The first and third lines are zero, because each contains the product $(g_{\alpha\varepsilon} + u_\alpha\ u_\varepsilon)\ u^\varepsilon$.

The prior equation reduces to:

$$0 = (g_{\alpha\varepsilon} + u_\alpha u_\varepsilon)\ (\mu + P)\ u^\varepsilon{}_{;\nu}\ u^\nu$$
$$+\ (g_{\alpha\varepsilon} + u_\alpha u_\varepsilon)\ g^{\varepsilon\nu}\ P_{,\nu}$$

Expanding the product terms yields:

$$0 = (\mu + P)\ g_{\alpha\varepsilon}\ u^\varepsilon{}_{;\nu}\ u^\nu + (\mu + P)\ u_\alpha\ u_\varepsilon\ u^\varepsilon{}_{;\nu}\ u^\nu$$
$$+\ g_{\alpha\varepsilon}\ g^{\varepsilon\nu}\ P_{,\nu} + u_\alpha u_\varepsilon\ g^{\varepsilon\nu}\ P_{,\nu}$$

$$0 = (\mu + P)\ (u_{\alpha;\nu})\ u^\nu + (\mu + P)\ u_\alpha\ (0)\ u^\nu$$
$$+\ (\delta_\alpha{}^\nu)\ P_{,\nu} + u_\alpha\ (u^\nu)\ P_{,\nu}$$

$$0 = (\mu + P)\ (u_{\alpha;\nu})\ u^\nu + P_{,\alpha} + u_\alpha\ u^\nu\ P_{,\nu}$$

Now examine the radial component ($\alpha = r$) of this equation in the rest frame of the ideal gas, where $u^\varepsilon = (1, 0, 0, 0)$. In a static star at equilibrium, only the radial derivative of pressure is nonzero.

$$-P_{,r} = (\mu + P)\ (u_{r;\nu})\ u^\nu + (0)\ (0)\ P_{,r}$$

We come now to an essential fact of curved spacetime calculus. In this case, u_r is zero. By the normal rules of calculus, the partial derivative $u_{r,\nu}$ is also zero. But the covariant derivative $u_{r;\nu}$ is *not* zero, in general, because it involves not just changes in u_r but also changes in the curved spacetime axes.

We must always expand covariant derivatives according to their rules, and then evaluate each resulting term.

$$-P_{,r} = (\mu + P)\ (u_{r,\nu})\ u^\nu - (\mu + P)\ \Gamma^\beta{}_{r\nu}\ u_\beta\ u^\nu$$
$$P_{,r} = (\mu + P)\ (0)\ u^\nu + (\mu + P)\ \Gamma^t{}_{rt}\ u_t\ u^t$$
$$P_{,r} = -\ (\mu + P)\ \Gamma^t{}_{rt}$$

Let's now calculate this connection coefficient. We will assume a spherically symmetric metric that has zero curvature in the polar and azimuthal directions, and whose curvature is not yet known along the time and radial axes. This is the most general metric for such a geometry. It is the starting point for the derivation of the Schwarzschild metric.

$$ds^2 = -\Lambda^2\ dt^2 + \Sigma^2\ dr^2 + r^2\ d\theta^2 + r^2\ \sin^2\theta\ d\phi^2$$

$$\Gamma^t{}_{rt} = g^{tt}\ (g_{tr,t} + g_{tt,r} - g_{rt,t})\ /\ 2$$
$$\Gamma^t{}_{rt} = (-1\ /\ \Lambda^2)\ (0 - 2\ \Lambda\ \Lambda{,}r - 0)\ /\ 2$$
$$\Gamma^t{}_{rt} = +\Lambda{,}r\ /\ \Lambda$$

With that, we obtain the equation for the derivative of pressure:

$$P_{,r} = -\ (\mu + P)\ \Lambda{,}r\ /\ \Lambda$$

§31.2 Employing Field Equations

Our derivation of the Schwarzschild metric in Appendix 7 applies here, up through the point of equating $G_{\sigma\sigma}$ with $8\pi\, T_{\sigma\sigma}$. We must redo the last step, because here, we have a different $T_{\varrho\varrho}$.

As in Appendix 7, let's employ an orthonormal non-coordinate basis, defined by:

$$d\sigma = \Lambda\, dt$$

$$d\varrho = \Sigma\, dr$$

$$d\Theta = r\, d\theta$$

$$d\Phi = r \sin\theta\, d\phi$$

The metric in this non-coordinate basis is:

$$ds^2 = -d\sigma^2 + d\varrho^2 + d\Theta^2 + d\Phi^2$$

The key results from Appendix 7 are:

$$G^{\varrho}{}_{\varrho} = -1 / r^2 + \{2\, \Lambda{,}r / r\, \Lambda + 1 / r^2\} / \Sigma^2$$

$$G_{\sigma\sigma} = (1 / r^2)\ \partial\, [r\, (1 - 1 / \Sigma^2)] / \partial r$$

$$1 / \Sigma^2 = 1 - 2\ m(r)\ / r$$

Where:

$$m(r) = 4\pi \int_0^r \mu\ \varrho^2\ d\varrho$$

The equation for Σ comes from equating the mass density terms:

$$G_{\sigma\sigma} = 8\pi\, T_{\sigma\sigma} = 8\pi\, \mu$$

What remains is to equate radial pressure terms:

$$G^{\varrho}{}_{\varrho} = G_{\varrho\varrho} = 8\pi\, T_{\varrho\varrho}$$

$$G^{\varrho}{}_{\varrho} = 8\pi\, P$$

$$8\pi\, P = -1 / r^2 + \{2\, \Lambda{,}r / r\, \Lambda + 1 / r^2\}\, (1 - 2\, m / r)$$

Multiplying out the { } and ( ) terms yields:

$$8\pi\, P = -1 / r^2 + 1 / r^2 - 2\, m / r^3$$
$$+ \{2\, \Lambda{,}r / r\, \Lambda\}\, (1 - 2\, m / r)$$

This simplifies slightly to:

$$8\pi\, P + 2\, m / r^3 = \{2\, \Lambda{,}r\ /\ r\ \Lambda\}\, (1 - 2\, m / r)$$

$$(8\pi\, P + 2\, m / r^3)\ /\ (1 - 2\, m / r) = \{2\, \Lambda{,}r\ /\ r\ \Lambda\}$$

$$(4\pi\, P\, r^3 + m)\ /\ [\, r^2\, (1 - 2\, m / r)\,] = \Lambda{,}r\ /\ \Lambda$$

§31.3 Oppenheimer-Volkoff Equation

Plugging the prior equation into the equation for the derivative of pressure, the last equation in Section §31.1, we obtain the Oppenheimer-Volkoff ("OV") equation of hydrostatic equilibrium:

$$dP/dr = -(\mu + P)(m + 4\pi P r^3) / [r^2 (1 - 2m/r)]$$

This contrasts with the much simpler Newtonian equivalent:

$$dP/dr = -\mu m / r^2$$

What changes does general relativity impose on stars?

Both terms in the numerator of the relativistic equation have an additional factor related to pressure, each of which increases the numerator.

Relativity also decreases the denominator by the typical curvature factor $(1 - 2m/r)$. The larger numerator and smaller denominator both increase the relativistic pressure gradient dP/dr.

Thus as one moves inward from the star's surface, where $P = 0$, pressure increases faster in the relativistic case than in the Newtonian case. And as the pressure increases, it accelerates the relativistic enhancement factors, further increasing the pressure in a positive feedback loop.

Therefore, the true self-gravity of massive stars is far greater than would be predicted by classical physics. Stars that would be stable in a Newtonian universe may collapse under their own gravity in our actual universe that is governed by general relativity.

§31.4 Neutron Stars

The Oppenheimer-Volkoff equation involves two unknown functions of radius: μ and P. Solving the OV equation requires knowing one, or the other, or a relationship between the two. Additionally, we must know the boundary conditions.

In some cases, we do know the relationship between μ and P, which is called the *equation of state*. For a gas of highly relativistic particles, such as photons, $P = \mu / 3$. But most stars are more complex.

While we cannot solve the most general case, let's consider a problem we can solve: an idealized neutron star, so severely compressed that all its particles are highly relativistic: for each particle, $E >> m$.

All stars, during the prime of their lives, are powered by nuclear fusion reactions in their cores. Nuclear fusion releases millions of times more energy per kg of fuel than burning gasoline. The immense amount of energy released by nuclear fusion creates an enormous pressure within the star that precisely balances its enormous self-gravity. Pressure seeks to blow the star apart, while gravity seeks to crush it.

Eventually, each star depletes its nuclear fuels. As its internal pressure drops, ever relentless gravity crushes the stellar core. The mass of that core defines its fate.

Neutron stars result from the gravitational collapse of stars somewhat more massive than our Sun — stars with core masses between 1.4 and 3 times our Sun's mass.

When subjected to extreme gravity, electrons and protons can combine to form neutrons. The space occupied by a normal atom shrinks to the size of a nucleus, whose volume is many trillion times smaller. The sum of the proton and electron masses is 0.08% less than the neutron mass. Hence, forming neutrons requires an energy source to make up that deficit. That source is the release of gravitational potential energy from the collapsing star, which is enabled by its collapsing atoms.

The predominant nuclear fusion reaction in stars produces one helium nucleus from four hydrogen nuclei; 0.69% of those hydrogen atoms' original mass is converted in energy, the star's light and heat.

Over the course of billions of years, a Sun-like star may burn 10% of its hydrogen, converting 0.05% of its original mass into energy. (Hydrogen accounts for 74% of the total atomic mass of the universe, and a similar percentage of most stars.)

During the core collapse that produces a neutron star, the core's gravitational potential energy decreases drastically, attaining an enormous negative value, and releasing as much as 15% of the core's original mass in other forms of energy.

In seconds, the core collapse releases hundreds of times more energy than that star released in billions of years.

For a comprehensive discussion of collapsed stars, see my ebook ***Black Holes, Supernovae & More*** or my print book ***Our Universe: The Known & The Unknown***.

Neutron stars are almost entirely comprised of neutrons in their interior, with proton, electrons, and possibly other particles in limited quantities. For simplicity, let's imagine an idealized neutron star that is comprised entirely of neutrons that are compacted to the normal nuclear density, or even higher.

Due to the Pauli exclusion principle, only two neutrons, one with spin up and the other with spin down, can occupy each tiny region of *phase space* of volume $(\hbar / 2)^3$. For a thorough discussion of phase space and the exclusion principle, see my ebook ***Quantum Mechanics 4: Identical Particles*** or my print book ***Quantum Mechanics Strange Theory of the Micro-World***.

Very briefly, the Heisenberg uncertainty principle states that the uncertainty Δx in the x-positon of any particle, and the uncertainty Δp_x in its x-momentum, cannot both be arbitrarily small — the product $(\Delta x\ \Delta p_x)$ must always be at least $(\hbar / 2)$. The same criterion applies separately to the y- and z-axes.

One can think of the product of the two quantities Δx and Δp_x as defining an area in a mathematical 2-D space. The y- and z-axes add two more dimensions each. In total, this 6-D space is called *phase space*. Thought of in this way, we interpret the Heisenberg uncertainty principle as saying every particle requires its own minimum 6-D volume of phase space in the amount of $(\hbar / 2)^3$.

Fermions, the particles of matter, including protons, neutrons, and electrons, will share their phase space with at most one other identical particle, and then only if their spins are opposite.

As the neutrons in a neutron star are squeezed into a very small 3D volume, their phase space must expand into the p_x-p_y-p_z dimensions. When phase space is fully occupied up to a maximum momentum F, the *Fermi level*, the gas is called *degenerate*.

White dwarfs are degenerate electron gases, and neutron stars are degenerate neutron gases.

The neutron number density is found by integrating phase space from momentum p = 0 to p = F:

$$n = 2\,(8\,\hbar^{-3})\ 4\pi \int_0^F p^2\ dp$$

$$n = 64\pi\ \hbar^{-3}\ F^3 / 3$$

In the first equation above, the factor of 2 accounts for the two spin states that can occupy each minimum unit of phase space. The equation says: the number density in x-y-z space equals the 3-D momentum space (the integral of $4\pi\, p^2\, dp$) divided by $(\hbar / 2)^3$ the minimum unit phase space for two neutrons.

The mass-energy density is similar, with the added factor of energy ($p \sim E$ for $E >> m$):

$$\mu = 2\,(8\,\hbar^{-3})\ 4\pi \int_0^F p^3\ dp$$

$$\mu = 16\pi\ \hbar^{-3}\ F^4$$

Note that:

$$\mu = n\,(\,3\ F / 4\,)$$

Employing a thermodynamic equation, we have:

$$dE + P\ dV = 0$$

$$P = -\ dE/dV$$

$$P = -\ d(\mu / n)\ /\ d(1 / n)$$

$$P = -\ d(3\ F / 4)\ /\ (-\ n^{-2}\ dn)$$

$$P = (3 / 4)\ dF\ /\ (n^{-2}\ 3n\ \ dF / F)$$

$$P = (1 / 4)\ n\ F$$

$$P = \mu / 3$$

This is also the equation of state of photons — not unexpected, since we assumed these neutrons are highly relativistic (E >> m). The OV equation can then be written:

$$d\mu/dr\ / 3 = -\ (4\ \mu / 3)\ (m + 4\pi\ \mu\ r^3 / 3)\ /\ [\ r\ (r - 2\ m)\ ]$$

$$3\ r\ (r - 2\ m)\ \ d\mu/dr = -\ 4\ \mu\ (3\ m + 4\pi\ \mu\ r^3\ )$$

Recall that:

$$m(r) = 4\pi \int_0^r \mu(\varrho)\ \varrho^2\ d\varrho$$

This equation seems daunting. There is nothing wrong with an experimental approach to solving equations — try a power-law solution. If that does not work, try an exponential. If you guess wrong, the math will quickly inform you. If you guess right, people will think you're brilliant. What have you got to lose?

With the benefit of experience, we assume:

For unknown constants a, x:

$$\mu(r) = a\ r^{-x}$$

$$d\mu/dr = -\ a\ x\ r^{-x-1}$$

$$m(r) = 4\pi \int_0^r a\ \varrho^{-x}\ \varrho^2\ d\varrho$$

$$m(r) = 4\pi\ a\ r^{3-x}\ /\ (3 - x)$$

The OV equation is then:

$$3\ r\ [\ r - 8\pi\ a\ r^{3-x}\ /\ (3 - x)\ ]\ (-\ a\ x\ r^{-x-1}\ )$$

$$= -\ 12\ (a\ r^{-x})\ [4\pi\ a\ r^{3-x}\ / (3 - x)] - 16\pi\ r^3\ a^2\ r^{-2x}$$

Multiply both sides by r^x / a:

$$-\ 3\ x\ [\ r - 8\ \pi\ a\ r^{3-x}\ /\ (3 - x)\ ]$$

$$= -\ 12\ [4\ \pi\ a\ r^{3-x} /\ (3 - x)] - 16\ \pi\ a\ r^{3-x}$$

For the above equation to be valid for all r, each term must contain r to the same power. This requires x = 2, leading to:

$$-\ 6\ [\ r - 8\ \pi\ a\ r\ ] = -\ 12\ (4\ \pi\ a\ r) - 16\ \pi\ a\ r$$

Divide by – r:

$$6 - 48\ \pi\ a = 64\ \pi\ a$$

$$a = 6 / 112\ \pi = 3 / 56\ \pi$$

$$m = 4\ \pi\ a\ r = 3\ r / 14$$

Putting this expression for m into the equations for mass density and pressure yields:

$$\mu = 3 / (56\,\pi\, r^2)$$

$$P = 1 / (56\,\pi\, r^2)$$

Do not be concerned that μ and P go to infinity as r goes to zero. All that is physically meaningful is the integral of μ or P over a nonzero volume. Those integrals are all finite and continuous, as shown by the following integral of P over an infinitesimal ball of radius ε centered at $r = 0$.

$$4\pi \int_0^{\varepsilon} P\ r^2\ dr = 4\pi \int_0^{\varepsilon} r^2\ dr / (56\,\pi\, r^2)$$

$$4\pi \int_0^{\varepsilon} P\ r^2\ dr = \int_0^{\varepsilon} dr\ / 14$$

$$4\pi \int_0^{\varepsilon} P\ r^2\ dr = \varepsilon / 14$$

The integral is clearly finite.

Chapter 32
Expansion Equation of the Universe

By now, it must be abundantly evident that solving the field equations of general relativity can be challenging. Indeed, in 100 years, the only discovered exact solutions are for highly symmetric situations.

§32.1 Cosmological Principle

Fortunately, it appears our universe is, in fact, highly symmetric. On a very large scale, it seems homogeneous and isotropic, having the same overall characteristics everywhere and in every direction. This is called the *cosmological principle* that we introduced in Section §19.1. This principle does not derive from the fundamental laws of nature, but appears consistent with observations. We thoroughly examine all aspects of cosmology in my ebook series ***Our Universe 1 thru 5*** and my print book ***Our Universe: The Known & The Unknown***.

Let's briefly examine two important observations that support the cosmological principle.

Figure 32-1 shows a map of the entire sky taken by the radio telescope in Arecibo, Puerto Rico. Each black dot in this image identifies the location of a radio galaxy, a galaxy that emits copious radiation at radio frequencies. Altogether, the image captures the 125,000 brightest radio galaxies visible from Earth.

The original image is spherical, just like the heavens surrounding Earth. To make the image printable, it has been transformed into an oval, which distorts the image less than would a rectangular shape. The curving white band, reminiscent of a samurai helmet, is the portion of our sky that is covered by the Milky Way. Since we wish to examine the distant universe, that portion of the data has been removed. Finally, the white bottom of this image corresponds to the portion of our sky's Southern Hemisphere that cannot be properly observed by a telescope at 18° North.

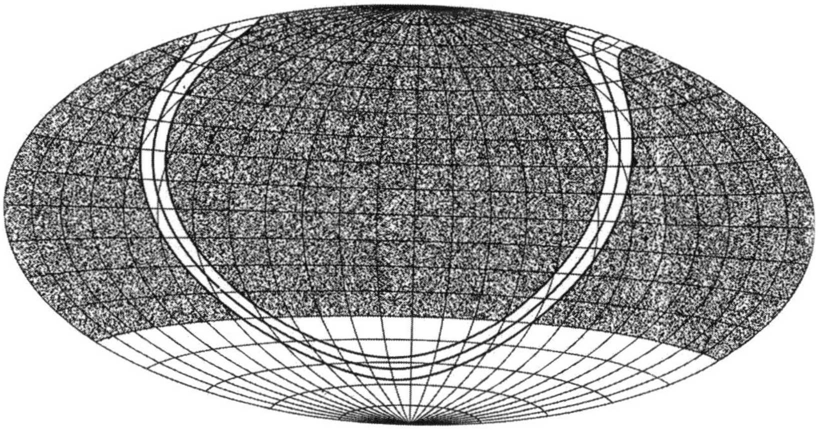

Figure 32-1 Universal Map of Radio Galaxies by Arecibo Radio Telescope

In the areas of Figure 32-1 that have not been whited-out, the radio galaxies seem quite evenly distributed — there are no large clumps or voids, and no apparent edges. This confirms the cosmological principle; our universe does indeed appear homogeneous on a large scale.

Figure 32-2 is also a map of the entire sky, again presented as an oval. This image, captured by NASA's WMAP satellite, shows the universal distribution of the Cosmic Microwave Background (CMB) radiation. The microwaves that our telescopes capture today come from the farthest reaches of our universe, having traveled through nearly 14 billion light-years of space to reach us. This light was emitted just 379,000 years after the Big Bang.

Figure 32-2 is therefore an image of our infant universe; it shows us what our universe was like when it was 1 / 37,000th of its current age.

The brighter areas in Figure 32-2 are where the universe was denser and hotter than the darker areas — but only slightly, the maximum difference being only 0.02%. Such tiny differences show that the universe, at nearly the beginning of time, was very nearly homogeneous throughout, again confirming the cosmological principle.

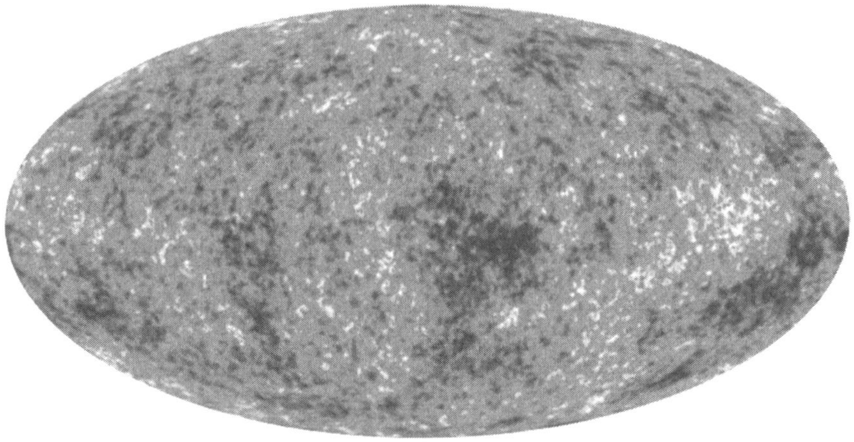

Figure 32-2 Universal CMB Map by WMAP Satellite

The cosmological principle pertains to uniformity throughout space, but not throughout time. The universe has certainly changed over time. It has, for example, grown larger and become colder. But as our universe changed, the same overall changes occurred everywhere, at least to a reasonable approximation.

§32.2 Constant Curvature Metrics

In accordance with the cosmological principle, let's assume our universe has the same spatial curvature everywhere, but allow that curvature to vary with time as the universe evolves.

What geometric shapes have the same curvature everywhere? A 2-D sphere, the surface of a 3-D ball, has the same positive (>0) curvature everywhere. A plane, a flat 2-D surface, has zero curvature everywhere. And, a hyperboloid, such as a potato chip, has the same negative curvature everywhere.

Mathematicians Howard Percy Robertson and Arthur Geoffrey Walker generalized these 2-D geometries to 3 spatial dimensions — 3-D *hypersurfaces* of constant curvature. Again, the curvature categories are: negative, zero, and positive, as we discussed in Section §15.9.

The most general metrics for these hypersurfaces are:

$$ds^2 = -dt^2 + A^2(t)\ \{ d\chi^2 + Q \}$$

Where Q depends on the curvature category as follows:

Curvature = 0: $\quad Q = \chi^2\, d\theta^2 + \chi^2 \sin^2\theta\, d\phi^2$

Curvature > 0: $\quad Q = \sin^2\chi\, d\theta^2 + \sin^2\chi\, \sin^2\theta\, d\phi^2$

Curvature < 0: $\quad Q = \sinh^2\chi\, d\theta^2 + \sinh^2\chi\, \sin^2\theta\, d\phi^2$

Where: $\quad \sinh x = (e^x - e^{-x}) / 2$

Above, A(t) is the scale factor of the universe at time t, and sinh is the hyperbolic sine function. The coefficient of dt^2 cannot depend on χ, θ, or ϕ, because that would violate spherical symmetry. Any constant coefficient of dt^2 merely rescales time; therefore we are free to set that constant equal to 1.

In 1922, Alexander Friedman solved Einstein's field equations for the Robertson-Walker metrics. His solutions are derived in Appendix 8. The primary Einstein tensor component for a hypersurface of constant curvature is:

$$G_{tt} = 3\ k\ A^{-2} + 3\ A^{-2}\ (A{,}t)^2$$

Here, k = +1, 0, −1 for positive, zero, and negative curvature. A(t) is the size scale of the universe at time t. If the universe has a finite spatial extent, A(t) can be its radius at time t. If the universe has an infinite spatial extent, A(t) can be the distance at time t between any chosen pair of "stationary" entities.

Both finite and infinite space are consistent with general relativity. In either alternative, the universe could be finite or infinite in temporal extent. Only measurements of cosmic parameters can establish which alternatives match our universe.

§32.3 FLRW equation

We next employ the field equations to relate geometric curvature, expressed by G, to mass and energy, encapsulated in T.

For the stress-energy-mass tensor T of our universe, let's assume the T tensor of an ideal gas. We treat the universe as a gas with isotropic pressure, comprised of idealized particles that do not exchange internal energy. For the universe, galaxies are particles.

In the rest frame of an ideal gas, with energy density μ and pressure P, the nonzero components of T are (see Sections §23.1 and §31.1):

$$T_{tt} = \mu$$

For α not equal t: $T_{\alpha\alpha} = P$

Equating G with 8π T yields the Friedman-Lemaître-Robertson-Walker (FLRW) equation for the evolution of the universe:

$$G_{tt} = 8\ \pi\ T_{tt}$$

$$3\ k\ A^{-2} + 3\ A^{-2}\ (A{,}t)^2 = 8\ \pi\ \mu$$

$$(A{,}t\ /\ A)^2 = (8\pi\ /\ 3)\ \mu - k\ A^{-2}$$

$$H^2 = (8\pi\ /\ 3)\ \mu - k\ A^{-2}$$

Here, k = +1, 0, −1 for positive, zero, and negative curvature. The last equation results from recognizing that (dA/dt) / A is the definition of H, the Hubble expansion rate, which may be a function of time. Hubble's observations showed that H is currently positive — the universe is now expanding.

The expansion of our universe is not due to galaxies flying apart. As explored more thoroughly in my ebooks and print books focused on cosmology, the proper explanation is that space is expanding uniformly in all directions in accordance with general relativity and the FLRW equation. As space expands, it carries with it all that it contains. The galaxies themselves are moving *through space* at much lower speeds than the large-scale expansion rate of space itself.

The expansion rate H has declined as the universe has aged. Based on the latest observations from the WMAP and Planck satellites, H is now about 74,300 km/h (or 46,400 mph) per million light-years. This means the distance between objects now separated by 100 million light-years is increasing at 7.43 million km/h (or 4.64 million mph).

The units of H are inverse time: 1 / H = 14.53 billion years. If H had never changed, 1 / H would be the age of the universe. However, H has changed — dramatically in the earliest epochs of our universe, but only gradually for many billion years. That is why 1 / H is close to, but not exactly equal to, the age of the universe, now estimated to be 13.817 ± 0.048 billion years.

§32.4 Energy Density

The energy density of the universe μ is comprised of three forms of energy: matter, radiation, and dark energy. The densities of these three forms vary differently with the expansion of the universe, and each has dominated during some epoch of its history.

The density of mass energy, including normal matter and dark matter, varies as $1 / r^3$ — a constant number of particles divided by an increasing volume, with r being the radius of the universe (or its scale factor if "radius" is inappropriate).

The second form of energy is radiation, whose density varies as $1 / r^4$. This is because the number of radiation particles does not change as the universe expands but their wavelengths stretch as space stretches. Hence, the density of radiation energy varies as $1 / r^3$ (particle density) multiplied by $1 / r$ for the decrease in each particle's energy ($E = h / \lambda$, and wavelength λ is proportional to r).

Finally, we have dark energy, whose density never changes. Dark energy remains mysterious, but is probably the result of virtual particles filling space, a phenomenon described by quantum mechanics (see Section §19.7). Everything we know about dark energy indicates that its density is constant, because every cubic foot of empty space is identical and contains the same virtual particle soup.

Separating our universe's energy density into its three distinct parts, the FLRW equation can be written:

$$H^2 = (8\,\pi / 3)\ (\mu_M\ A^{-3}\ +\ \mu_R\ A^{-4}\ +\ \mu_\Lambda)\ -\ k\,A^{-2}$$

Here, μ_M is the present density of matter (normal plus dark), μ_R is the present density of radiation, μ_Λ is the unchanging density of dark energy, and A is the radius / scale factor that we define to be 1 at the present time. A was less than 1 at all earlier times, and approaches 0 as t approaches 0.

The dark energy term is traditionally called Λ, in homage to Einstein's ill-fated *cosmological constant*.

Let's examine scenarios for the evolution of some possible universes that are consistent with general relativity.

§32.5 Negative Curvature

Let's first assume that the dark energy density μ_Λ is not negative.

If a universe has negative spatial curvature ($k = -1$), the right hand side of the FLRW equation is always greater than zero. This means H^2 is always strictly greater than zero, and H cannot change sign, since it cannot pass through zero.

If this universe is expanding ($H > 0$) at any particular moment, it must have been expanding since its beginning and must continue expanding forever. As A grows larger without limit, the impacts of matter, radiation, and the curvature term $k\,A^{-2}$ eventually become negligible compared with the constant term μ_Λ. The resulting equation would then become:

$$\partial A/\partial t\ =\ H\ A\ =\ A\ \sqrt{(\,8\,\pi\ \mu_\Lambda / 3\,)}$$

Such a universe grows exponentially, without limit. This is called an *open universe*.

Conversely, if this universe is contracting ($H < 0$), it will continue contracting until it is consumed in a *Big Crunch*.

If the dark energy density μ_Λ is negative, the right hand side of the FLRW equation may be positive initially, when A is small, and this universe may be expanding. But H will eventually become zero when A becomes large enough to sufficiently reduce the radiation, matter, and curvature terms. This universe will then stop expanding, and begin contracting. This is a *closed universe* whose ultimate fate is a *Big Crunch*.

§32.6 Positive Curvature

If a universe has positive (>0) curvature (k = + 1), the FLRW equation becomes:

$$H^2 = (8\pi / 3)\ \mu - A^{-2}$$

$$\mu = \mu_M\ A^{-3} + \mu_R\ A^{-4} + \mu_\Lambda$$

If this universe originated in a Big Bang akin to ours, μ would be extremely large initially, making the initial expansion rate also extremely large ($H >> 0$).

As this universe expands, A increases and μ decreases rapidly. The matter term drops with the third power of A, and the radiation term drops with the fourth power of A.

If the dark energy term is but small, but greater than zero, eventually $(8\pi / 3)\ \mu = A^{-2}$, making $H = 0$. This universe will then stop expanding and begin contracting, eventually collapsing in a *Big Crunch*. This is called a *closed universe*.

If, however, the dark energy term is positive and sufficiently large, it might keep $H^2 > 0$ forever, and thereby prevent a collapse. As A continues increasing, eventually this universe would be dominated by dark energy, and would expand exponentially, as in an *open universe*.

§32.7 Zero Curvature

If a universe has zero curvature (k = 0), the FLRW equation becomes:

$$H^2 = (8\pi / 3)\ (\mu_M\ A^{-3} + \mu_R\ A^{-4} + \mu_\Lambda)$$

If $\mu_\Lambda > 0$, the right hand side is greater than 0 for all values of A, and H cannot change sign. Hence, an expanding universe continues expanding forever and never collapses.

If the dark energy term μ_Λ is zero, the mass term eventually dominates, and the FLRW equation becomes:

$$(A,t / A)^2 = (8\pi / 3)\ \mu_M\ A^{-3}$$

$$A\ (A,t)^2 = (8\pi / 3)\ \mu_M$$

$$(\sqrt{A})\ dA = dt\ \sqrt{(8\pi\ \mu_M / 3)}$$

$$(2 / 3)\ A^{3/2} = t\ \sqrt{(8\pi\ \mu_M / 3)}$$

$$A = t^{2/3}\ (6\pi\ \mu_M)^{1/3}$$

This universe expands forever with the two-thirds power of time, a rather sedate expansion compared with the exponential growth of other scenarios. This is called a *flat universe*.

§32.8 Our Universe

Our universe appears to be flat (k = 0). The latest measurements show its total energy density is 100.1 ± 0.2% of the critical density required to make the universe flat (zero global curvature). These measurements also show that our universe's energy density is dominated by dark energy, with $\mu_\Lambda > 0$ and accounting for about 69% of total energy.

As shown in Figure 32-3, we believe dark energy began to dominate our universe's energy density about 6 billion years ago (6 Gya), and that it will remain dominant forevermore.

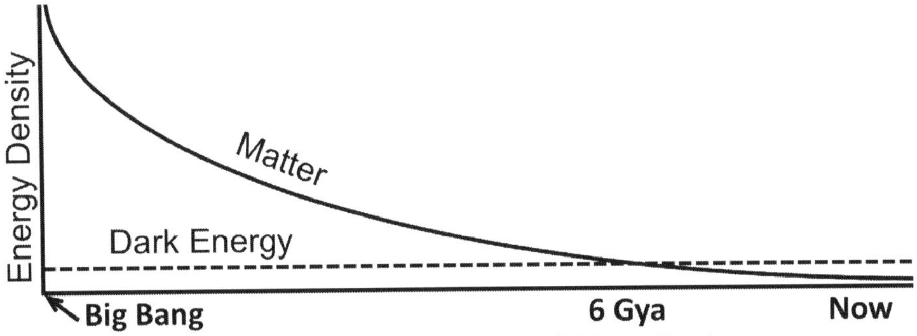

Figure 32-3 Evolution of Our Universe's Energy Density

On very large scales, this means our universe will expand exponentially, like an *open universe*.

However, on "small" scales — galaxies, solar systems, and planets — the attractive self-gravity of matter in these objects is strong enough to keep these structures intact.

Galaxy clusters seem to be near the dividing line. Denser galaxy clusters have enough self-gravity to retain their structure. However, galaxies on the periphery of less-dense clusters may be swept away by the exponentially expanding space.

While very remote objects will fly away ever faster, Earth will not recede from the Sun or be torn apart by cosmic expansion.

This presumes our understanding and measurements of dark energy are reasonably accurate.

Chapter 33
Linearized Theory in Weak Gravity

§33.1 Why Linearize?

This chapter provides the mathematical tools required to explore one of the most exciting discoveries of 21st Century physics: gravitational waves, the subject of the next chapter.

As discussed more thoroughly in Chapter 34, gravitational waves are small variations in the curvature of spacetime that propagate through space. In 1916, Einstein predicted that changes in the distribution of mass and energy would produce changes in spacetime curvature that spread outward through space in waves that gradually diminish, eventually becoming tiny ripples. While the proper technical term is gravitational waves, I will sometimes use their more common name: gravity waves.

Due to the complexity of the equations of general relativity, no one has ever derived comprehensive, mathematically precise equations for gravitational waves — the equations are just too difficult to solve analytically. By employing supercomputers, one can numerically integrate the second-order differential equations of general relativity, and thereby analyze these waves in strong gravity, where non-linearity dominates.

What Einstein did, and what we will do, is explore weak gravitational waves by approximating the true equations with much simpler linear equations. Linearized general relativity is not adequate to analyze the production of gravitational waves in the merger of two black holes, but it is adequate to analyze the tiny ripples that ultimately reach our detectors.

§33.2 Linearized General Relativity

We therefore assume a spacetime metric that is nearly flat, as defined by:

$$g_{\alpha\beta} = \eta_{\alpha\beta} + h_{\alpha\beta}$$

Here, $\eta_{\alpha\beta}$ is the metric of zero-curvature (flat) 4-D spacetime, and $h_{\alpha\beta}$ describes tiny curvature variations in that spacetime. Assuming weak gravity ensures the absolute values of all $h_{\alpha\beta}$ are $<< 1$. $\eta_{\alpha\beta}$ is generally called the Minkowski metric that is defined by:

In flat spacetime: $ds^2 = \eta_{\alpha\beta}\ dx^\alpha\ dx^\beta$

In rectilinear coordinates: $ds^2 = -\ dt^2 + dx^2 + dy^2 + dz^2$

In spherical coordinates: $ds^2 = -\ dt^2 + dr^2 + r^2\ d\theta^2 + r^2 \sin^2\theta\ d\phi^2$

With the nearly flat metric, we can proceed to calculate the usual curvature tensors, and now drop terms proportional to $(h_{\alpha\beta})^2$ and higher powers of these small quantities.

The connection coefficients in a coordinate basis are:

$$\Gamma^\mu{}_{\alpha\beta} = g^{\mu\sigma}\ (\ h_{\sigma\alpha,\beta} + h_{\sigma\beta,\alpha} + h_{\alpha\beta,\sigma}\ )\ /\ 2$$

Since all $\eta^{\mu\sigma}$ are constants, the metric derivatives involve only h. Our first approximation is replacing $g^{\mu\sigma}$ with $\eta^{\mu\sigma}$ whenever raising or lowering an index, since these quantities differ only by $h^{\mu\sigma}$, which leads to second and higher order terms of h. The prior equation reduces to:

$$\Gamma^\mu{}_{\alpha\beta} = (\ h^\mu{}_{\alpha,\beta} + h^\mu{}_{\beta,\alpha} - h_{\alpha\beta}{}^{,\mu}\ )\ /\ 2$$

Evidently, each $\Gamma^\mu{}_{\alpha\beta}$ is proportional to first derivatives of various components of h.

The Riemann components in a coordinate basis are:

$$R^{\alpha}{}_{\beta\mu\varepsilon} = \Gamma^{\alpha}{}_{\beta\varepsilon,\mu} - \Gamma^{\alpha}{}_{\beta\mu,\varepsilon} + \Gamma^{\alpha}{}_{\mu\delta}\,\Gamma^{\delta}{}_{\beta\varepsilon} - \Gamma^{\alpha}{}_{\varepsilon\delta}\,\Gamma^{\delta}{}_{\beta\mu}$$

We drop the last two terms, because they are second order, yielding:

$$R^{\alpha}{}_{\beta\mu\varepsilon} = (\,h^{\alpha}{}_{\beta,\varepsilon\mu} + h^{\alpha}{}_{\varepsilon,\beta\mu} - h_{\beta\varepsilon}{}^{,\alpha}{}_{,\mu}\,)\,/\,2$$
$$-\,(\,h^{\alpha}{}_{\beta,\mu\varepsilon} + h^{\alpha}{}_{\mu,\beta\varepsilon} - h_{\beta\mu}{}^{,\alpha}{}_{,\varepsilon}\,)\,/\,2$$

Since partial derivatives commute, the first terms in each line cancel one another, leaving:

$$R^{\alpha}{}_{\beta\mu\varepsilon} = (\,+\,h^{\alpha}{}_{\varepsilon,\beta\mu} - h_{\beta\varepsilon}{}^{,\alpha}{}_{,\mu} - h^{\alpha}{}_{\mu,\beta\varepsilon} + h_{\beta\mu}{}^{,\alpha}{}_{,\varepsilon}\,)\,/\,2$$

Set $\mu = \alpha$, and contract: $$R^{\alpha}{}_{\beta\alpha\varepsilon} = (\,+\,h^{\alpha}{}_{\varepsilon,\beta\alpha} + h_{\beta\alpha,\varepsilon}{}^{,\alpha} - h_{\beta\varepsilon}{}^{,\alpha}{}_{,\alpha} - h^{\alpha}{}_{\alpha,\beta\varepsilon}\,)\,/\,2$$

Raise & lower α's: $$R_{\beta\varepsilon} = (\,+\,h^{\alpha}{}_{\beta,\varepsilon\alpha} + h^{\alpha}{}_{\varepsilon,\beta\alpha} - h_{\beta\varepsilon,\alpha}{}^{,\alpha} - h_{,\beta\varepsilon}\,)\,/\,2$$

Where we define: $$h = h^{\alpha}{}_{\alpha}$$

With the Ricci components, we next calculate the curvature scalar R.

$$R = g^{\beta\varepsilon}\,R_{\beta\varepsilon}$$
$$R = (\,+\,h^{\alpha\beta}{}_{,\beta\alpha} + h^{\alpha\beta}{}_{,\beta\alpha} - h^{\beta}{}_{\beta,\alpha}{}^{,\alpha} - h_{,\beta}{}^{,\beta}\,)\,/\,2$$
$$R = h^{\alpha\mu}{}_{,\alpha\mu} - h_{,\alpha}{}^{,\alpha}$$

Next is the Einstein tensor G.

$$2\,G_{\beta\varepsilon} = 2\,R_{\beta\varepsilon} - \eta_{\beta\varepsilon}\,R$$
$$2\,G_{\beta\varepsilon} = +\,h^{\alpha}{}_{\beta,\varepsilon\alpha} + h^{\alpha}{}_{\varepsilon,\alpha\beta} - h_{\beta\varepsilon,\alpha}{}^{,\alpha} - h_{,\beta\varepsilon}$$
$$-\,\eta_{\beta\varepsilon}\,(\,h^{\alpha\mu}{}_{,\alpha\mu} - h_{,\alpha}{}^{,\alpha}\,)$$
$$2\,G_{\beta\varepsilon} = +\,h_{\alpha\beta,\varepsilon}{}^{,\alpha} + h_{\alpha\varepsilon,\beta}{}^{,\alpha}$$
$$-\,h_{,\beta\varepsilon} - \eta_{\beta\varepsilon}\,h^{\alpha\mu}{}_{,\alpha\mu} - h_{\beta\varepsilon,\alpha}{}^{,\alpha} + \eta_{\beta\varepsilon}\,h_{,\alpha}{}^{,\alpha}$$

For brevity, define: $$\hbar_{\beta\varepsilon} = h_{\beta\varepsilon} - \eta_{\beta\varepsilon}\,h\,/\,2$$
$$h_{\beta\varepsilon} = \hbar_{\beta\varepsilon} + \eta_{\beta\varepsilon}\,h\,/\,2$$

Note that: $$\hbar = \hbar_{\alpha}{}^{\alpha} = h_{\alpha}{}^{\alpha} - \eta_{\alpha}{}^{\alpha}\,h\,/\,2$$
$$\hbar = h - 2\,h = -\,h$$

Let's make this substitution:

$$2\,G_{\beta\varepsilon} = \hbar_{\alpha\beta,\varepsilon}{}^{,\alpha} + \eta_{\alpha\beta}\,h_{,\varepsilon}{}^{,\alpha}\,/\,2 + \hbar_{\alpha\varepsilon,\beta}{}^{,\alpha} + \eta_{\alpha\varepsilon}\,h_{,\beta}{}^{,\alpha}\,/\,2$$
$$-\,h_{,\beta\varepsilon} - \eta_{\beta\varepsilon}\,(\,\hbar^{\alpha\mu}{}_{,\alpha\mu} + \eta^{\alpha\mu}\,h_{,\alpha\mu}\,/\,2\,)$$
$$-\,\hbar_{\beta\varepsilon,\alpha}{}^{,\alpha} + \eta_{\beta\varepsilon}\,h_{,\alpha}{}^{,\alpha}\,/2$$
$$2\,G_{\beta\varepsilon} = \hbar_{\alpha\beta,\varepsilon}{}^{,\alpha} + h_{,\varepsilon\beta}\,/\,2 + \hbar_{\alpha\varepsilon,\beta}{}^{,\alpha} + h_{,\beta\varepsilon}\,/\,2$$
$$-\,h_{,\beta\varepsilon} - \eta_{\beta\varepsilon}\,(\,\hbar^{\alpha\mu}{}_{,\alpha\mu} + h_{,\alpha}{}^{,\alpha}\,/\,2\,)$$
$$-\,\hbar_{\beta\varepsilon,\alpha}{}^{,\alpha} + \eta_{\beta\varepsilon}\,h_{,\alpha}{}^{,\alpha}\,/2$$
$$2\,G_{\beta\varepsilon} = \hbar_{\alpha\beta,\varepsilon}{}^{,\alpha} + \hbar_{\alpha\varepsilon,\beta}{}^{,\alpha} - \eta_{\beta\varepsilon}\,\hbar^{\alpha\mu}{}_{,\alpha\mu} - \hbar_{\beta\varepsilon,\alpha}{}^{,\alpha}$$

With the prior equation for the components of the Einstein tensor, we arrive at the final step: employing Einstein's field equations to relate curvature to mass-energy:

$$8\,\pi\;T_{\beta\varepsilon} \;=\; G_{\beta\varepsilon}$$

$$16\,\pi\;T_{\beta\varepsilon} \;=\; -\;\hbar_{\beta\varepsilon,\alpha}{}^{,\alpha} \;+\; \hbar_{\alpha\beta,\varepsilon}{}^{,\alpha} \;+\; \hbar_{\alpha\varepsilon,\beta}{}^{,\alpha} \;-\; \eta_{\beta\varepsilon}\;\hbar^{\alpha\mu}{}_{,\alpha\mu}$$

Now, note that each of the last three terms has a divergence factor of the form $h_{\alpha\beta}{}^{,\alpha}$. Because the h components are very small and we are dropping quantities proportional to second and higher powers of h, we can eliminate all these $h_{\alpha\beta}{}^{,\alpha}$ terms with a *gauge transformation*, basically a change in coordinate system (see Section §33.3). We therefore choose a gauge in which:

Gauge condition for all μ: $$\hbar^{\mu\alpha}{}_{,\alpha} \;=\; 0$$

$$16\,\pi\;T_{\beta\varepsilon} \;=\; -\;\hbar_{\beta\varepsilon,\alpha}{}^{,\alpha}$$

The last remaining term on the right side of this equation is the sum of the second derivatives of $\hbar_{\beta\varepsilon}$ with respect to each coordinate — this is the sourced-wave equation for a wave moving at velocity c = 1, with $T_{\beta\varepsilon}$ being the source. Often, one writes wave equations using the symbol $\square$, as in:

$$16\,\pi\;T_{\beta\varepsilon} \;=\; -\;\square\;\hbar_{\beta\varepsilon}$$

In rectilinear coordinates: $$\square \;=\; -\;\partial^2/\partial t^2 \;+\; \partial^2/\partial x^2 \;+\; \partial^2/\partial y^2 \;+\; \partial^2/\partial z^2$$

§33.3 Gauge Transformations

This Section discusses gauge transformations in linearized general relativity. Such transformations exemplify the power of *general covariance*, a central theme of general relativity that we discussed in Section §15.1. General covariance ensures that properly formulated physical laws and equations (those written in tensor calculus) that are valid in any one coordinate system or reference frame are automatically valid in all coordinate systems and references frames without modification. While employing tensor calculus may seem burdensome, the reward can be spectacular.

You may join me in exploring gauge transformations, or choose to skip on to the next Section.

Consider a gauge transformation that changes the coordinates with which we are analyzing gravity waves by adding an arbitrary function to each coordinate. For the remainder of this section, I will denote transformed quantities in bold italics. The transformation equations are:

$$\boldsymbol{x}^\mu \;=\; x^\mu \;+\; \xi^\mu$$

$$x^\mu \;=\; \boldsymbol{x}^\mu \;-\; \xi^\mu$$

In keeping with the requirements of linearized theory, the absolute value of all ξ^μ must be << 1.

Let's first examine the effect of such a transformation on a simple function Q(P) at a point $P = x^\mu$.

If we shift the x-axis by an amount $+\Delta$, the x-coordinate of all physical points P decreases by Δ.

Since physical points do not move when the x-axis shifts, the value of $\boldsymbol{Q}$ at the shifted $\boldsymbol{x}^\mu$ equals the value of Q at the original x^μ. For small Δ, we have:

$$\boldsymbol{Q}(\mathrm{P}) \;=\; \mathrm{Q}(\mathrm{P})$$

$$\boldsymbol{Q}(\boldsymbol{x}^\mu) \;=\; Q(x^\mu) \;=\; Q(\boldsymbol{x}^\mu - \Delta) \;=\; Q(\boldsymbol{x}^\mu) \;-\; \Delta\;\partial Q/\partial \boldsymbol{x}$$

We next need to identify the effect of this transformation on the metric components $g_{\beta\varepsilon}$.

The general equation for changing the coordinates of tensors is:

$$\boldsymbol{g}_{\beta\varepsilon}(\boldsymbol{x}^\alpha) = g_{\mu\nu}(x^\alpha)\ \partial x^\mu/\partial \boldsymbol{x}^\beta\ \partial x^\nu/\partial \boldsymbol{x}^\varepsilon$$

In the present case, this is:

$$\boldsymbol{g}_{\beta\varepsilon}(\boldsymbol{x}^\alpha) = g_{\mu\nu}(x^\alpha)\ (\delta^\mu{}_\beta - \xi^\mu{}_{,\beta})\ (\delta^\nu{}_\varepsilon - \xi^\nu{}_{,\varepsilon})$$

We now drop second and higher powers of small quantities, such as h ξ and ξ ξ.

$$\boldsymbol{g}_{\beta\varepsilon}(\boldsymbol{x}^\alpha) = g_{\mu\nu}(x^\alpha)\ (\delta^\mu{}_\beta\ \delta^\nu{}_\varepsilon - \delta^\nu{}_\varepsilon\ \xi^\mu{}_{,\beta} - \delta^\mu{}_\beta\ \xi^\nu{}_{,\varepsilon})$$

$$\boldsymbol{g}_{\beta\varepsilon} = (\eta_{\mu\nu} + h_{\mu\nu})\ (\delta^\mu{}_\beta\ \delta^\nu{}_\varepsilon - \delta^\nu{}_\varepsilon\ \xi^\mu{}_{,\beta} - \delta^\mu{}_\beta\ \xi^\nu{}_{,\varepsilon})$$

$$\boldsymbol{g}_{\beta\varepsilon} = \eta_{\mu\nu}\ (\delta^\mu{}_\beta\ \delta^\nu{}_\varepsilon - \delta^\nu{}_\varepsilon\ \xi^\mu{}_{,\beta} - \delta^\mu{}_\beta\ \xi^\nu{}_{,\varepsilon}) + h_{\mu\nu}\ (\delta^\mu{}_\beta\ \delta^\nu{}_\varepsilon - \delta^\nu{}_\varepsilon\ \xi^\mu{}_{,\beta} - \delta^\mu{}_\beta\ \xi^\nu{}_{,\varepsilon})$$

$$\boldsymbol{g}_{\beta\varepsilon} = \eta_{\beta\varepsilon} + h_{\beta\varepsilon} - \delta_{\mu\varepsilon}\ \xi^\mu{}_{,\beta} - \delta^\nu{}_\beta\ \xi^\nu{}_{,\varepsilon}$$

In the new gauge, the metric has the form:

$$\boldsymbol{g}_{\beta\varepsilon} = \eta_{\beta\varepsilon} + \boldsymbol{h}_{\beta\varepsilon}$$

Hence h is transformed to ***h*** as:

$$\boldsymbol{h}_{\beta\varepsilon} = h_{\beta\varepsilon} - \xi_{\beta,\varepsilon} - \xi_{\varepsilon,\beta}$$

Let's now calculate the Riemann tensor ***R*** resulting from the gauge-transformed ***h***:

$$2\,R_{\alpha\beta\mu\varepsilon} = +\,h_{\alpha\varepsilon,\beta\mu} + h_{\beta\mu,\alpha\varepsilon} - h_{\beta\varepsilon,\alpha\mu} - h_{\alpha\mu,\beta\varepsilon}$$

$$2\,\boldsymbol{R}_{\alpha\beta\mu\varepsilon} = +\,\boldsymbol{h}_{\alpha\varepsilon,\beta\mu} + \boldsymbol{h}_{\beta\mu,\alpha\varepsilon} - \boldsymbol{h}_{\beta\varepsilon,\alpha\mu} - \boldsymbol{h}_{\alpha\mu,\beta\varepsilon}$$

$$\begin{aligned}2\,\boldsymbol{R}_{\alpha\beta\mu\varepsilon} = &+ h_{\alpha\varepsilon,\beta\mu} - \xi_{\alpha,\varepsilon\beta\mu} - \xi_{\varepsilon,\alpha\beta\mu}\\ &+ h_{\beta\mu,\alpha\varepsilon} - \xi_{\beta,\mu\alpha\varepsilon} - \xi_{\mu,\beta\alpha\varepsilon}\\ &- h_{\beta\varepsilon,\alpha\mu} + \xi_{\beta,\varepsilon\alpha\mu} + \xi_{\varepsilon,\beta\alpha\mu}\\ &- h_{\alpha\mu,\beta\varepsilon} + \xi_{\alpha,\mu\beta\varepsilon} + \xi_{\mu,\alpha\beta\varepsilon}\end{aligned}$$

Because partial derivatives commute, four pairs of terms in the prior equation cancel one another. To show this more graphically, in the equation below, these pairs are designated A, B, C, and D.

$$\begin{aligned}2\,\boldsymbol{R}_{\alpha\beta\mu\varepsilon} = &+ h_{\alpha\varepsilon,\beta\mu} - A - B\\ &+ h_{\beta\mu,\alpha\varepsilon} - C - D\\ &- h_{\beta\varepsilon,\alpha\mu} + C + B\\ &- h_{\alpha\mu,\beta\varepsilon} + A + D\end{aligned}$$

$$2\,\boldsymbol{R}_{\alpha\beta\mu\varepsilon} = +\,h_{\alpha\varepsilon,\beta\mu} + h_{\beta\mu,\alpha\varepsilon} - h_{\beta\varepsilon,\alpha\mu} - h_{\alpha\mu,\beta\varepsilon}$$

$$\boldsymbol{R}_{\alpha\beta\mu\varepsilon} = R_{\alpha\beta\mu\varepsilon}$$

After the cancellations, we discover that the gauge transformation leaves the Riemann tensor unchanged, and therefore leaves unchanged all tensors derived from Riemann. We can therefore employ any such transformation we wish without changing the physical meaning of our results.

Our final step is deriving the gauge transformation that most simplifies the wave equation.

Recall two prior equations:

$$\hbar_{\beta\varepsilon} = h_{\beta\varepsilon} - \eta_{\beta\varepsilon}\ h\ /\ 2$$

$$\boldsymbol{h}_{\beta\varepsilon} = h_{\beta\varepsilon} - \xi_{\beta,\varepsilon} - \xi_{\varepsilon,\beta}$$

Now consider:

$$\boldsymbol{\hbar}_{\alpha\beta}{}^{,\alpha} = \boldsymbol{h}_{\alpha\beta}{}^{,\alpha} - \eta_{\alpha\beta}\ \boldsymbol{h}_{\mu}{}^{\mu,\alpha}\ /\ 2$$

$$\boldsymbol{\hbar}_{\alpha\beta}{}^{,\alpha} = h_{\alpha\beta}{}^{,\alpha} - \xi_{\alpha,\beta}{}^{,\alpha} - \xi_{\beta,\alpha}{}^{,\alpha}$$

$$- \eta_{\alpha\beta}\ (h_{\mu}{}^{\mu,\alpha} - \xi_{\mu}{}^{,\mu\alpha} - \xi_{\mu}{}^{,\mu\alpha})\ /\ 2$$

$$\boldsymbol{\hbar}_{\alpha\beta}{}^{,\alpha} = h_{\alpha\beta}{}^{,\alpha} - \eta_{\alpha\beta}\ h_{\mu}{}^{\mu,\alpha} / 2 - \xi_{\alpha,\beta}{}^{,\alpha} - \xi_{\beta,\alpha}{}^{,\alpha} + \xi_{\mu}{}^{,\mu}{}_{,\beta}$$

$$\boldsymbol{\hbar}_{\alpha\beta}{}^{,\alpha} = \hbar_{\alpha\beta}{}^{,\alpha} - \xi_{\beta,\alpha}{}^{,\alpha}$$

If we set:

$$\xi_{\beta,\alpha} = \hbar_{\alpha\beta}$$

We obtain the desired result:

$$\boldsymbol{\hbar}_{\alpha\beta}{}^{,\alpha} = 0$$

Note ξ is not completely constrained; adding anything not a function of α to $\xi_{\beta,\alpha}$ yields the same result.

§33.4 Gravity Waves in Vacuum

In vacuum — in the absence of matter, energy, and pressure — the gauge-transformed wave equation is:

$$T_{\beta\varepsilon} = 0$$

$$\Box\ \hbar_{\beta\varepsilon} = 0$$

Let's discover what types of waves are possible.

The simplest waves are plane waves. The amplitude of a plane wave moving in the +z-direction is the same throughout each x,y plane — nothing is a function of x or y. The amplitude is different in different x,y planes, and all amplitudes change over time. In vacuum, a plane wave is described by:

$$\hbar_{\beta\varepsilon} = \mathbf{Re}\ \{\ A_{\beta\varepsilon}\ \exp(\ i\ k_{\mu}\ x^{\mu}\ )\ \}$$

Here, **Re** { } denotes the real part of { }, A is the wave amplitude, and k_{μ} is the wave number 4-vector. Also, exp(w) denotes e^{w}; in print, the former allows subscripted and superscripted exponents that the latter does not. The amplitude tensor $A_{\beta\varepsilon}$ and wave number 4-vector k_{μ} are governed by these equations:

$$k_{\mu}\ k^{\mu} = 0$$

$$A_{\beta\mu}\ k^{\mu} = 0$$

The analogy with electromagnetic waves may be helpful. A plane wave of electromagnetic radiation (light), with velocity c shown explicitly, is described by:

In 4-D spacetime:

$$E_{\beta} = \mathbf{Re}\ \{\ A_{\beta}\ \exp(\ i\ k_{\mu}\ x^{\mu}\ )\ \}$$

In classical 3-D:

$$E_{\beta} = \mathbf{Re}\ \{\ A_{\beta}\ \exp(\ i\ [\ -\omega\ t\ +\ \boldsymbol{k} \bullet \boldsymbol{r}\ ]\ )\ \}$$

Where:

$$\omega = k_{t}\ c$$

$$\boldsymbol{k} = (k_{x}\ ,\ k_{y}\ ,\ k_{z})$$

$$\boldsymbol{r} = (x,\ y,\ z\ )$$

Key properties of this wave of electromagnetic radiation are:

$$k_\mu k^\mu = 0 = -\omega^2 + k_x k^x + k_y k^y + k_z k^z = -\omega^2 + \boldsymbol{k} \bullet \boldsymbol{k}$$

Direction of wave motion: $\boldsymbol{k} / k_t$

Wave speed: $\omega / k_t = c$

At first, it is quite startling that $k_\mu k^\mu = 0$ *does not* imply $k_\mu = 0$, as it would with 3-vectors. In 4-D, a nonzero 4-vector can be perpendicular to itself due to the minus sign in the t-component of all subscripted 4-vectors.

Returning to gravitational waves, our next step is defining additional gauge conditions to constrain mathematical degrees of freedom that are physically irrelevant. These additional gauge conditions are:

$$A_\mu{}^\mu = 0$$

$$A_{\beta\mu} u^\mu = 0$$

Here, u^μ is any 4-vector field that we might wish. Let's select a reference frame in which:

$$u^0 = 1$$

$$u^1 = u^2 = u^3 = 0$$

In this frame:

$$A_{0\mu} u^\mu = A_{00} = 0$$

For n = 1, 2, 3: $A_\mu{}^\mu = A_n{}^n = 0$

For $\beta = 0$: $A_{0\mu} k^\mu = 0$

Since the last equation must be valid for any k^μ, it must be that:

$$A_{0\mu} = 0 = \bar{h}_{0\mu}$$

Also:

$$\bar{h} = \bar{h}_\beta{}^\beta = \textbf{Re}\{ A_\beta{}^\beta \exp(i\, k_\mu x^\mu) \} = 0$$

$$\bar{h} = 0 = -h$$

$$\bar{h}_{\beta\varepsilon} = h_{\beta\varepsilon} - \eta_{\beta\varepsilon} h / 2 = h_{\beta\varepsilon}$$

Since there is no longer any distinction between h and $\bar{h}$, we will hereafter drop the bar.

Recapping the collection of all these constraints on $h_{\beta\varepsilon}$, we have:

$$h_{0\mu} = 0$$

For n = 1, 2, 3: $h_n{}^n = 0$

$$h_{\mu n}{}^{,n} = 0$$

In the prior three equations, the upper one states only the spatial components of h are nonzero. The middle equation states the spatial components of h have zero trace, and the lower equation states the spatial components of h have zero divergence.

Any tensor satisfying these three equations is called a *transverse-traceless* (TT) tensor. The combination of all our constraints specifies the *transverse-traceless gauge*.

Finally, because these equations are linear, we know from classical physics that a wave of any type is comprised of a linear superposition of plane waves. Hence, everything we derived above for plane waves, in the transverse-traceless gauge, is also valid for any type of wave.

In linearized gravitational theory, all gravitational waves in vacuum are governed by the above constraints and by:

$$\Box\, h_{\beta\varepsilon} = \hbar_{\beta\varepsilon,\alpha}{}^{,\alpha} = 0$$

Let's now calculate the curvature tensors for these waves in the TT gauge. Our gauge manipulations will simplify these calculations.

We do this first for a wave moving in any direction.

First is the Riemann tensor. From Section §33.2:

$$2\,R_{\alpha\beta\mu\varepsilon} = +\,h_{\alpha\varepsilon,\beta\mu} + h_{\beta\mu,\alpha\varepsilon} - h_{\beta\varepsilon,\alpha\mu} - h_{\alpha\mu,\beta\varepsilon}$$

For $\alpha = \mu = t$:

$$2\,R_{t\beta t\varepsilon} = +\,h_{t\varepsilon,\beta t} + h_{\beta t,t\varepsilon} - h_{\beta\varepsilon,tt} - h_{tt,\beta\varepsilon}$$

Since all $h_{t\sigma} = 0$:

$$2\,R_{t\beta t\varepsilon} = -\,h_{\beta\varepsilon,tt}$$
$$2\,R_{txtx} = -\,h_{xx,tt}$$
$$2\,R_{txty} = -\,h_{xy,tt}$$
$$2\,R_{txtz} = -\,h_{xz,tt}$$
$$2\,R_{tyty} = -\,h_{yy,tt}$$
$$2\,R_{tytz} = -\,h_{yz,tt}$$
$$2\,R_{tztz} = -\,h_{zz,tt}$$

For $\alpha = t$:

$$2\,R_{t\beta\mu\varepsilon} = +\,h_{\beta\mu,t\varepsilon} - h_{\beta\varepsilon,t\mu}$$
$$2\,R_{txxy} = +\,h_{xx,ty}$$
$$2\,R_{txxz} = +\,h_{xx,tz}$$
$$2\,R_{txyz} = +\,h_{xy,tz} - h_{xz,ty}$$
$$2\,R_{tyxy} = -\,h_{yy,tx}$$
$$2\,R_{tyxz} = +\,h_{yx,tz} - h_{yz,tx}$$
$$2\,R_{tyyz} = +\,h_{yy,tz}$$
$$2\,R_{tzxy} = +\,h_{zx,ty} - h_{zy,tx}$$
$$2\,R_{tzxz} = -\,h_{zz,tx}$$
$$2\,R_{tzyz} = -\,h_{zz,ty}$$

Components with no t index:

$$2\,R_{xyxy} = -\,h_{yy,xx} - h_{xx,yy}$$
$$2\,R_{xyxz} = -\,h_{yz,xx} - h_{xx,yz}$$
$$2\,R_{xyyz} = +\,h_{xz,yy} + h_{yy,xz}$$
$$2\,R_{xzxz} = -\,h_{zz,xx} - h_{xx,zz}$$
$$2\,R_{xzyz} = -\,h_{zz,xy} - h_{xy,zz}$$
$$2\,R_{yzyz} = -\,h_{zz,yy} - h_{yy,zz}$$

Next is the Ricci tensor. From Section §33.2:

$$2\,R_{\beta\varepsilon} = +\,h^{\alpha}{}_{\beta,\varepsilon\alpha} + h^{\alpha}{}_{\varepsilon,\beta\alpha} - h_{\beta\varepsilon,\alpha}{}^{,\alpha} - h_{,\beta\varepsilon}$$

$$2\,R_{\beta\varepsilon} = -\,h_{\beta\varepsilon,\alpha}{}^{,\alpha}$$

Now comes the curvature scalar. From Section §33.2:

$$R = h^{\alpha\mu}{}_{,\alpha\mu} - h_{,\alpha}{}^{,\alpha}$$

$$R = 0$$

Finally, we have the Einstein tensor. From Section §33.2:

$$2\,G_{\beta\varepsilon} = \hbar_{\alpha\beta,\varepsilon}{}^{,\alpha} + \hbar_{\alpha\varepsilon,\beta}{}^{,\alpha} - \eta_{\beta\varepsilon}\,\hbar^{\alpha\mu}{}_{,\alpha\mu} - \hbar_{\beta\varepsilon,\alpha}{}^{,\alpha}$$

$$2\,G_{\beta\varepsilon} = -\,\hbar_{\beta\varepsilon,\alpha}{}^{,\alpha}$$

Now, let's further simplify these tensors by considering only plane waves moving in the z-direction. In this case, all x and y derivatives are zero, and any component of h with either a t or z index is zero.

The only nonzero Reimann components are:

$$R_{txtx} = -\,h_{xx,tt}\;/\;2$$

$$R_{txty} = -\,h_{xy,tt}\;/\;2$$

$$R_{tyty} = -\,h_{yy,tt}\;/\;2$$

$$R_{txxz} = +\,h_{xx,tz}\;/\;2$$

$$R_{txyz} = +\,h_{xy,tz}\;/\;2$$

$$R_{tyxz} = +\,h_{yx,tz}\;/\;2$$

$$R_{tyyz} = +\,h_{yy,tz}\;/\;2$$

$$R_{xzxz} = -\,h_{xx,zz}\;/\;2$$

$$R_{xzyz} = -\,h_{xy,zz}\;/\;2$$

$$R_{yzyz} = -\,h_{yy,zz}\;/\;2$$

Recall, from the start of this section, the equation for a plane wave moving in the +z-direction.

$$h_{\beta\varepsilon} = \mathbf{Re}\,\{\,A_{\beta\varepsilon}\,\exp(\,i\,k_{\mu}\,x^{\mu}\,)\,\}$$

$$h_{\beta\varepsilon} = \mathbf{Re}\,\{\,A_{\beta\varepsilon}\,\exp(-i\,\omega\,t + i\,k_z\,z)\,\}$$

$$k_{\mu}\,k^{\mu} = 0 = -\,\omega^2 + \boldsymbol{k}\bullet\boldsymbol{k} = -\,\omega^2 + k_z^{\,2}$$

Hence, $\omega = k_z$, and we can rewrite the wave equation as:

$$h_{\beta\varepsilon} = \mathbf{Re}\,\{\,A_{\beta\varepsilon}\,\exp(-i\,\omega\,[t - z\,]\,)\,\}$$

This means the t and z derivatives of every component of h are related by:

$$h_{\beta\varepsilon,t} = -\,h_{\beta\varepsilon,z}$$

Also, recall that the spatial components of h are traceless in this gauge. For a plane wave moving in the +z-direction, all h components with a z index are zero. Therefore:

$$h_n{}^n = 0 = h_x{}^x + h_y{}^y$$

$$h_{xx} = -\ h_{yy}$$

Lastly, since $h_{\beta\varepsilon} = g_{\beta\varepsilon} - \eta_{\beta\varepsilon}$, and both g and η are symmetric tensors, h must also be symmetric:

$$h_{xy} = +\ h_{yx}$$

These many special conditions greatly simply the Riemann components of a plane wave moving in the +z-direction. We can rearrange Riemann in a more symmetric form.

$$R_{txtx} = -\ h_{xx,tt}\ /\ 2$$

$$R_{txtx} = -\ (\ h_{xx} + h_{xx}\ )_{,tt}\ /\ 4$$

$$R_{txtx} = -\ (\ h_{xx} - h_{yy}\ )_{,tt}\ /\ 4$$

Similarly:

$$R_{xtxt} = R_{ytyz} = R_{xzxz} = -\ (\ h_{xx} - h_{yy}\ )_{,tt} / \ 4$$

$$R_{ytyt} = R_{xtxz} = R_{yzyz} = +\ (\ h_{xx} - h_{yy}\ )_{,tt} / \ 4$$

$$R_{xtyt} = R_{xzyx} = R_{xzyz} = -\ (\ h_{xy} + h_{yx}\ )_{,tt} / \ 4$$

$$R_{xtyz} = R_{xtyx} = R_{xzyt} = +\ (\ h_{xy} + h_{yx}\ )_{,tt} / \ 4$$

The only nonzero Ricci and Einstein tensor components are:

$$G_{\beta\varepsilon} = R_{\beta\varepsilon} = -\ h_{\beta\varepsilon,\alpha}{}^{,\alpha}\ /\ 2$$

$$G_{\beta\varepsilon} = R_{\beta\varepsilon} = -\ (\ h_{\beta\varepsilon,t}{}^{,t} + h_{\beta\varepsilon,z}{}^{,z}\ )\ /\ 2$$

$$G_{\beta\varepsilon} = R_{\beta\varepsilon} = +\ h_{\beta\varepsilon,tt}$$

$$G_{xx} = R_{xx} = +\ h_{xx,tt} = +\ (\ h_{xx} - h_{yy}\ )_{,tt}\ /\ 2$$

$$G_{yy} = R_{yy} = +\ h_{yy,tt} = -\ (\ h_{xx} - h_{yy}\ )_{,tt}\ /\ 2$$

$$G_{xy} = R_{xy} = +\ h_{xy,tt} = +\ (\ h_{xy} + h_{yx}\ )_{,tt} / \ 2$$

$$G_{yx} = R_{yx} = +\ h_{yx,tt} = +\ (\ h_{xy} + h_{yx}\ )_{,tt} / \ 2$$

We see that there are only two independent components of h. We can rewrite these, resulting in a more revealing expression for the metric g:

$$h_+ = (\ h_{xx} - h_{yy}\ )\ /\ 2$$

$$h_\times = (\ h_{xy} + h_{yx}\ )\ /\ 2$$

$$ds^2 = g_{\alpha\beta}\ dx^\alpha\ dx^\beta$$

$$ds^2 = (\ \eta_{\alpha\beta} + h_{\alpha\beta})\ dx^\alpha\ dx^\beta$$

$$ds^2 = -\ dt^2 + (1 + h_+)\ dx^2 + (1 - h_+)\ dy^2$$

$$+\ 2\ h_\times\ dx\ dy + dz^2$$

Here, the factor of 2 arises because the sum over α and β includes the two terms dx dy and dy dx.

Yet another version of the metric is:

$$ds^2 = \eta_{\alpha\beta}\ dx^\alpha\ dx^\beta\ +\ h_+\ (dx^2 - dy^2)\ +\ 2\ h_\times\ dx\ dy$$

On the right side of the last equation, the first term is the metric of flat space, and the last two terms describe the transverse spatial curvature due to a gravitational wave. Such waves have two independent polarization states: h_+ and $h_\times$ are the amplitudes of what are called + and × polarizations, respectively. We explore these two polarizations further, with illustrative graphics, in Section §34.2.

While we employed a transverse-traceless gauge to prove that gravitational waves have exactly two polarization states, that conclusion is valid in any gauge, reference frame, or coordinate system. The number of polarizations is a real physical entity that is the same in all coordinate systems.

Since h_+ and $h_\times$ are both very small, the x-spatial deformation of + polarization is given by:

$$\sqrt{\{ (1 + h_+)\ dx^2 \}} = (1 + h_+ / 2)\ dx$$

Similar expressions can be found for y-deformations, and for × polarization.

The key point is: spacetime curvature is proportional to a gravitational wave's amplitude. This is in sharp contrast to electromagnetism, where the force exerted by an electromagnetic wave on a charged particle is proportional to the wave amplitude squared.

Therefore, deformations from gravitational waves decrease with distance r from the source by the factor $1/r$, rather than $1/r^2$, as is the case for electromagnetic waves.

This is extremely fortunate for physicists. Detecting gravity waves that create strains of only 10^{-22} already requires heroic efforts — if these strains were 10^{-44}, dropping as $1/r^2$, they would be impossible to detect.

§33.5 Gravity Waves from Compact Binaries

The most interesting sources of gravitational waves, given our current instrumental technologies, are compact binaries, systems of two massive bodies orbiting one another in close proximity.

The orbital mechanics of two bodies of comparable mass is somewhat different from the more common situation of a small body orbiting a much more massive body.

Before proceeding, let's recall the equations we derived in Section §10.5 for circular orbits of two bodies of comparable mass.

Let's define the masses, 3-vector accelerations, and 3-vector positions of the bodies to be m_1 and m_2, $\boldsymbol{a}_1$ and $\boldsymbol{a}_2$, and $\boldsymbol{r}_1$ and $\boldsymbol{r}_2$, respectively. We also define the relative acceleration $\boldsymbol{a}$, the reduced mass μ, and separation r as:

Relative acceleration: $\boldsymbol{a} = \boldsymbol{a}_1 - \boldsymbol{a}_2$

Reduced mass: $\mu = m_1\ m_2\ /\ (m_1 + m_2)$

Separation r: $r^2 = (\boldsymbol{r}_2 - \boldsymbol{r}_1)^2$

Let's also define M to be the total mass of both orbiting bodies:

$$M = m_1 + m_2$$

From the definition of reduced mass, note that:

$$\mu\ M = m_1\ m_2$$

The force that each body exerts on the other can now be written:

$$F = \mu \, \boldsymbol{a} = G \, \mu \, M \, / \, r^2$$

This is the equation of a body of mass μ orbiting a stationary body of mass M in a circular orbit with radius r. Our analysis reduces a problem with two moving bodies to a much simpler problem with only one moving body.

Now, after all this math, what can we learn about the radiation of gravitational waves by a compact binary?

In their book ***Gravitation***, on pages 988 and following, authors Charles W. Misner, Kip S. Thorne, and John Archibald Wheeler analyze the decaying orbits of two celestial bodies in a compact binary system. They employ techniques beyond the scope of this book, and refer to an even more complex research paper by Peters and Mathews (1963). Even those analyses are approximations. As yet, no one has derived an exact analytical solution of Einstein's field equations of general relativity for the generation of gravitational waves. I wish to thank Misner and Thorne for graciously addressing my many naïve questions.

These approximate analyses state the gravitational waves radiated by a compact binary reduce the orbiting bodies' total energy E at the rate:

$$dE/dt = -\, 32 \, f(\varepsilon) \, G^4 \, \mu^2 \, M^3 \, / \, 5 \, c^5 \, r^5$$

Where: $$f(\varepsilon) = (\, 1 + 73\,\varepsilon^2 / 24 + 37\,\varepsilon^4 / 96 \,)\,(\, 1 - \varepsilon^2 \,)^{-7/2}$$

Here, μ is the reduced mass, M is the sum of the two bodies' masses, r is the distance between the bodies' centers, and ε is the eccentricity of the orbit. We are assuming the bodies are spherically symmetric. As Figure 33-1 shows, the eccentricity function $f(\varepsilon)$ equals 1 for a circular orbit ($\varepsilon = 0$), and grows rapidly with increasing ε; $f(\varepsilon)$ approaches infinity as ε approaches 1, which is physically unrealistic in any case.

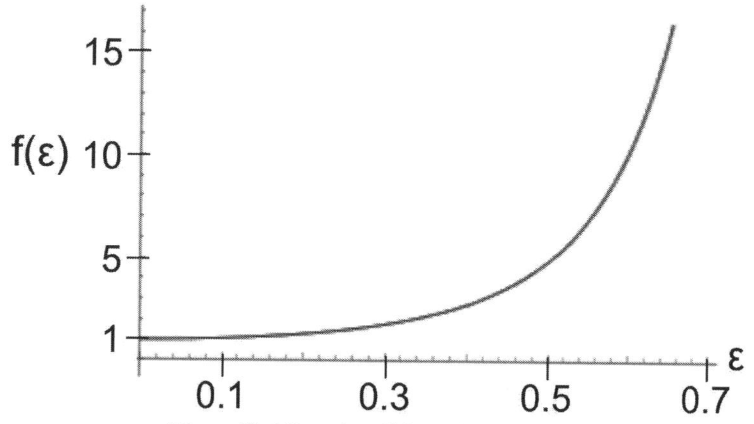

Figure 33-1 Function $f(\varepsilon)$ vs. Eccentricity ε

Let's avoid this elliptical complexity and focus on circular orbits.

Assuming non-relativistic velocities and Newtonian orbits, the gravitational potential energy U, and the total energy E of two such orbiting bodies is:

$$U = -\, G \, \mu \, M \, / \, r$$

The virial theorem states: $$E = T + U = -\,U \, / \, 2$$

Here, T is the sum of the bodies' kinetic energy.

The time derivative of the last equation is:

$$dE/dt = -\, dU/dt \, / \, 2 = -\,(dU/dr)\,(dr/dt) \, / \, 2$$

Evaluating the radial derivative of gravitational potential U yields:

$$dE/dt = (+ G \mu M / 2 r^2) (dr/dt)$$

Inserting the energy loss rate due to Misner, Thorne, and Wheeler yields:

$$dr/dt = (- 32 G^4 \mu^2 M^3 / 5 c^5 r^5) (2 r^2 / G \mu M)$$

$$dr/dt = - 64 G^3 \mu M^2 / 5 c^5 r^3$$

$$5 c^5 r^3 dr = - 64 G^3 \mu M^2 dt$$

Let's now integrate this equation from r = R, the separation at time t = 0, to r = 0, when the bodies collide at time τ.

$$5 c^5 \int_R^0 r^3 dr = - 64 G^3 \mu M^2 \int_0^\tau dt$$

$$- 5 c^5 R^4 / 4 = - 64 G^3 \mu M^2 \tau$$

$$\tau = 5 c^5 R^4 / 256 G^3 \mu M^2$$

Using G Msun = 1.327×10^{20} m^3 / s^2, let's restate our equations so that masses μ and M are measured in multiples of Msun, the mass of our Sun, and R is measured in meters.

We obtain:

$$dr/dt = - [1.23 \times 10^{19} \mu M^2 / R^3] \text{ meters / second}$$

$$\tau = [6.45 \times 10^{-28} R^4 / \mu M^2] \text{ years}$$

$$\tau = [2.03 \times 10^{-20} R^4 / \mu M^2] \text{ seconds}$$

$$dE/dt = - [1.63 \times 10^{69} \mu^2 M^3 / R^5] \text{ watts}$$

$$T = 5.44 \times 10^{-13} \sqrt{(R^3 / M)} \text{ years}$$

$$T = 1.715 \times 10^{-5} \sqrt{(R^3 / M)} \text{ seconds}$$

Here, T is the orbital period. Recall that dE/dt is the rate of change of the orbiting bodies' total energy. That energy is carried away by gravitational waves. Hence, the power radiated in gravitational waves equals minus dE/dt.

For Earth orbiting the Sun, $\mu = 3.003 \times 10^{-6}$, M = 1, and R = 1.496×10^{11} meters. These yield:

For Earth's orbit:

$$dr/dt = - 0.94 \times 10^{-15} \text{ meters / day}$$

$$\tau = 1.1 \times 10^{23} \text{ years}$$

$$dE/dt = - 192 \text{ watts}$$

These data show that gravitational wave radiation decreases Earth's orbital radius by about the width of a single proton each day, and Earth's orbital energy loss is only enough to power two incandescent light bulbs.

The time τ for Earth to fall into the Sun due to gravitational wave radiation alone, 10^{23} years, is astronomically greater than the current age of the universe (14 billion years), or Sun's remaining lifetime (about 5 billion years).

§33.6 Free-Falling Bodies in Linear Theory

In many applications, gravity is sufficiently weak that the linearized theory of general relativity is adequate. Our Solar System is a prime example.

The Schwarzschild metric outside a body of mass M is:

In spherical coordinates: $$ds^2 = -\Omega\, dt^2 + dr^2/\Omega + r^2\, d\theta^2 + r^2 \sin^2\theta\, d\phi^2$$

In rectilinear coordinates: $$ds^2 = -\Omega\, dt^2 + (dx^2 + dy^2 + dz^2)/\Omega$$

Where: $$\Omega = 1 - 2M/r$$

From the surface of the Sun outward, half way to the next star, $2M/r$ is less than 5×10^{-5}.

We can therefore approximate the metric as:

$$ds^2 = -(1 - 2M/r)\, dt^2 + (1 + 2M/r)\, dr^2 + r^2\, d\theta^2 + r^2 \sin^2\theta\, d\phi^2$$

Since the Newtonian gravitational potential is $\Phi = -M/r$, we can think of the metric in those terms:

$$ds^2 = -(1 - 2\Phi)\, dt^2 + (1 + 2\Phi)\, dr^2 + r^2\, d\theta^2 + r^2 \sin^2\theta\, d\phi^2$$

Let's confirm some simple cases for an object orbiting a stationary mass M, within the $\theta = \pi/2$ plane. The metric then reduces to:

$$ds^2 = -(1 - 2M/r)\, dt^2 + (1 + 2M/r)\, dr^2 + r^2\, d\phi^2$$

Following Section §33.2, we define:

$$h_{tt} = h_{rr} = +2M/r$$

The nonzero connection coefficients are:

$$\Gamma^{\mu}{}_{\alpha\beta} = \eta^{\mu\mu}\,(h_{\mu\alpha,\beta} + h_{\mu\beta,\alpha} - h_{\alpha\beta,\mu})\,/\,2$$

$$\Gamma^{r}{}_{tt} = \eta^{rr}\,(h_{rt,t} + h_{rt,t} - h_{tt,r})\,/\,2 = (1)(-h_{tt,r})/2$$

$$\Gamma^{r}{}_{tt} = +M/r^2$$

$$\Gamma^{r}{}_{rr} = \eta^{rr}\,(h_{rr,r} + h_{rr,r} - h_{rr,r})\,/\,2 = (1)(+h_{rr,r})/2$$

$$\Gamma^{r}{}_{rr} = -M/r^2$$

$$\Gamma^{r}{}_{\phi\phi} = \eta^{rr}\,(h_{r\phi,\phi} + h_{r\phi,\phi} - h_{\phi\phi,r})\,/\,2 = (1)(-h_{\phi\phi,r})/2$$

$$\Gamma^{r}{}_{\phi\phi} = -r$$

With these connection coefficients, let's evaluate the geodesic equation.

The geodesic equation for radial motion is:

$$0 = \partial^2 r/\partial t^2 + \Gamma^r_{\alpha\beta}\ \partial x^\alpha/dt\ \partial x^\beta/\partial t$$

$$\partial^2 r/\partial t^2 = -\Gamma^r_{tt}\ (\partial t/\partial t)^2 - \Gamma^r_{rr}\ (\partial r/\partial t)^2 - \Gamma^r_{\phi\phi}\ (\partial\phi/\partial t)^2$$

Let's substitute the connection coefficients, and define radial velocity $v_r = \partial r/\partial t$, and azimuthal velocity $v_\phi = r\ \partial\phi/\partial t$. The result is:

$$\partial^2 r/\partial t^2 = -M/r^2 + M\,v_r^2/r^2 + v_\phi^2/r$$

Let's first consider an initially stationary object, with $v_r = v_\phi = 0$. This reduces the metric to the normal Newtonian equation (with G = 1).

$$\partial^2 r/\partial t^2 = -M\ /\ r^2$$

Next consider an object in a circular orbit, with $v_r = \partial r/dt = 0$. The metric equation becomes the normal Newtonian equation for the centripetal acceleration required to maintain a circular orbit of radius r around a stationary mass M.

$$M / r^2 = v_\phi^2 / r$$

§33.7 Redshift of Light in Linear Theory

Now, let's consider another weak-field scenario. Imagine a photon emitted at spacetime event x^β, where gravity is weak, that propagates to spacetime event y^β, where gravity is also weak. Let the photon's frequency be f_x at event x and f_y at event y. What is the equation relating f_x and f_y?

To be specific, let's consider the experiments of Pound and Rebka in 1960 and Pound and Snider in 1965. I will simplify the description of these experiment slightly for clarity. From decays of radioactive Fe^{57} from a source in the basement of the Jefferson Laboratory at Harvard University, they produced photons of energy 14.4 keV that traveled through a helium-filled tube up to the 6th floor attic, an elevation change of 22.5 meters.

The metric tensor tells us how time changes in gravitational fields. In weak gravity, a clock in its own rest frame (dx = dy = dz = 0), is described by this metric:

$$ds^2 = -(1 - 2M/r)\,dt^2$$

$$d\tau^2 = -ds^2 = +(1 - 2M/r)\,dt^2$$

$$d\tau = (1 - M/r)\,dt$$

Here, τ is proper time, the time that a perfect clock keeps if no forces (other than gravity) act upon it, and r is the clock's distance from the center of mass M. This equation describes weak-field gravitational time dilation — clocks run slower in stronger gravity — as we discovered in Chapter 17.

We next need to know how gravitational time dilation affects a photon's frequency f.

Imagine that, at event x, we generate a pulse of light comprised of N full wave cycles, and that we later detect that pulse at event y. Many things might be different at x and y, but N must be the same. The number of wave cycles is an absolute quantity, not a relative one. In principle, we can precisely count wave cycles — each cycle of an electromagnetic wave accelerates electrons back and forth, which is easily measurable (this is what radios and TV's do). If electrons accelerate back and forth N times in one reference frame, they must do so in every other frame as well.

The frequency of a wave with N full cycles is $f = N / \Delta\tau$, where $\Delta\tau$ is the time interval during which the N cycles occur — 10 cycles in 2 seconds means f = 5 Hz.

Let's define a reference frame and coordinate system centered on mass M, and define f(r) to be the frequency of this particular light pulse as measured at distance r from mass M. From the equation for gravitational time dilation, the equation for f(r) is:

$$f(r) = N / [(1 - M/r)\ \Delta t]$$

At $r = \infty$, where there is no gravitational time dilation ($M / r = 0$), f has its minimum value of:

$$f(r = \infty) = N / \Delta t$$

Close to mass M, M / r is large, and time is substantially dilated. Here, clocks run slower, the N wave cycles seem to pass by in less clock time. That makes f appear larger closer to M than it does far away.

Evidently, for any light wave, with any number of cycles, the expression on the left side of the following equation has the same value for any value of r:

$$f(r)\ (1 - M/r) = N / \Delta t$$

We can now compare photon frequencies at any two radii, call them r_x and r_y, with this equation:

$$f(r_x)\ (1 - M/r_x) = f(r_y)\ (1 - M/r_y)$$

For the experiments of Pound et. al., in weak gravity, we have:

$$f(r_{up})\ (1 - M/r_{up}) = f(r_{down})\ (1 - M/r_{down})$$

Here, r_{up} and r_{down} are the distances from Earth's center to the attic and to the basement, respectively.

Let's rearrange this equation.

$$f(r_{up}) / f(r_{down}) = (1 - M/r_{down}) / (1 - M/r_{up})$$

$$f(r_{up}) / f(r_{down}) = (1 - M/r_{down})\ (1 + M/r_{up})$$

$$f(r_{up}) / f(r_{down}) = 1 - M/r_{down} + M/r_{up}$$

$$f(r_{up}) / f(r_{down}) = 1 - M\ (1/r_{down} - 1/r_{up})$$

$$f(r_{up}) / f(r_{down}) = 1 - M\ (r_{up} - r_{down}) / r_{up}\ r_{down}$$

$$f(r_{up}) / f(r_{down}) = 1 - (r_{up} - r_{down})\ G\ M / r^2\ c^2$$

Near Earth's surface, $G\ M / r^2 = 1\ g = 9.807\ m / s^2$.

$$f(r_{up}) / f(r_{down}) = 1 - (22.5\ m)\ (9.807\ m/s^2) / c^2$$

$$f(r_{up}) / f(r_{down}) = 1 - 2.45 \times 10^{-15}$$

Employing the Mossbauer effect, Pound, Rebka, and Snider succeeded in measuring photon frequency changes to a few parts per 10^{17}. They confirmed Einstein's prediction of gravitational time dilation to a precision of 1%.

In 1980, a space-borne maser experiment confirmed gravitational time dilation to a precision of 0.01%.

Here is another, similar problem: what is the frequency shift of a photon emitted at the surface of the Sun and detected far away? Let's assume far away is at $r = \infty$. From above, the frequency equation is:

$$f(r_{SUN})\ (1\ -\ M_{SUN} / r_{SUN})\ =\ f(r = \infty)\ (1\ -\ M / \infty\ )$$

$$f(r_{SUN})\ -\ f(r_{SUN})\ M_{SUN} / r_{SUN}\ =\ f(r = \infty)$$

$$f(r = \infty)\ -\ f(r_{SUN})\ =\ -\ f(r_{SUN})\ M_{SUN} / r_{SUN}$$

$$[\ f(r = \infty)\ -\ f(r_{SUN})\ ]\ /\ f(r_{SUN})\ =\ -\ M_{SUN} / r_{SUN}$$

$$[\ f(r = \infty)\ -\ f(r_{SUN})\ ]\ /\ f(r_{SUN})\ =\ -\ 1.47\text{ km} / 696{,}392\text{ km}$$

$$[\ f(r = \infty)\ -\ f(r_{SUN})\ ]\ /\ f(r_{SUN})\ =\ -\ 2.11 \times 10^{-6}$$

Sunlight is therefore redshifted — its frequency decreases — as it moves away from the Sun. Its energy decreases as it climbs out of the Sun's gravitational potential well.

Chapter 34
Gravitational Waves

§34.1 Gravity Wave Overview

Gravity waves are small variations in the curvature of spacetime that propagate through the cosmos, somewhat like ripples across the surface of a lake. These ripples in spacetime are properly called *gravitational waves* to distinguish them from various phenomena in fluid dynamics that are driven by normal gravity, which have long been described as *gravity waves*. However, most people, either unaware or uninterested in fluid dynamic nomenclature, refer to waves of spacetime curvature as *gravity waves*. In this book, for brevity or ease of communication, I sometimes use *gravity waves* to describe waves of spacetime curvature.

In 1916, one year after completing the final version of his theory of general relativity, Einstein predicted gravitational waves would result from a variety of apocalyptic cosmic events — events that drastically alter the local distribution of energy, and thereby create enormous changes in local spacetime curvature. As these curvature disturbances propagate outward, their amplitudes diminish and eventually become tiny ripples. An analogy is a breaching humpback whale creating enormous waves in its immediate vicinity that spread across the ocean and are barely noticeable miles away.

It took theorists, including Einstein, over 40 years to understand what the nature of gravity waves might be. Experimentalists then began building ever more elaborate gravity wave detectors, but failed to find them for another half-century.

In 2015, 99 years after Einstein's original prediction, gravitational waves were detected for the first time by LIGO, the Laser Interferometer Gravitational-wave Observatory. For this momentous discovery, the leaders of LIGO — Rainer Weiss, Kip Thorne, and Barry Barish — received the 2017 Nobel Prize.

By late 2017, a total of five gravity wave events had been detected by the twin LIGO observatories in the U.S. VIRGO, a similar observatory in Italy, also detected the fourth and fifth events. Four events are thought to have originated from the merger of two black holes, while the other is attributed to the merger of two neutron stars. The data show that each of these mergers released a fabulous amount of energy, some more than all the stars in the universe at that instant. Most of that energy was carried away by gravitational waves. As those great waves spread out through the vastness of space, their intensity progressively diminished. By the time these waves reached Earth, they were tiny ripples that distorted our spacetime by only about one part in 10^{21}, one in a billion trillion.

LIGO monitors the lengths of two orthogonal 4-km-long arms, and measures changes in those lengths with a precision of about 1/10,000th of the width of a proton. It is incredible that mere mortals can detect such infinitesimal changes.

LIGO detected 5 events in two years, in sharp contrast to the LHC (Large Hadronic Collider) that produces up to 600 million events per second. But, it is LIGO that fascinates the physics world.

§34.2 Polarization States

In Sections §33.2 and §33.3, we showed that gravitational waves have two possible polarization states. Electromagnetic radiation (light) also has two polarization states; the two phenomena have similarities and also differences. The polarizations of both gravity waves and light can be described as either linear or circular. Let's first consider linear polarizations, and turn to circular polarization later.

Electromagnetic radiation has a dipole effect, exerting opposite forces in directions that are 180° apart. The linear polarization states of photons are superpositions of two orthogonal basis states, such as horizontal and vertical polarization, that are orthogonal to the wave's direction of motion.

In contrast to light, gravitational waves have a quadrupole effect, acting in directions that are 90° apart. These waves stretch space in one direction, while compressing space in an orthogonal direction, and later alternating stretching and compressing directions. Linear polarization states are superpositions of two orthogonal basis states called + and ×, as described below and illustrated in Figure 34-1. The deformation directions are always orthogonal to the wave's direction of motion.

+ Polarization

First half-cycle:	Stretch Horizontally and Compress Vertically
Second half-cycle:	Stretch Vertically and Compress Horizontally

× Polarization

First half-cycle:	Stretch at + 45° and Compress at – 45°
Second half-cycle:	Stretch at – 45° and Compress at + 45°

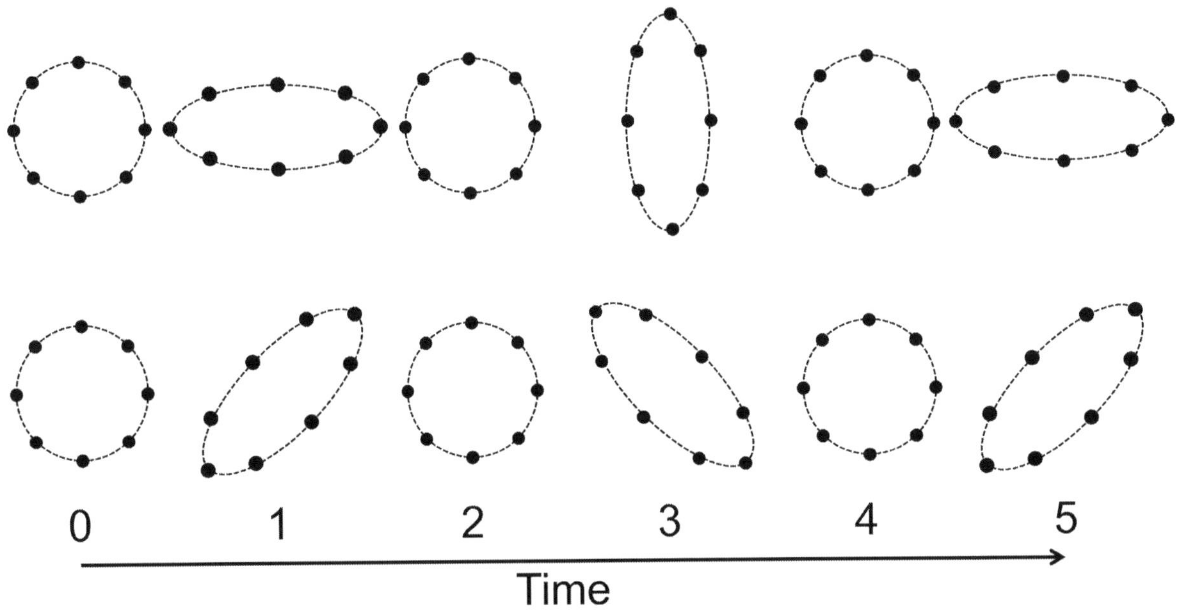

Figure 34-1 Ring Deformed by + Wave (upper) and × Wave (lower)

Figure 34-1 shows a circular array of beads being deformed by a gravity wave moving perpendicular to and into the page. As the wave propagates, the deformation axes alternate. The upper row shows the effect of + polarization, with deformation alternating between horizontal and vertical directions. The lower row shows the effect of × polarization, with the deformation axes at + 45° and – 45°.

In each row, the leftmost image, labeled time 0, shows the start of a wave cycle, when its amplitude is zero. The next image in each row, labeled time 1, shows the wave at maximum positive amplitude. At time 2, the amplitude has returned to zero. At time 3, the amplitude reaches its greatest negative value. At time 4, one cycle ends and the next begins. The wave at time 5 is again at maximum positive amplitude.

The two polarization basis states are named for the patterns they trace out as the wave propagates — the upper polarization traces a "+", while the lower polarization traces a "×".

Gravitational waves reaching us from remote cosmic events are too feeble to deform bodies made rigid by electric forces, such as Earth, people or beads. Gravitational waves deform space and thereby the distance between free-floating objects.

Photon polarizations can also be described as superpositions of basis states with right hand circular (RHC) and left hand circular (LHC) polarizations, which are themselves superpositions of horizontal and vertical polarizations. Gravitational waves can also be described as superpositions of basis states with right hand circular (RHC) and left hand circular (LHC) polarizations, as illustrated in Figure 34-2.

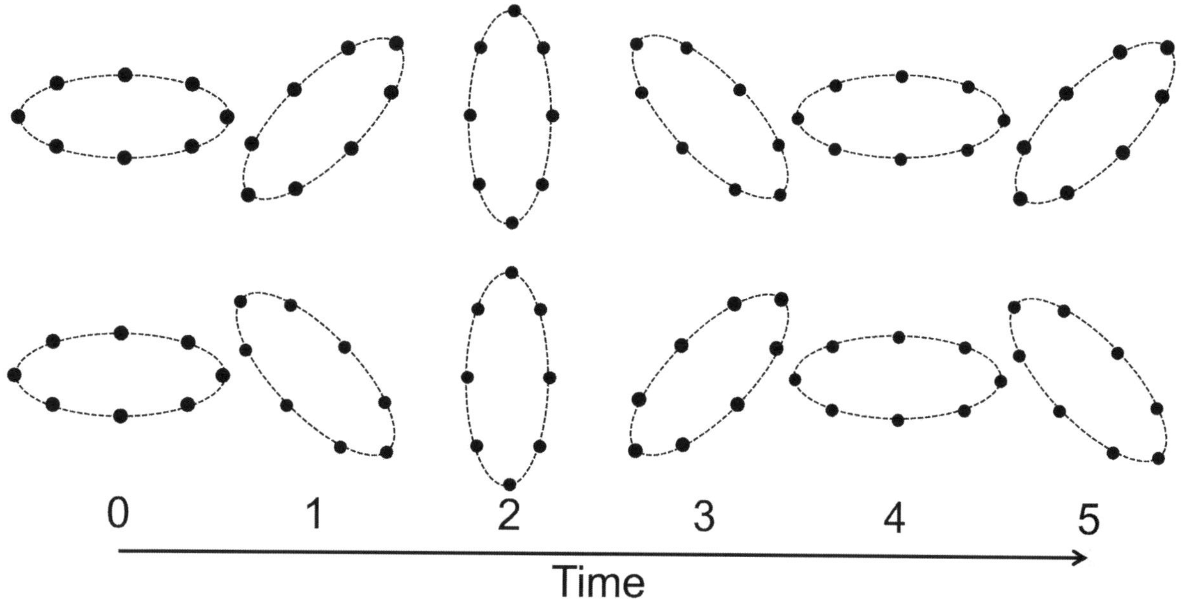

Figure 34-2 Ring Deformed by RHC Wave (upper) and LHC Wave (lower)

Figure 34-2 shows a circular array of beads being deformed by a gravitational wave moving into and perpendicular to the page. As the wave propagates, the deformation axes rotate. In the upper row, for RHC polarization, the deformation turns counterclockwise. In the lower row, for LHC polarization, the deformation turns clockwise. Note that neither the array of beads nor the individual beads themselves rotate, each bead moves slightly and then returns to its original position; only the spatial deformation pattern rotates.

As in Figure 34-1, the leftmost image in each row, labeled time 0, shows the start of a wave cycle. Time increases to the right, and at time 4, one cycle ends and the next begins.

§34.3 Generation of Gravity Waves

Gravitational waves are generated whenever spacetime curvature changes abruptly — when the "fabric" of spacetime is shaken — when black holes collide, when supernovae explode, when volcanoes erupt, when rockets take off, when lasers turn on, when popping champagne releases pressure, or when you pick up this book.

However, very few curvature changes produce detectable waves, because gravity is intrinsically very weak (Newton's constant G is 6.67×10^{-11}). I can easily stand up, overcoming the gravitational force of each of the 10^{51} particles in the Earth pulling on each of the 10^{28} particles in my body.

Appreciable gravitational waves are produced only by cataclysmic events: the destruction of mammoth stars, or the Big Bang at the birth of our universe.

To be a bit more precise, gravitational waves are generated when massive bodies accelerate asymmetrically. A spherically symmetric star does not radiate gravity waves as it spins on its axis, or even if it pulsates, alternately expanding and contracting.

To be much more technical, gravity waves are generated when the third time derivative of a system's quadrupole moment is nonzero. Rather than explain what that means, I will instead provide examples.

Two close massive bodies orbiting one another *do* radiate gravitational waves, because the spacetime curvature they create oscillates asymmetrically as the bodies continuously move.

The best sources of gravitational waves are compact binary systems. Binary stellar systems — two stars orbiting one another — are very common in the cosmos, but most have wide separations with orbital periods of many years. Those do not produce appreciable gravity waves. However, some pairs of cosmic bodies orbit in close proximity; these are called *compact binaries*.

As compact binaries orbit one another, and eventually merge, they radiate gravitational waves primarily within their orbital plane. Much less energy is radiated normal to that plane. This is illustrated in the *embedding diagram* in Figure 34-3. Be aware that the vertical direction in this image is only intended to suggest spatial curvature; it is *not* the third dimension of 3-D space. Nonetheless, it suggests action that is primarily within the orbital plane.

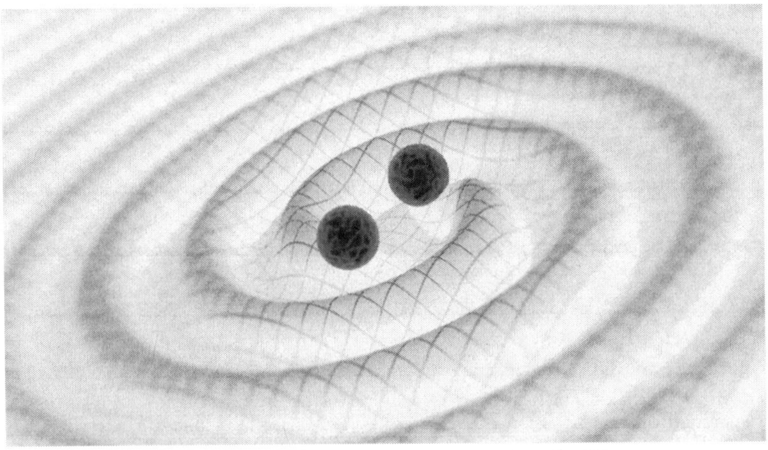

Figure 34-3 Compact Binary Generating Gravity Waves

Compact binaries contain two stellar bodies that might include collapsed stars, of which there are three types: white dwarfs, neutron stars, and black holes. Some compact binaries contain two collapsed stars of the same type: two white dwarfs, two neutron stars, or two black holes. These end-stage stars are described in Chapter 18 and Sections §25.2 and §31.4. For our present purposes, here are some highlights.

White dwarfs are the collapsed cores of low-mass stars that have exhausted their nuclear fuels. White dwarfs have masses up to 1.39 Msun (the Chandrashekar limit), and are typically 10,000 km wide, about the size of Earth. About 95% of all stars end their lives as white dwarfs.

Neutron stars are the collapsed cores of medium-mass stars that have exhausted their nuclear fuels. Neutron stars have masses above the Chandrashekar limit, and radii of about 12 km. All neutron stars discovered to date have masses between 1.4 and 2.01 Msun, although masses up to 3.7 Msun may be theoretically possible. About 5% of all stars evolve into neutron stars.

Black holes produced in the collapse of high-mass stars with core masses above 3 Msun; we know of no upper limit, but the most massive black hole discovered as yet has about 66 billion times our Sun's mass. Anything that enters a black hole's event horizon is consumed for eternity in its central singularity. A non-rotating black hole of mass M has an event horizon radius of 2.94 km multiplied by (M / Msun).

In this chapter, let's focus on compact binaries in circular orbits. Recall these equations derived in Section §33.5 for τ, the time for orbital decay, and the power radiated in gravity waves by a compact binary with separation R, total mass M, and reduced mass μ:

$$\tau = 2.03 \times 10^{-20} \ R^4 \ / \ \mu \ M^2 \text{ seconds}$$

$$\text{GW power} = 1.63 \times 10^{69} \ \mu^2 \ M^3 \ / \ R^5 \text{ watts}$$

Let's first consider a compact binary comprised of two white dwarfs, each of mass 1 Msun, in a tight circular orbit about one another.

If the binary white dwarfs are initially separated by 150,000 km (1/1000th of our distance to the Sun), their orbital period T would be 700 seconds. The time τ to before their collision is about 160,000 years — when their centers are separated by 10,000 km and T is only 12 seconds.

Consider next two neutron stars, each of mass 1.4 Msun, with an initial separation of 10,000 km. Their orbital period T is initially 10 seconds, but T drops to 0.001 seconds just before their collision, which occurs in about 400 days.

Substantive gravitational radiation is expected from neutron star mergers, but only for a few seconds before impact, a tiny fraction of the lifetime of such binary systems that can extend to billions of years. This means only a miniscule fraction of all existing compact neutron star binaries produce detectable gravitational waves each year.

The most powerful generators of gravitational waves, those easiest to detect, are compact binary black holes (BBH). Figure 34-4 is a supercomputer simulation by NASA/Caltech of gravitational waves radiating from two black holes orbiting one another in the vertical plane at the center of this image.

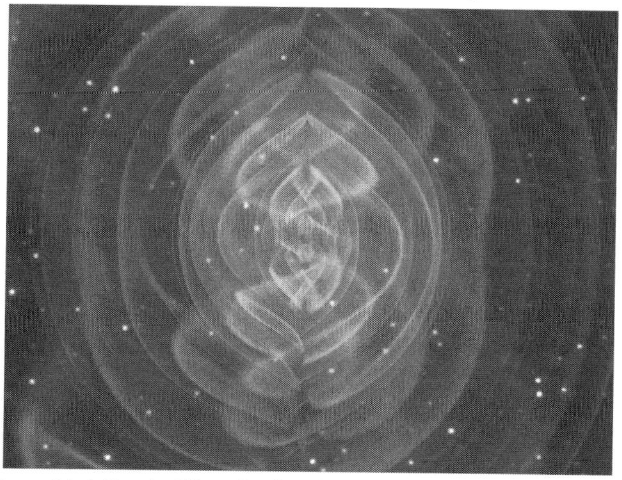

Figure 34-4 Gravity Wave Radiated by Compact Binary Black Holes

Two black holes, each of mass 8 Msun, with an initial separation of r = 10,000 km, have an orbital period T of only 4.3 seconds. They rapidly spiral toward one another, and collide in 2.3 days, forming a more massive black hole. When r drops to 1,000 km, their period T = 0.14 seconds, and the remaining time before collision is τ = 200 seconds. When r = 40 km, τ = 50 microseconds, and T = 0.0011 seconds. When r = 24 km, the black holes collide (their event horizons contact).

The table below details the demise of two black holes, of masses 36 and 29 Msun.

r radius (km)	τ time to collision	T orbit (sec.)	P power radiated
10,000	50 min	2.1	1.2E+42 watts
3,000	24 sec	0.35	4.7E+44
1,800	3.1 sec	0.16	6.1E+45
1,000	0.30	0.067	1.2E+47
510	0.020	0.025	3.3E+48
360	0.0050	0.015	1.9E+49
240	0.0010	0.008	1.5E+50
180	0.0003	0.005	6.1E+50
135	0.0001	0.003	2.6E+51
95	0	0.002	1.5E+52

Above, for better clarity, I use the computer notation 1.5E+52 = $1.5 \times 10^{+52}$. As the top line shows, the initial conditions are: separation r = 10,000 km, orbital period T = 2.1 seconds, and time before merger is τ = 50 minutes. These are the parameters of LIGO's first detected gravitational wave.

Since I calculated the data in the above table using the weak-gravity linearized theory, we cannot expect numerical precision in this strong-gravity scenario. Nonetheless, the trends are interesting.

LIGO says this gravitational wave's peak power, as estimated by numerical integration of Einstein's nonlinear equations of general relativity, was actually 3E49 watts, more in an instant than is continually radiated by all the stars in our observable universe.

Note that the radiated power, carried away by gravity waves, pikes sharply in the end, increasing by a factor of 100,000 in the final 0.1 seconds. An event billions of years in the making ends in less than a heartbeat.

After the two event horizons merge, there is a brief *ring-down* period during which the event horizon of the merging pair becomes spherical.

§34.4 Detection of Gravity Waves

Forty years after Einstein predicted gravitational waves, theorists were still struggling to understand even the simplest aspects of these waves.

In 1922, Sir Arthur Eddington, whose solar eclipse observations provided the first experimental confirmation of general relativity, wrote a paper calling gravity waves a mere mathematical artifact of Einstein's poor choice of coordinate systems. Eddington said Einstein's waves were effectively ripples in the coordinates rather than in space itself.

Indeed, in 1936, with his assistant Nathan Rosen, Einstein prepared a paper refuting the existence of gravity waves. He said his prior analysis predicting gravity waves was wrong, because he had employed a linearized version of general relativity that failed to properly describe an inherently nonlinear reality. After subsequent consultations with colleagues, Einstein (without Rosen) rewrote this paper before its publication and reaffirmed his belief in gravitational waves.

Along the way, several theorists "proved" gravity waves could not transport energy or momentum. If that were true, gravity waves would be even more elusive than neutrinos — they would be incapable of exerting any type of force on any material object, thus nullifying all conceivable detection schemes. Indeed, one might well argue that anything with zero energy has no physical reality.

Then, in 1955, Rosen demonstrated that two well-known pseudo-tensors — the Landau-Lifschitz and the Einstein pseudo-tensors — were both identically zero for any form of gravity wave. Rosen claimed that proved gravity waves must have zero energy. (Pseudo-tensors are not covariant, and are not reliable descriptions of nature; hence the name "pseudo".)

In 1957, at a worldwide conference on gravitational waves, physicists bemoaned the possibility that they would never be able to detect zero-energy gravity waves, even if they did exist. But then, British theoretical physicist Felix Pirani presented the most convincing argument to date for the central role of the Riemann tensor in general relativity, particularly in relation to the tidal forces that produce relative accelerations of geodesics (see Section §A5.6).

At that same conference, based on Pirani's presentation on geodesic deviation, a young American physicist, Richard P. Feynman, made a compelling argument: a thought-experiment in which gravity waves pass a rod that threads through the centers of movable beads, as shown in Figure 34-5.

Feynman said the electric forces between the rod's atoms are much stronger than the forces exerted by gravity waves; hence the rod's length does not change as the wave passes. However, the gravity waves slide the beads back and forth along the rod. If there is any slight friction between the rod and the "sticky" beads, both will heat up as the beads move back and forth, thereby dissipating energy. The ultimate source of this "sticky bead" heat energy must be the gravitational wave itself. Hence, these waves must transport energy.

While some theorists continued arguing about pseudo-tensors, most physicists thought Feynman's "sticky bead" thought-experiment quashed all doubt that gravitational waves carry energy.

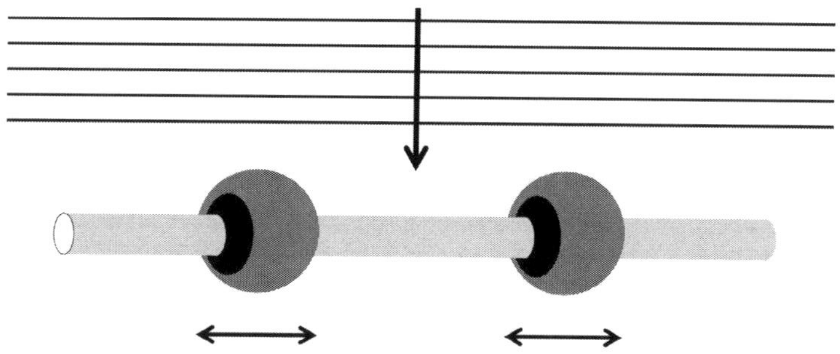

Figure 34-5 Gravity Waves Coming downward onto Feynman's Rod with "Sticky Beads"

American physicist Joseph Weber then started building the first gravity wave detectors. His *Weber bars*, massive aluminum cylinders covered with piezoelectric sensors, were delicately suspended in a vacuum chamber to minimize sensitivity to vibrations from terrestrial sources, such as earthquakes and jackhammers. The bars were up to 153 cm long, 66 cm in diameter, and weighed 3 tons. Weber imagined that passing gravity waves would excite the bars' resonant frequency, making them vibrate, which the piezoelectric sensors would detect. While Weber claimed to have recorded numerous gravity waves, others who replicated his experiments found none. His bars were not sensitive enough, and his vibration isolation was inadequate.

Yet, Weber inspired others to try.

Starting in the 1960's, physicists began planning more advanced experiments.

In 1988, the U.S. National Science Foundation (NSF) agreed to fund the initial design phase of a gravitational wave detector employing an interferometer, the most precise of all optical instruments.

In 1994, NSF approved $395 million for the construction of the Laser Interferometer Gravitational-wave Observatory (LIGO), the largest grant ever awarded by NSF. LIGO began construction in 1998, and started observing in 2002.

The LIGO interferometer is the ultimate refinement of the Michelson interferometer that proved the constancy of the speed of light in 1887, as in Section §1.4. Michelson used his interferometer to explore how electromagnetic radiation moves through space; 130 years later, today's physicists use interferometers to explore how gravitational radiation moves through spacetime.

Like Michelson's interferometer, LIGO has two orthogonal arms along which light beams travel back and forth. However, LIGO's arms are 400 times longer; each arm is 4 km (2.5 miles) long. Furthermore, the entire interferometer assembly is temperature-stabilized and enclosed in an enormous vacuum chamber.

Like Michelson, LIGO uses the interference method to search for tiny variations in the difference in the transit times of the two light beams. When two coherent light beams combine, they interfere. If the two beams are *in-phase* (if they crest and trough together), they combine with *constructive interference*, producing a combined light beam of high-intensity. If the two beams are *out-of-phase* (one crests when the other troughs), they combine with *destructive interference*, producing a combined light beam of very low, or even zero, intensity.

As long as an interferometer's arm lengths do not change, its light beams continue combining in the same way, and their interference remains constant. But if arm lengths do change, if one shortens and/or the other lengthens, the beams will interfere differently, changing their combined light intensity.

One of LIGO's greatest challenges is isolating the interferometer from terrestrial vibration that would mask changes in spacetime curvature. LIGO is so sensitive that it detects ocean waves breaking on shores hundreds of miles away, and trucks driving on distant highways.

LIGO is so sensitive that it is even affected by the thermal vibrations of atoms in its key components.

To minimize spurious arm-length changes, LIGO employs the finest vibration-isolating system ever made, including a 4-stage mirror suspension system — itself a triumph of engineering and science.

LIGO has over 200,000 sensors that monitor the interferometer, its support systems, and its external environment to detect any potential malfunction or external interference. Scientific data collection is suspended whenever a significant concern is detected.

As ultimate protection from spurious signals, the LIGO team built two identical observatories, one at Hanford, in Washington State, and the other at Livingston, Louisiana. A spurious signal, perhaps from a passing truck, might trigger one interferometer, but not the other 3000 km away.

Figure 34-6 is a LIGO schematic. A *beam splitter* divides a laser beam into two coherent beams that proceed in orthogonal directions, each along a 4-km-arm. The light beams bounce back and forth between mirrors hanging near the start and at the end of each arm. The beams traverse 150 round trips along the arms, making the effective path length along each arm nearly 1200 km (720 miles). The two beams then return to the beam splitter, where they recombine and interfere. A photo detector monitors the combined light beam, whose intensity is critically dependent on the light transit time difference in the two arms.

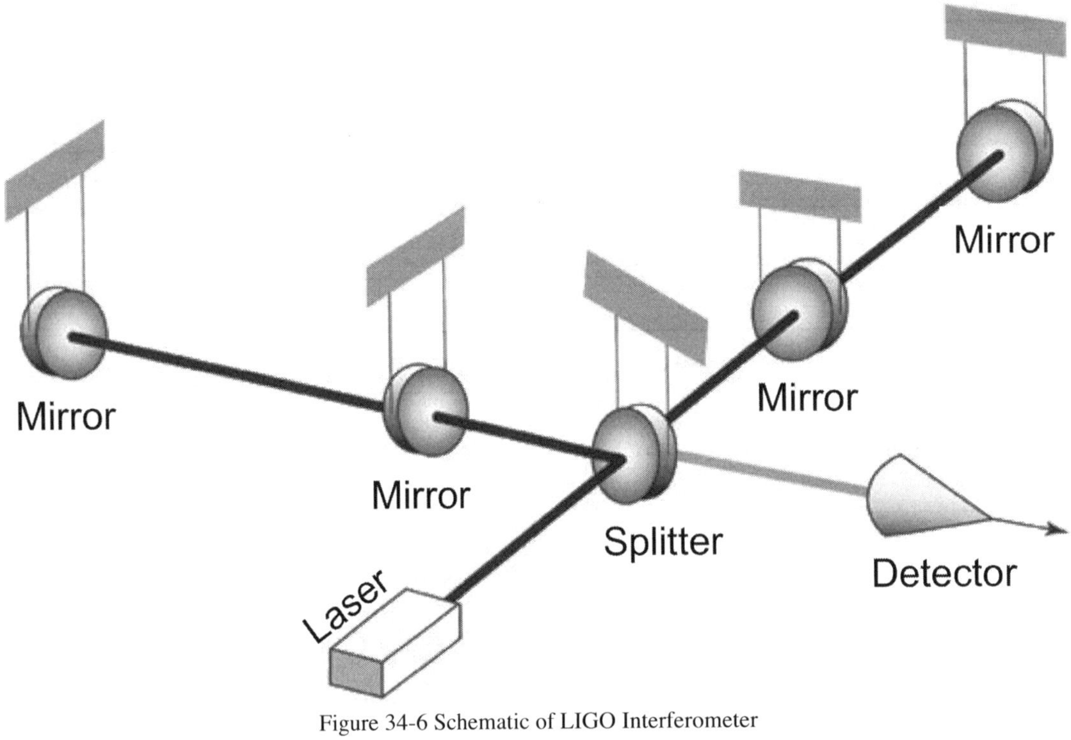

Figure 34-6 Schematic of LIGO Interferometer

LIGO can detect interference changes resulting from path length changes as small as a few parts in 10^{22} — that is 10,000, 000,000, 000,000, 000,000.

LIGO's buildings, tunnels, and vacuum chambers are rigid objects bound by strong atomic-level, electric forces; their dimensions are not altered by gravity waves. The mirrors, however, are delicately suspended with 0.4 mm diameter glass fibers, and thus move freely as space expands or contracts.

Lasers operate according to a principle of quantum mechanics discovered by Albert Einstein. It is amusing that LIGO employs one Einstein discovery to test another Einstein discovery. This actually is not unusual. Every current physics experiment I can think of relies on a technology or a theory that originated with Einstein.

Figure 34-7 illustrates how changes in path lengths and wavelengths alter the interference of LIGO's two light beams. The figure shows the slanted beam splitter, and the photo detector depicted by a circle on the right of the beam splitter. Plus signs mark some of the places where light waves are at their maximum amplitude, at one instant in time.

For clarity, Figure 34-7 shows each beam making only one round trip along its arm, and shows the beams offset. The actual scheme employed by LIGO is more sophisticated.

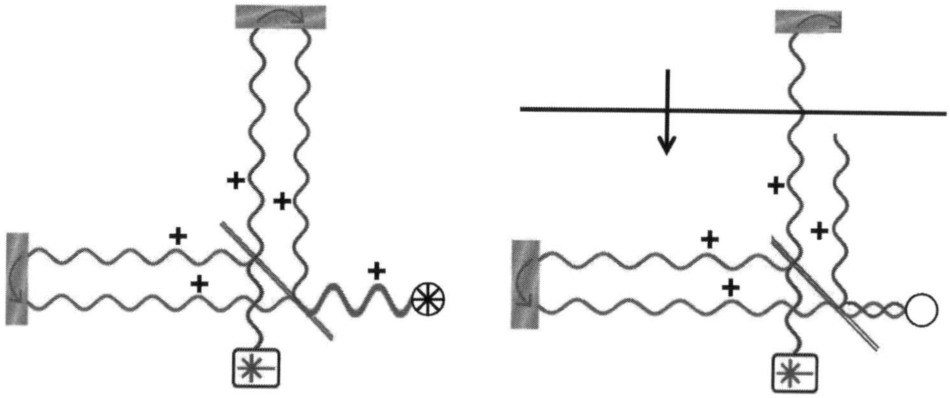

Figure 34-7 Length Change Impacts Interference

The left image shows the quiescent condition, the beams returning to the beam splitter in-phase — the amplitudes of both light beams crest at the same time — their sum is large, and the photo detector registers a strong signal.

The right image shows a gravity wave moving downward within the plane of this page; its wavefront is indicated by the bold horizontal line, and its direction of motion by the arrow. This gravitational wave stretches the horizontal arm and the wavelength of the horizontal light beam, while leaving the vertical arm and light beam unchanged.

For clarity, the image shows the horizontal arm length and light beam wavelength both increasing by 15.8%. As shown here, the horizontal round trip path length is 9.5 wavelengths of the horizontal light beam, both with and without the gravity wave. The 15.8% length increase corresponds to 1.5 wavelengths of the vertical light beam, the original wavelength of both beams.

The horizontal light beam returns to the beam splitter with exactly the same phase, with or without the gravity wave. But with the gravity wave, it is delayed by its increased path length. Therefore, by the time the horizontal beam arrives at the beam splitter, the vertical beam has oscillated through 1.5 additional cycles (in the right image). The returning beams are therefore out-of-phase — they combine destructively, their sum is zero, and the photo detector registers no signal.

If a gravity wave arrives moving perpendicular to the page, one light path may stretch while the other shrinks, thus doubling the interference shift.

Figure 34-8 shows the photo detector signal at the magical moment of the first-ever detection of a gravitational wave.

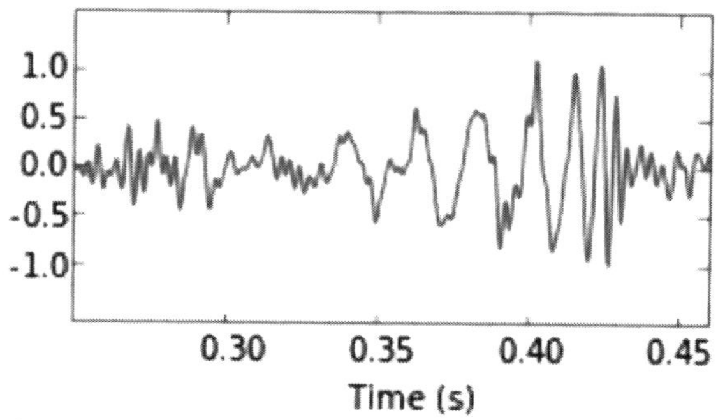

Figure 34-8 Plot of Strain vs. Time of First Detected Gravitational Wave

Here, the vertical scale is the fractional change ("strain") in light path length in units of 10^{-21} — "+1" means the path length increased by 1 part in a billion trillion, equivalent to changing the distance to Alpha Centauri, the nearest star system, by the width of one human hair. The horizontal scale is time in seconds from the nominal start of the signal. The signal stands out clearly from about 0.35 to 0.43 seconds.

The waveform in Figure 34-8 reflects the motion of LIGO's mirrors, which is affected by the gravity wave and the harmonic resonance characteristics of the suspended mirrors.

For a compact binary, with two comparably massive bodies, the radiated gravitational waves are at their greatest intensity when the wave frequency f = 2 / T, where T is the binary's orbital period. This is because, during each orbital period, two bodies pass through each point on the orbital path. The table in the prior section demonstrates how sharply frequency f increases just before the collision.

Although gravity waves are nothing like sound waves, gravitational wave physicists fondly say they are "listening to the whispers of the universe". In that vein, LIGO converts photo detector signals to audio. Figure 34-9 shows a spectrum of the *Chirp Heard 'Round the World* — LIGO's first event — with frequency in Hz plotted vertically, and time in seconds horizontally. At each frequency and time, the intensity of black denotes the intensity of the "sound" of two black holes merging.

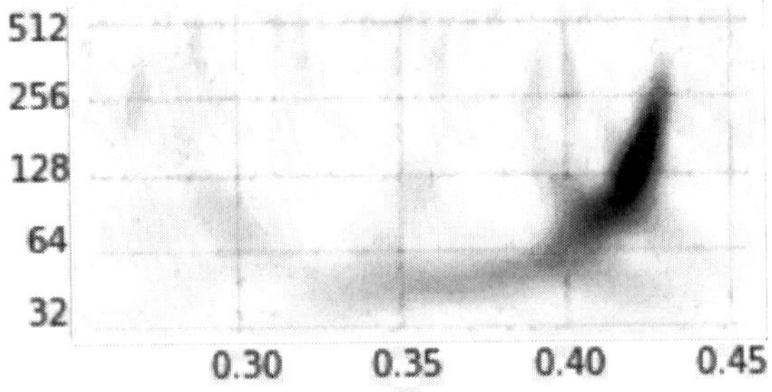

Figure 34-9 "Chirp" of Merging Black Holes.

This amazing discovery comes with an amusing story. LIGO shut down for 5 years, from October 2010 to September 2015, for upgrades. After replacing the most critical devices with improved versions, the final series of engineering tests were running behind schedule. With only four days before the official start of scientific observation, engineers and scientists worked until 4:00 a.m. before deciding to go home for a few hours sleep. They just happened to leave LIGO running by itself.

At 4:50 a.m., the first discovered gravitational wave hit both detectors — Livingston first, followed by Hanford 0.007 seconds later. A wave that had been traveling for a billion years arrived whether LIGO was ready or not. Fortunately, LIGO was ready.

LIGO proved that gravitational waves really do exist, confirming the last outstanding prediction of Albert Einstein, and opening a new chapter in astronomy.

In May 2017, the VIRGO gravitational wave observatory near Pisa, Italy, completed its performance enhancement program and joined the LIGO twins in the search for gravitational waves.

§34.5 Observational Results

LIGO has reported 74 gravitational wave detections in a total of about 24 months of science-grade operation over three observing runs denoted O1, O2 and O3, and ending on March 2020 due to COVID. O4 began in May 2023 and is expected to conclude in January 2025. Events are denoted "GWyymmdd", GW for gravity wave, and "yymmdd" for detection date.

The wiggling interference signal of each event, such as the one shown in Figure 34-8, is compared to an ever-increasing library of wave forms predicted by general relativity for a variety of masses and spins of each orbiting body and their orbit size, eccentricity, and orientation. Each prediction requires numerical integration by supercomputers of general relativity's complex equations. We are entering a new field of science, and still have a lot to learn. The parameters of each event and their total number are still being refined as we climb this learning curve.

Of the 74 reported events, 67 are classified as binary black hole (BBH) mergers, 5 as neutron star-black hole mergers (NSBH), and 2 as binary neutron star (BNS) mergers. The black hole masses range from 5 to 95 Msun, while neutron star masses range from about 1.3 to 2.6 Msun.

Here are some highlights of a few special events.

GW150914, detected on 14 Sept. 2015, is believed due to a BBH merger at a distance of 1.3 billion light-years. LIGO estimates the initial black hole masses were 36 Msun and 29 Msun, and the post-merger black hole mass was 62 Msun, with 3 Msun of mass converted into energy. Estimated peak power was 3E49 watts, about 1000 times the power emitted by all the stars in the observable universe.

GW170817 is believed due to a BNS merger 130 million light-years away. Estimated initial masses were 1.46 Msun and 1.27 Msun. The energy released in this merger was at least 0.025 Msun. The result of this merger is not firmly established — the post-merger body might have been a very massive neutron star that rapidly collapsed to a black hole. VIRGO detected GW170817 first, followed 0.022 seconds later by LIGO-Livingston, and 0.003 seconds after that by LIGO-Hanford — the first triple-observatory gravitational wave detection.

The FGST (Fermi Gamma-Ray Space Telescope) detected a short gamma ray burst (GRB) arriving 1.7 seconds after GW170817. The close time proximity suggests neutron star mergers are an important progenitor of GRB's. This gamma ray burst is denoted GRB170817A. Within 14 seconds, FGST alerted the astronomical community. LIGO-VIRGO's announcement came 40 minutes later. (Gamma-ray physicists have optimized their automated alert systems over many years; gravitational wave physicists are at the start of this learning curve.)

These announcements triggered an astronomical feeding frenzy. Within 24 hours, nearly 100 preprints were submitted, coauthored by almost 4000 astronomers (about 1/3 of the world population) representing 900 institutions. All this activity may have created its own gravity wave.

This event, called both GW170817 and GRB170817A, fits models of BNS mergers that produce *kilonovae*, stellar explosions that are prodigious producers of heavy elements. It is believed this kilonova produced 16,000 Earth masses of heavy elements (elements heavier than iron), including 10 Earth masses of gold and platinum. Too bad it's so far away. Over 70 optical observatories studied this kilonova. Perhaps strangely, no neutrinos were detected from this event.

Figure 34-10 shows two Hubble space telescope images taken 5 and 9 days after GW/GRB170817's detection. The boxed area enclosing the kilonova is enlarged in the upper right of each image.

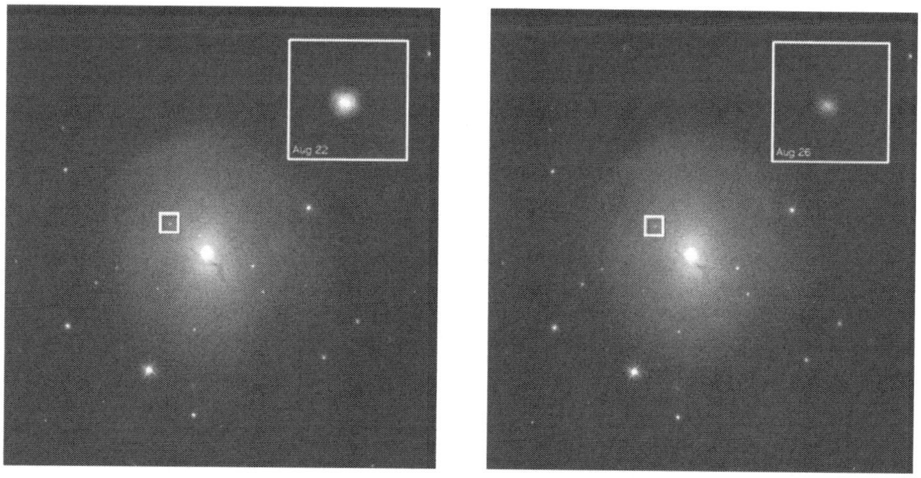

Figure 34-10 Hubble Images of Kilonova: 22Aug. (left) and 26Aug. (right)

GW170817 and GRB170817A most likely share the same source, but need not have been generated at exactly the same time. If we assume simultaneous origins, a 1.7 second arrival time difference confirms to 1 part in 1000 trillion Einstein's prediction that gravitational waves travel at the speed of light.

GW190521 is the biggest bang detected so far: a BBH merger with initial black holes masses of about 90 and 69 Msun, producing a roughly 150 Msun black hole and releasing 9 Msun of energy.

To date, every observed GW event is consistent with general relativity. This early data set limits on proposed modifications of general relativity. LIGO sets an upper limit of 1.3E–23 eV on the mass of the hypothetical *graviton* proposed by field theory (about 7000 times the photon mass upper limit). LIGO sets an upper limit on the scale of Lorentz invariance violation at 4E–16 m, about the radius of a proton.

LIGO data also seem to exclude the possibility of more than 3 large spatial dimensions.

LIGO's most important discovery is that gravitational waves really do exist and we can detect them.

A key consequence of the constant speed of light is that as we look farther out into the cosmos we are also looking farther back in time. If we detect a compact binary black hole merger 1 billion light-years away, we will be observing an event that happened 1 billion years ago. As the sensitivities of our gravity wave observatories improve, enabling the detection of ever-more distant mergers, we will be observing mergers that occurred at ever-earlier epochs in the evolution of our universe.

The universe has changed throughout its nearly 14-billion-year life. Billions of years ago, there may have been many more mergers of smaller black holes than have occurred more recently. Another possibility is that 10 billion years ago, there may have been far more mergers of supermassive black holes in the centers of galaxies than occurred 1 billion years ago. Due to the constancy of the speed of light, we can only detect the former if they are 10 billion light-years away, while we can detect the latter only if they are 1 billion light-years away.

We cannot sample the temporal and spatial distributions separately; time and space are linked by light, as Einstein's theories of relativity reveal.

§34.6 Future of Gravity Wave Astronomy

As an observational modality, gravitational waves feature unique characteristics that astronomers confidently hope will lead to a continuing sequence of important new discoveries.

Gravitational waves provide an entirely new way of observing the universe. Many of the greatest advances in science were enabled by new methods and new devices that opened our eyes to previously unimagined phenomena. Some key instrumental examples in astronomy are Galileo's telescope, Edwin Hubble's Hooker telescope, the Hubble space telescope, and the WMAP microwave space telescope.

WMAP discovered crucial facts about our universe — its age, geometry, and energy composition — by collecting light in a way that no prior instrument could. WMAP examined the first light released by the cosmos when the primordial fireball cooled enough to became a transparent gas, about 379,000 years after the Big Bang. No earlier image in any form of light will ever be possible since no earlier light exists.

Gravitational waves allow astronomers to observe phenomena that do not produce electromagnetic radiation, making them undetectable by conventional means. Gravitational wave astronomy offers at least the possibility of one day detecting gravitational waves produced by the Big Bang itself. This will require a giant leap in technology, as did LIGO. I would not bet against human determination and creativity.

Furthermore, unlike electromagnetic radiation, gravity waves are not attenuated by cosmic dust, plasma clouds, or even massive bodies — they simply pass straight through everything, including objects that are opaque to light. Gravity waves can probe obscure regions, like the center of our galaxy, in ways that light cannot. We may learn more about the formation of supermassive black holes with gravitational wave observatories than we ever could with optical observatories.

Fortunately, gravitational waves and light are not mutually exclusive; we can now choose both. Astronomers relish the opportunity to observe cosmic phenomena with both gravity waves and light, as they did with GW/GRB170817.

Figure 34-11 shows LIGO's past and future performance. The vertical axis plots detector noise level, its response to external terrestrial sources and its own internal vibrations, including the vibrations of its mirrors due to the thermal energy of their atoms. The horizontal axis is wave frequency. Note this is a log-log plot; each major division corresponds to a change by a factor of ten. The upper right box lists the estimated maximum detection range in millions of light-years for BNS (binary neutron stars) mergers.

To convincingly detect a gravitational wave, the strain amplitude induced by the gravitational wave must be significantly larger than the detector's strain noise level, so designers strive to minimize strain noise. Note that LIGO's detection sensitivity is strongly dependent on the incoming gravitational wave frequency — LIGO's sensitivity is 100 times better at 200 Hz than it is at 10Hz.

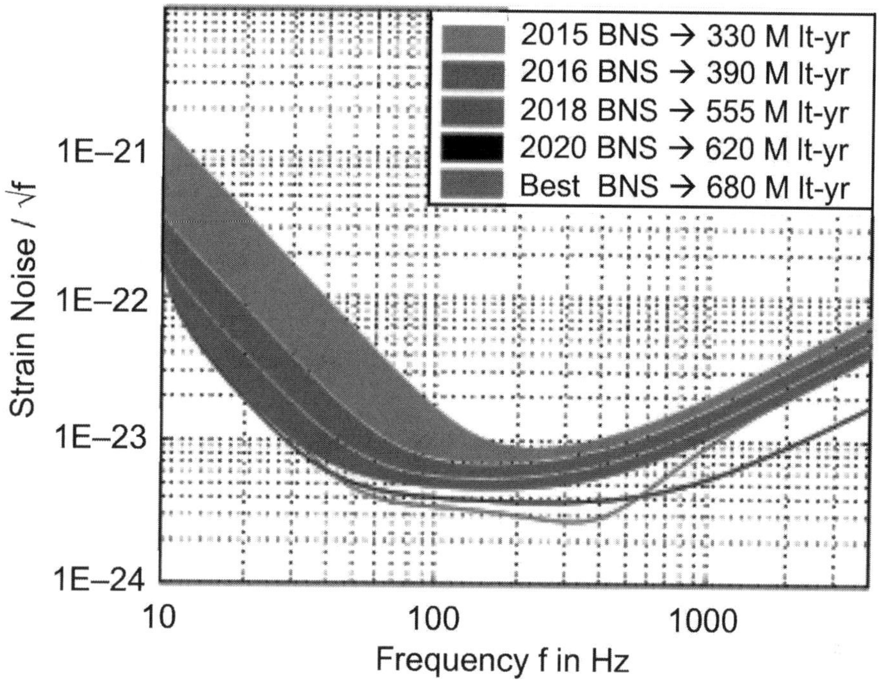

Figure 34-11 Detection Limits of LIGO Through 2020

The lowest curve (thin gray) is the estimated performance in a mode specifically optimized to detect BNS mergers. The next lowest curve (thin black) is LIGO's ultimate design goal, which has excellent response over a broader frequency range.

The general shape of LIGO's sensitivity-frequency plot is typical of any interferometer-based gravitational wave observatory on Earth's surface. Several other interferometer-based Earth-bound observatories are active, being built, or being planned. The most promising of these are:

VIGRO, originally a French-Italian collaboration, began construction in 1996. Advanced VIRGO, which promised to be 10× more sensitive, became operational in May 2017, in time to join LIGO in detecting GW170817. VIRGO is very similar to LIGO but with 3-km arms. VIRGO and LIGO collaborate in both system design and data analysis.

GEO 600 near Hanover in Germany is a U.K.-Germany collaboration, with a similar interferometer design. As its name suggests, GEO 600 has 600-m arms.

KAGRA, near Kamioka, Japan, is the successor to two smaller Japanese pilot programs. It has an interferometer with 3-km arms and became operational in February 2020.

LIGO and the government of India have agreed in principle to install a duplicate LIGO observatory in India that would be called IndIGO. Multiple distant observatories improve locating the origin of detected gravitational waves.

Figure 34-12 plots the past and planned performance of three gravitational wave observatories. Recall that LIGO has twin observatories separated by 3000 km. The upper chart shows each observatory's BNS merger detection range — the maximum distance of reliable detection — in millions of light-years. The lower chart shows each observatory's BBH merger detection range.

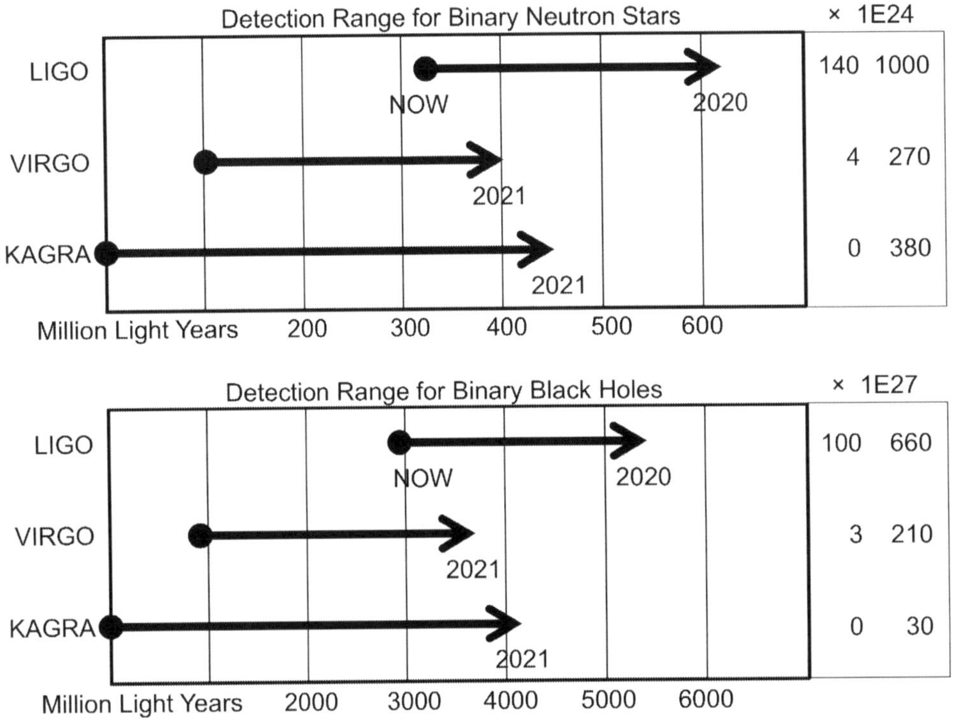

Figure 34-12 Past (black circle) & Planned (arrow tip) Detection Ranges

The boxed numbers at the right are the corresponding detection volumes in cubic light-years, with the current volume on the left and the planned volume on the right. Each number must be multiplied by the scale factor above the box. For example, in 2020, LIGO's expected detection range for BBH mergers is 5400 million light-years, for a detection volume of 660E27 cubic light-years. Detection rate should grow linearly with detection volume — tripling volume should triple event rate. Since the volume V of a ball is proportional to its radius R cubed, nearly half of V is farther from the center than 80% of R. Thus, most of any gravitational wave observatory's detected events will be near the limit of its range. Here, we assumed merger rates are uniformly distributed throughout space and time.

A very different type of gravitational wave observatory employs millisecond pulsars. The European Pulsar Timing Array (EPTA), the Parkes Pulsar Timing Array (PPTA) in Australia, and the North American Nanohertz Observatory for Gravitational Waves (NANOGrav), and are searching for variations in pulsar timing due to passing gravitational waves, and they are just starting to get interesting results.

Pulsars are rapidly spinning neutron stars with intense magnetic fields that sweep radio waves across the cosmos, like lighthouse beacons. Some beacons sweep past Earth. Pulsars spin at highly stable rates, due to enormous moments of inertia. For GW observatories, each pulsar provides a very long arm with Earth at one end. Anything that changes the apparent length of that arm alters the pulse arrival time. By monitoring many different pulsars in many different locations, we hope to observe correlated changes in pulse arrival times — to see gravitational waves sweep across the cosmos.

Let's be clear about the term "nanohertz" — 1 nanohertz means each full wave oscillation takes 1 billion seconds, or about 32 years. Imagine doing an experiment searching for a wave that oscillates once in a career. How does one convince graduate students to do their Ph.D. theses on that time scale?

One great advantage of the pulsar timing technique is that nature provides most of the wave detector.

The North American group employed the world's two largest single-dish radio telescopes, Arecibo and Green Banks. In 2012, they reported monitoring 17 pulsars over a five-year period ending in 2010. These pulsars have periods ranging from 3 to 16 milliseconds. The team found pulsar timing variations of less than 1 microsecond for 15 of the 17 sources, with the best two sources having timing variations between 30 and 50 nanoseconds — they were stable to about 1 millionth of their rotational periods.

One thing these astronomers hope to discover is how supermassive holes form — by gradual accretion, or by occasional mergers of behemoths, or some combination thereof.

Another approach is the proposed launch of an interferometer-based gravitational wave observatory. LISA, the Laser Interferometer Space Antenna, a project of the European Space Agency, would have three satellites orbiting the Sun at Earth's orbital radius. The array would follow Earth around the Sun, trailing our planet by about 3 weeks.

Each LISA satellite would form the corner of one interferometer, with the other two satellites being the ends of that interferometer's arms. Thus, LISA would have three independent interferometers, whose arm lengths would each be 2.5 million km.

The larger arms make LISA sensitive to longer wavelength, lower frequency gravitational waves. LISA's optimal sensitivity is projected to be 50× worse than LIGO, but LISA would explore a different frequency band. Its peak performance would be centered at 0.003 Hz, about 30,000 times lower than LIGO's 100 Hz. At that frequency, it is hoped that LISA would detect the inspirals of million-solar-mass binaries. As of now, LISA is approved for launch sometime in the 2030's.

Figure 34-13 shows the sensitivities of various observatories (black lines) and predicted gravity wave signals (gray bands) from a variety of different sources. In this log-log plot, the vertical axis is strain and the horizontal axis is wave frequency. Christopher Moore, Robert Cole and Christopher Berry created this image.

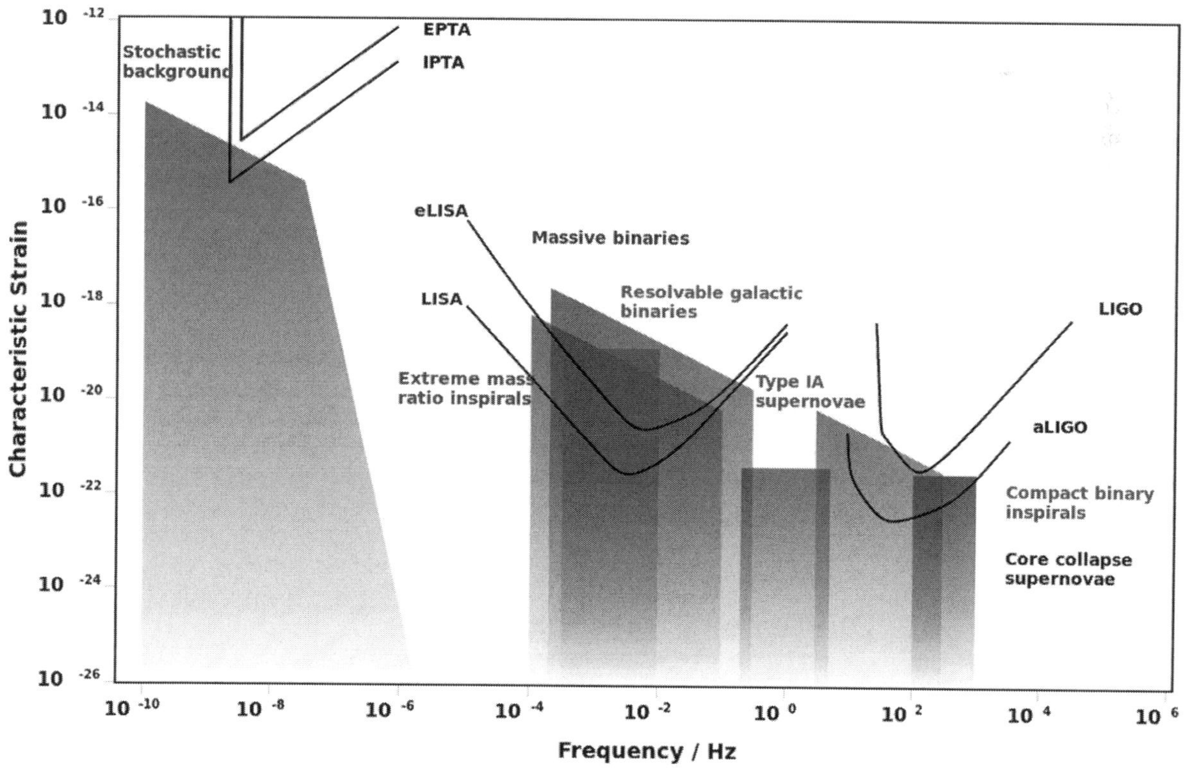

Figure 34-13 Performance of Leading Present and Future Observatories and Possible Sources

§34.7 Validations of General Relativity

All of the major predictions of Einstein's theory of general relativity have now been confirmed by the ultimate arbiter of scientific truth: direct observation of nature.

The list is impressive:

Gravitational bending of light: confirmed to 0.05% — see Section §24.4.

Precession of Mercury: confirmed to 2% — see Section §25.3.

Gravitational redshift of light: confirmed to 0.01% — see Section §33.7.

Precession of orbiting satellite: confirmed to 0.3% — see Section §29.5.

Frame-dragging by Earth: confirmed to 18% — see Section §29.5.

Orbital decay of binary pulsar: confirmed to 0.2% — see Section §34.8.

Speed of Gravitational Waves: confirmed to 1 in 10^{15} by LIGO

Gravitational waves: confirmed to 4% by LIGO.

To be clear, none of these experimental tests found any deviations from general relativity. The quoted levels of precision correspond only to the experimental uncertainties.

LIGO's observation of gravitational waves validates the last previously-unproven prediction of Einstein's theory of general relativity — the prediction of propagating spacetime curvature waves resulting from cataclysmic cosmic events — the strong field limit where gravity is immense. The events detected so far are a 96% match to Einstein's predictions, as perfect a match as our instruments allow.

This is the beginning of a new field of science with limitless potential: gravitational astronomy.

§34.8 Precession of Binary Pulsar

Pulsars are neutron stars that emit intense beams of radiation at radio frequencies. The beams shoot out from the north and south poles of the pulsar's magnetic field, which can be trillions of times greater than Earth's magnetic field.

Pulsars typically spin very rapidly about their rotational axes, which, like Earth, may not be aligned with their magnetic axes. If a Pulsar's axes are not aligned, its radio beam sweeps across the heavens like a lighthouse beacon. We can only detect those few pulsars whose beams happen to sweep by Earth.

A *binary pulsar* is a system of two neutron stars in a mutual orbit about their common center of mass, with one or both neutron stars being pulsars.

American astronomers Russell Hulse and Joseph Taylor discovered the first binary pulsar in 1974. Still the most famous and most important of its class, PSR B1913+16 is comprised of a pulsar and a silent companion. Hulse and Taylor received the 1993 Nobel Prize for their extraordinarily extensive and precise observations of this important system.

Some key measurements of PSR B1913+16:

Pulsar mass:	1.4398 Msun
Companion mass:	1.3886 Msun
Orbital Period:	0.322,997,448,930 Earth-days
Pulsar Period:	0.059,029,997,929,88 seconds
Periastron, d:	765,000 km
Apastron, d:	3,340,000 km
Periastron, v:	300 km / sec
Apastron, v:	75 km / sec

Here, Msun is the mass of our Sun. The pulsar's orbital period is about 8 hours, and it spins on its axis nearly 17 times per second. I included all these digits to demonstrate just how precisely these quantities have been measured. Periastron is the point in their mutual orbit of minimum distance d between the two stars, and maximum velocity v. Apastron the point of maximum distance d, and minimum velocity v.

In Figure 34-14, the two neutron stars are depicted as solid circles at periastron, and as open circles at apastron. The gray diamond marks the central focal point of both ellipses. The orbits are clearly highly eccentric — from Section §25.4, their orbital eccentricity $\varepsilon = 0.627$. The two arrows indicate the direction of motion of each star at periastron.

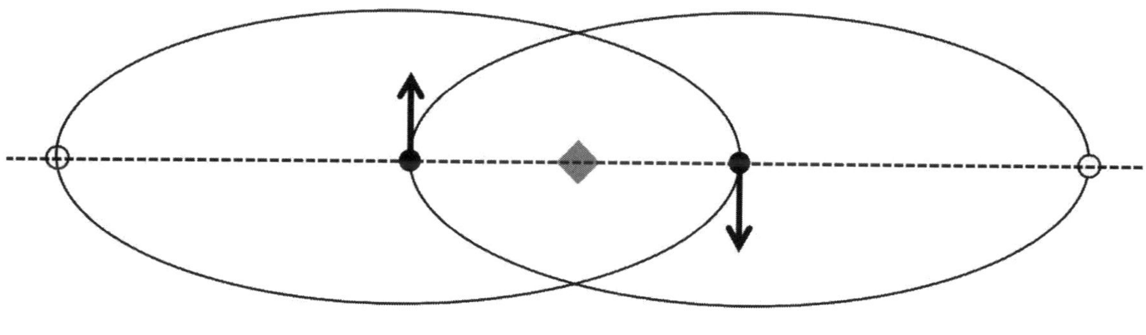

Figure 34-14 Binary Pulsar System Orbits

For the measured properties of PSR B1913+16, the prediction of general relativity for the fractional decrease in orbital period resulting from the emission of gravity waves is:

Predicted by GR: $2.40253 \times 10^{-12} = 75.680$ microseconds per year

A 2010 paper by Taylor et. al., based on 35 years of observations, reports a fractional decrease in the orbital period of:

Measured: $(2.396 \pm 0.005) \times 10^{-12} = 75.45 \pm 0.16$ microseconds per year

This corresponds to 0.997 ± 0.002 of what Einstein's theory predicts.

Figure 34-15 is a plot of this system's cumulative periastron precession, comparing measurements (black dots) with the prediction of general relativity (curve). In 30 years, the orbital ellipse has precessed about one-half degree.

The prediction and measurements match spectacularly.

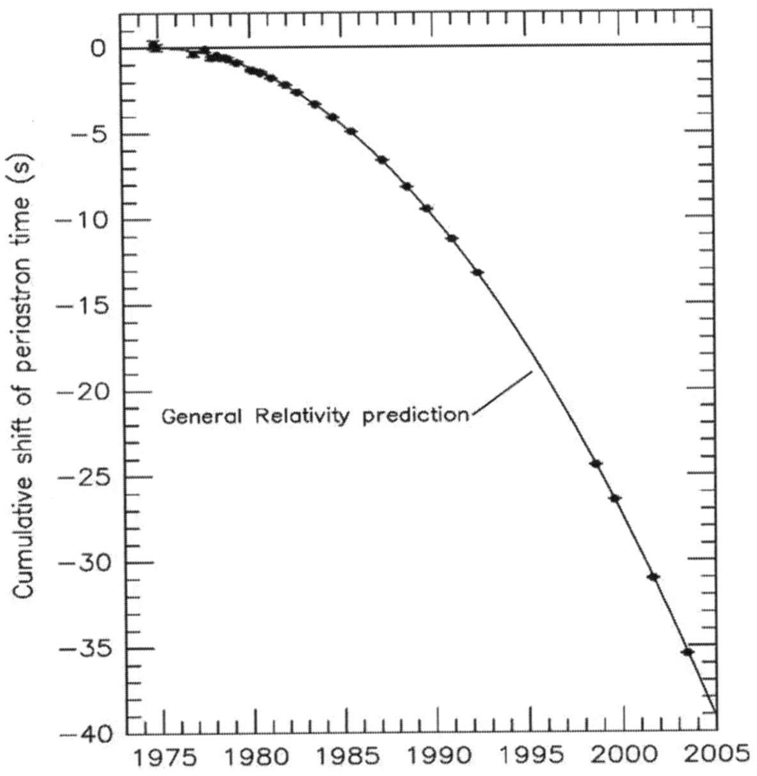

Figure 34-15 Cumulative Precession of Binary Pulsar over 30 Years

I am amazed that Einstein could sit at his desk, with only paper and pen, and precisely predict the behavior of a phenomenon that no one had ever seen or even imagined. At that time, no one knew what a neutron was, let alone a neutron star or pulsar.

The first binary pulsar was discovered 60 years after general relativity was published. It is 100,000 trillion miles from Earth.

Yet this incredible object behaves exactly as Einstein predicted.

Is there any greater testimony to the power of the human mind?

Appendix 1
Symbols, Units & Dimensional Analysis

§A1.1 Common Symbols

The following symbols are commonly used in physics:

$A = B$	Equality, A and B have the same value
$A > B$	A is greater than B
$A >> B$	A is much greater than B
$A < B$	A is less than B
$A << B$	A is much less than B
$A <= B$	A is less than or equal to B
$A >= B$	A is greater than or equal to B
Re { }	The real part of the complex quantity { }
dq	A small change in the variable q, see Appendix §A2.1
dx/dt	Common derivative, ratio of dx to dt, see Appendix §A2.1
$\partial x/\partial t$	Partial derivative, like dx/dt, except all other variables are fixed, see Appendix §A2.2
$A^{\alpha}{}_{\beta}$	Rank 2 tensor with contravariant index α and covariant index β, see Section §20.2
A,b	Tensor notation for $\partial A/\partial b$, see Section §22.2
A;b	Tensor notation for the covariant derivative, see Section §22.2
$\int$	Integration symbol, see Appendix §A2.3
γ	Relativistic Factor: $\gamma = 1 / \sqrt{(1 - u^2 / c^2)}$
$g_{\alpha\beta}$	Spacetime Metric, see Chapter 21
$\Gamma_{\alpha\beta\gamma}$	Connection Coefficient, see Section §22.3
$R_{\alpha\beta\gamma\delta}$	Riemann Tensor, see Section §22.3
$R_{\alpha\beta}$	Ricci Tensor, see Section §22.3
$G_{\alpha\beta}$	Einstein Tensor, see Section §22.3
$T_{\alpha\beta}$	Mass-Energy Tensor, see Section §22.3

§A1.2 Units of Measure

Quantities of interest in physics commonly are expressed as some number of units appropriate to that type of quantity. For example, distance can be stated as being X microns, Y meters, or Z light-years. Each of those units is a standard measure of distance, although their sizes are very different. Clearly, each is appropriate in quite different circumstances.

Time is typically measured in seconds or years. Electric charge is typically measured in coulombs or multiples of the charge of a single proton. Mass is typically measured in grams or kilograms; pounds are a poor choice, because pounds are used for both mass and force, two very different physical entities.

Speed, the distance moved divided by the time to move, can be stated as X microns per nanosecond, Y meters per second, or Z light-years per year. In the latter case, we expect Z to be less than or equal to 1.

The set of units that scientists most commonly use has long been called the *mks* system, which stands for meter-kilogram-second. This metric system of units has evolved into the *Système International d'unités*, which is abbreviated SI. Most fields of science define additional units appropriate to their unique needs.

No choice of units is wrong per se. You could even use miles per month, or fathoms per fortnight. But following standard conventions will probably minimize mistakes and facilitate communication with colleagues. We generally believe the choices of units and coordinate systems is arbitrary: we are free to choose whatever we please, because nature has no intrinsic units or coordinates. However, there may be some exceptions, some instances in which nature hints at a "right" choice.

One example is the speed of light. Einstein's theory of special relativity, the most precisely and extensively validated concept of all human thought, insists that the speed of light is invariant, that it has the same value c in every circumstance, in every location, and at every moment in time. The speed c is a fundamental property of our universe that establishes a natural unit of velocity.

§A1.3 Matching Units

Checking an equation's units is a great way to gain insight and minimize mistakes. This is often called *dimensional analysis*. In every valid equation, both sides of the equal sign **must** have **exactly** the same units. Consider a simple example:

$$s = g\ t^2 / 2$$

The left side of this equation is a snap; we know that s is a distance, so measuring it in meters makes sense. On the right side, the units are $(m / sec^2) \times (time)^2$. If we measure time in seconds, the units on the right side will be meters, just like the left side. If the right side had t instead of t^2, the units on the right would be $(m / sec^2) \times (sec) = m / sec$, which does not match the left side and is therefore **wrong**.

There is no partial credit for units that are only slightly wrong; only a perfect match is acceptable.

Note that we could choose to measure time in minutes. But, that would give the wrong answer:

$$1\ g = 9.8\ m / sec^2$$

$$1\ g = 35{,}280\ m / min^2$$

These are very different numbers. The units must match *exactly* to get the right answer.

Units also tell us something about the physics: m / sec^2 means meters are changing with the square of time — that becomes very fast, very quickly.

§A1.4 Dimensional Analysis

For brevity, physicists sometimes adopt units for distance and time that make the speed of light c equal to 1. One can do this by measuring time in seconds and distance in light-seconds (the distance light travels in one second, about 300,000 km). Astronomers prefer to measure time in years and distances in light-years. Some experimental high-energy physicists measure time in nanoseconds and distance in feet. Either way, c = 1. Adopting this convention simplifies equations, allowing us to better focus on the key physics, rather than mathematical bookkeeping.

If you find an equation that was derived assuming c = 1, and you wish to calculate the real value of some physical quantity, you will first need to restore the missing c's.

To do restore the missing c's in an equation:

	replace each "t" with "c t"
	replace each "v" with "v / c"
	replace each "E" with "E / c"
For electrostatic potential ø:	replace each "ø" with "ø / c"
For electric current density:	replace each "ϱ" with "c ϱ"

After that, check the units on both sides of each equation and add c's as necessary to dimensionally balance the equation.

Let me demonstrate this process with an example.

Consider an equation relating the magnetic vector field ***B*** to changes in the electric vector field ***E*** (assuming no electric currents). For now, there is no need to worry about what this equation means — just foucs on the units.

With c = 1:
$$(dB_z / dy) - (dB_y / dz) = (dE_x / dt)$$

Here the notation dQ / dP means the small change in Q due to a small change in P. Using the replacement list above, the equation becomes:

$$(dB_z / dy) - (dB_y / dz) = ([dE_x / c] / [c\, dt])$$
$$c^2 [(dB_z / dy) - (dB_y / dz)] = (dE_x / dt)$$

The replacement list correctly restored the missing c's.

Alternatively, if you forget the replacement list, you can go directly to dimensional analysis. Consider the Lorentz force equation:

$$F_x = q (E_x + v_y B_z - v_z B_y)$$

From this, we know that the units of E must be the same as the units of v B; let's write that [E] = [v B]. In our original equation, the units are:

$$[B] / [x] = [E] / [t]$$

The units on the left are magnetic field B divided by distance. The units on the right are electric field E divided by time t. Substituting [E] = [v B] yields:

$$[B] / [x] = [v\, B] / [t]$$
$$[B] = [v\, B] [x] / [t]$$
$$[B] = [v]^2 [B]$$

The units of distance [x] divided by time [t] are the units of velocity [v]. To balance this equation, we must multiply the left side by a velocity squared. Since we left out the c's, this means the restored equation is:

$$c^2 [(dB_z / dy) - (dB_y / dz)] = (dE_x / dt)$$

Try this on a few equations that you already know. With a little practice, you will find that dimensional analysis is easy and highly effective.

If you want to be doubly sure, keep all the c's, *and* do dimensional analysis on your results.

Appendix 2
Introduction to Calculus

Appendix 2 provides a reminder / brief introduction to the calculus of derivatives and integrals.

My book, ***Math For Physicists***, in ebook and paperback, presents a more comprehensive exploration of all the math scientists need to know.

For physicists, knowing how to differentiate and integrate is a survival skill. Calculus will not only help you pass exams and succeed in your careers, but it will also change your worldview. After learning calculus, you will appreciate the majesty of the universe more profoundly than ever before.

For those discovering calculus for the first time, this may be exciting, but perhaps a bit frightening. Do not expect to fully appreciate calculus after just one reading. Changing your worldview will take some time. For me, and for others I know, calculus was a dense fog that, in a single instant, seemed to turn into a bright beautiful blue sky. I still vividly recall that one moment.

§A2.1 Common Derivatives

Derivatives describe rate of change.

Since everything in the universe changes, derivatives are the foundation upon which science strives to describe all natural phenomena.

Most physical laws describe how quantities change over time. Consider, for example, a ball dropped from the *Leaning Tower of Pisa*. Newton's law of gravity does not tell us the ball's height or speed, but it does tell us how rapidly its speed changes. To make use of these laws, we must be able to quantify rate of change, and that requires differential calculus.

Let's talk about speed: what exactly do physicists mean by "speed" ?

Speed is distance traveled divided by travel time. The speedometer in your car tells you the car's speed. If you are driving along a highway and the speedometer reads 100 kilometers per hour (km / hr), it means you would travel 100 km in the next hour if you constantly maintained that speed. How does the speedometer know how far you would travel in one hour if you just started driving? If the highway ends in 2 km, you probably are not going to go 100 km in the next hour? Clearly, "100 km / hr" cannot be the whole story.

What the speedometer actually measures is how far the car's wheels turn in some small interval of time, such as one second. The number of wheel rotations, combined with the known tire circumference, determines how many meters your car travels in one second. If the speedometer reads 100 km / hr, what it really means is that your car is moving, at this instant, at the rate of about 28 meters per second (m / sec).

Averaging over one second might be good enough for a car, but what about a ball falling from a great height? It turns out that 9 seconds after being dropped, the ball's speed is 88 m / sec. This increases to 98 m / sec one second later. Since that is a substantial change, we should probably compute its speed over a time interval that is even less than one second.

The ultimate answer is to define speed in terms of the *infinitesimal distance* traveled during an *infinitesimal time* interval. This concept, developed independently by Isaac Newton and Gottfried Leibniz is the basis of *differential calculus*, the first "new math."

Calculus is a rich and beautiful branch of mathematics that was essential to the development of physics. It allows us to deal sensibly with infinitesimal quantities and also to properly accumulate them to describe macroscopic results.

By convention, we use the symbol *ds* for an infinitesimal change in distance s, and *dt* for an infinitesimal change in time. Calculus provides a precise definition of velocity denoted by the letter v:

$$v = \text{limit as } dt \rightarrow 0 \text{ of } \{ ds / dt \}$$

This equation reads: v equals the *limit* of the ratio ds/dt as *dt goes to zero*.

A ratio of two infinitesimals, such as (ds/dt), is well-behaved if the ratio comes closer and closer to a final value of v as we compute the ratio for smaller and smaller time intervals dt. In which case, v is called the *asymptotic limit* of (ds/dt).

Not all ratios are well behaved. The ratio 1/x is not well behaved as x approaches 0, because it grows ever larger and is infinite at x = 0. But the ratio (sinx)/x is well behaved as x approaches 0, because the numerator and denominator both approach 0 *at the same rate*. For every infinitesimal but nonzero value of x, (sinx)/x equals 1. As we get closer and closer to x = 0, but not at x = 0, the ratio remains 1. This means the *limit* of this ratio equals 1.

Let's consider another example of this concept of limits: what is the speed of a falling ball 8 seconds after its release? From Newton's laws, we know that the equation relating s, the distance dropped, and t, the time lapsed since release, is:

$$s(t) = g\ t^2 / 2$$

Here, g is the acceleration of gravity near Earth's surface, which equals 9.8 m / sec^2. To compute speed, we calculate the distance dropped at time t, and also at an infinitesimally later time t + dt.

$$s(t) = g\ t^2 / 2$$

$$s(t + dt) = g\ (t + dt)^2 / 2$$

The infinitesimal distance traveled, ds, is the change in distance during dt.

$$ds = s(t + dt) - s(t)$$

$$ds = g\ [\ (t + dt)^2 - t^2\ ] / 2$$

$$ds = g\ [\ t^2 + 2\,t\,dt + dt^2 - t^2\ ] / 2$$

$$ds/dt = g\ [\ t + dt / 2]$$

Now, we take advantage of dt being extremely small, dt << 1. We can discard the dt term in the [ ]'s in the last equation because dt << 2 t, and the limit as dt goes to zero of [t + dt / 2] equals t. Now, we can compute v(t), the speed as a function of time, and evaluate it at t = 8 sec:

$$v(t) = ds/dt = g\ t$$

$$v(8 \text{ sec}) = [9.8 \text{ m} / \text{sec}^2]\ [8 \text{ sec}] = 78 \text{ m} / \text{sec}$$

The procedure performed above on s(t) is called *differentiation* or more specifically *taking the derivative of* s *with respect to* t. Differentiation is important enough to merit an entire symbology: dq is not d times q, but is rather a single symbol denoting a *differential*, a tiny increment in the variable q or a tiny range of values of q.

The d-symbology applies to any variable. The ones that you will see most commonly are coordinate differentials, such as dt, dx, dy, dz.

The expression ds/dt, the *derivative* of s with respect to t, is the ratio of two differentials. The two d's in ds/dt **do not** cancel one another leaving s/t — ds/dt and s/t are two entirely different expressions.

Any normal function or equation in physics can be differentiated. To differentiate X with respect to z, calculate:

$$dX/dz = [\ X(z + dz) - X(z)\ ] / dz$$

and take the limit as dz goes to zero.

Derivatives have graphic significance. Figure A2-1 shows a plot of y(x) = sinx. The derivative of y(x) at each value of x is the slope of sinx at that x.

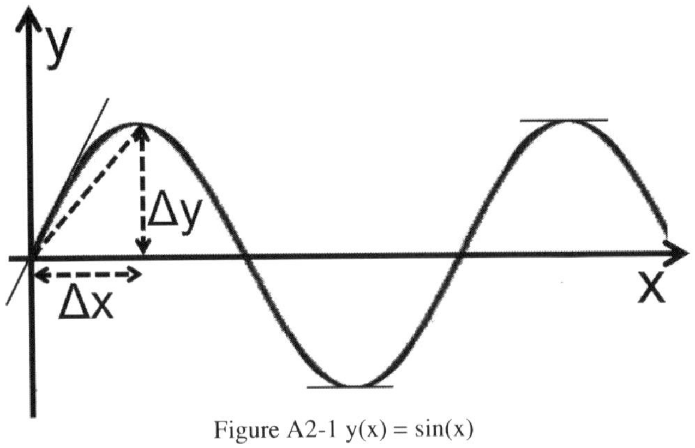

Figure A2-1 y(x) = sin(x)

Above, the slope of y(x) is zero at the *minimum* value of y(x) about halfway along the x-axis, and also at the *maximum* value of y(x) near the right end of the x-axis. At both points, the tangent lines drawn above are horizontal, the slope of the curve is zero, and the derivative dy/dx equals 0.

Figure A2-1 also shows two large distances, Δy and Δx, that are indicated by dashed lines. These are increments from the origin (x = y = 0) to some point on the curve y(x). We see that for a substantial value of Δx, the ratio Δy/Δx, the slanted dashed line, differs from the tangent line at x = 0. We show next that Δy/Δx comes closer and closer to the tangent line as Δx gets closer and closer to zero.

The sine function can be expressed as an infinite series of which the first three terms are:

$$\sin(x) = x - x^3/6 + x^5/120 - \ldots$$

The derivative with respect to x of sin(x) at x = 0 is the limit as Δx goes to zero of:

$$d\sin(0)/dx = \{\sin(\Delta x) - \sin(0)\}/\Delta x$$

$$d\sin(0)/dx = \sin(\Delta x)/\Delta x$$

$$d\sin(0)/dx = \{\Delta x - \Delta x^3/6 + \Delta x^5/120 - \ldots\}/\Delta x$$

$$d\sin(0)/dx = 1 - \Delta x^2/6 + \Delta x^4/120 - \ldots$$

We see that for moderate values of Δx, the right side of the last equation is less than 1. But as Δx goes to zero, Δx^2 and higher powers of Δx become negligible (much less than the leading term 1), and the expression approaches 1 — the limit as Δx goes to zero of { sin(Δx) / Δx } equals 1, the true derivative.

The maxima and minima of any function are always at points at which the derivative of that function is zero.

Now, let's step this up a notch, and consider acceleration a, the derivative of speed with respect to time. Continuing from above:

$$v(t) = g\ t$$

$$v(t+dt) = g\ [t+dt]$$

$$dv = g\ [t+dt] - g\ t = g\ dt$$

$$a = dv/dt = g = 9.8\ \text{m/sec}^2$$

Since acceleration is the derivative of speed, and speed is the derivative of distance, we say acceleration is the *second derivative* of distance. Here, "second" means we differentiate twice. The symbology denoting the second derivative of s with respect to t is:

$$a = dv/dt = d^2s/dt^2$$

First derivatives, like ds/dt, are ubiquitous in physics, and we often omit the word "first." Second derivatives, like d^2s/dt^2, are less common; third and higher order derivatives are rare.

The third derivative of distance, d^3s/dt^3, is called *jerk*. If your car speeds up with constant acceleration, you will be pressed back in your seat, but will not be unduly uncomfortable. But if the driver alternately floors and releases the gas pedal, the resulting *changes* in acceleration will toss you back and forth, which is very uncomfortable. That discomfort is due to jerk. Smooth rides are all about minimizing jerk.

§A2.2 Partial Derivatives

Physicists often deal with functions that vary with two or more variables. Temperature is a simple example: it is a function of time and the three dimensions of space.

The highest recorded air temperature on Earth is 134°F (56.7°C) at mid-afternoon, on July 10, 1913 at Furnace Creek, in Death Valley, California. But, we might be more interested in knowing *where* was the highest-ever noon temperature, or we might wish to know *when* was the highest-ever temperature at the North Pole. The highest-ever temperature yields neither. To get what we want, we need to find the maximum temperature at a fixed position or at a fixed time.

This is what partial derivatives do.

Next, consider an example from general relativity shown in Figure A2-2: the curved space near a massive body. The curvature here, shown as a downward deflection, is clearly a function of both x and y.

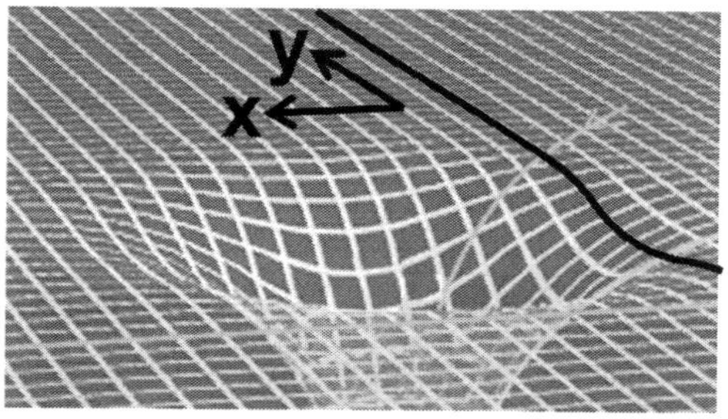

Figure A2-2 Spacetime Curvature Near Mass

The maximum curvature is deep down within the chasm. But we might be interested in the maximum curvature along the indicated black line parallel to the y-axis. We can obtain that by taking the partial derivative of the curvature function with respect to y, *while holding x fixed.* Holding x fixed ensures finding the greatest curvature for that value of x only, thus avoiding the even greater curvature deep within the chasm.

Mathematically, taking a partial derivative is identical to taking the derivative of a function of only one variable. Said another way: for functions of only one variable, there is no difference between the normal derivative and the partial derivative.

Consider a function f of n + 1 variables:

$$f(x,\ y_1,\ y_2, \ldots,\ y_n)$$

Here, x is one independent variable, and y_1, y_2, …, y_n are n other independent variables. The partial derivative of f with respect to x is defined to be:

$$\partial f/\partial x \ = \ \partial f(x,\ y_1,\ y_2, \ldots,\ y_n)/\partial x$$

$$\partial f/\partial x \ = \ \text{limit of } \{Q \,/\, dx\} \text{ as } dx \to 0$$

$$Q \ = \ f(x + dx,\ y_1,\ y_2, \ldots,\ y_n) \ - \ f(x,\ y_1,\ y_2, \ldots,\ y_n)$$

Note the notation: ∂/∂x rather than d/dx. Also note that none of the y's change.

§A2.3 Integrals

Integrals and derivatives are inverse operations like division and multiplication.

An example is:

If: $X = dY/dt$

Then: $\int X \, dt = Y + C$

Here, X and Y are functions of variable t, dt is an infinitesimal change in t, and the symbols { $\int X \, dt$ } denote the *integral of* X *with respect to t* — X is called the *integrand*.

This integral equals Y + C, because X is the derivative of Y with respect to t. C is any constant; it can have any arbitrary value, because the derivative of any constant is zero.

There are two types of integrals. The one we just examined is an *indefinite integral*; it finds the function Y(t) whose derivative is X(t).

The other type is the *definite integral*. Definite integral are the same as indefinite integrals, but with one additional step. Definite integrals are performed between *limits*, two values of the integration variable, t in this case. We write this:

$$\int_A^B X \, dt = Y \,|_A^B = Y(B) - Y(A)$$

Here, we calculate Y just as in an indefinite integral, then we evaluate Y(t) at each limit t = A and t = B, and lastly, we subtract Y(t = A) from Y(t = B). The symbols $Y \,|_A^B$ denote: Y(B) – Y(A). In a definite integral, the arbitrary constant C cancels since it is the same at both limits.

A definite integral effectively calculates the area under the curve X(t), as illustrated in Figure A2-3. The quantity { X dt } equals the area of a rectangle of height X and width dt, like one of the rectangles in Figure A2-3. For each value of t, we have one rectangle. The integral sums the rectangles for all values of t between whatever limits we specify; in this figure, we sum from t = 1 to t = 7.

As you see, these rectangles cover most, but not all, of the area under the curve X(t). Making dt smaller, reduces the gaps between the curve X(t) and the rectangles. The essential trick of calculus is taking the limit as dt, the width of each rectangle, goes to zero. In that limit, the sum of an infinite number of infinitesimally narrow rectangles exactly matches the area under the curve.

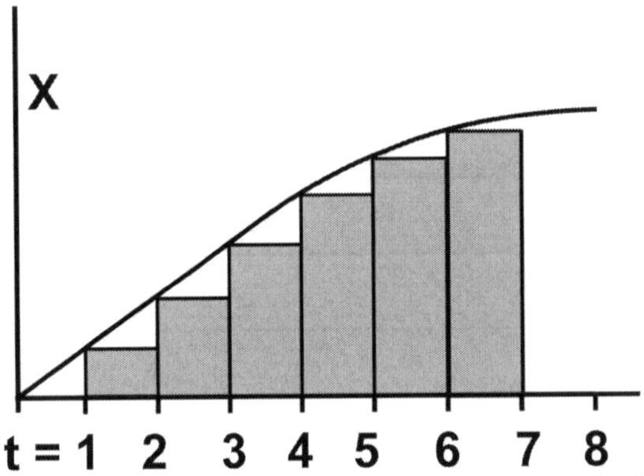

Figure A2-3 Summing Rectangles to Calculate Area under Curve

Many physics equations, including many in general relativity, determine the values of derivatives of quantities of interest.

For example, Newton's equation $\boldsymbol{F} = m\,\boldsymbol{a}$, combined with a knowledge of $\boldsymbol{F}$ and m determines the acceleration $\boldsymbol{a}$, the time derivative of velocity. Integrating $\boldsymbol{a}$ with respect to time yields velocity, which in turn is the time derivative of position. Integrating again, we finally get the position as a function of time for an object of mass m subject to force $\boldsymbol{F}$.

§A2.4 Integrals in N-Dimensions

Multidimensional integrals are employed to integrate quantities over a specified area or volume.

In N-dimensions, we need to integrate over N variables, typically coordinate axes. We evaluate multiple integrals sequentially: we first integrate over one variable, then integrate that result over a second variable, and continue until all integrals are done. The order of integration is arbitrary; regardless of which variables are integrated first or last, one obtains the same ultimate result.

For ease of discussion, let the word "volume" describe any defined region with any number of dimensions, whether it is a 2-D area, a 3-D volume, or a 4-D hyper-volume.

In rectilinear coordinate systems, integrating over an N-dimensional volume simply requires doing N one-dimensional integrals. For example, consider an integral over x, over y, and over z with these limits:

$$x: 0 \text{ to } X$$

$$y: 0 \text{ to } Y$$

$$z: 0 \text{ to } Z$$

This integral is over the entire volume V of a cuboid whose opposite corners are at (0, 0, 0) and (X, Y, Z). To make this as simple as possible, let the integrand be f, a constant.

$$Q = \int_0^Z \int_0^Y \int_0^X f \, dx \, dy \, dz$$

$$Q = \int_0^Z \int_0^Y (x f |_0^X) \, dy \, dz = X \int_0^Z \int_0^Y f \, dy \, dz$$

$$Q = X \int_0^Z (y f |_0^Y) \, dz = X \, Y \int_0^Z f \, dz$$

$$Q = f \, X \, Y \, Z = f \, V$$

If we now let f be a function of the coordinates, Q becomes:

$$Q = \langle f \rangle V$$

Here, $\langle f \rangle$ represents the average value of f throughout the cuboid.

This procedure works for any rectilinear coordinate system with any number of dimensions.

Let's next examine the 2-D integral of f across a disk of radius R, centered at the origin. Assume for now that f is constant. Figure A2-4 shows how we will do this: for each y value, from y = – R to y = + R, we sum a rectangle of height dy and width 2 xlim, where at each value of y:

$$\text{xlim} = \sqrt{(R^2 - y^2)}$$

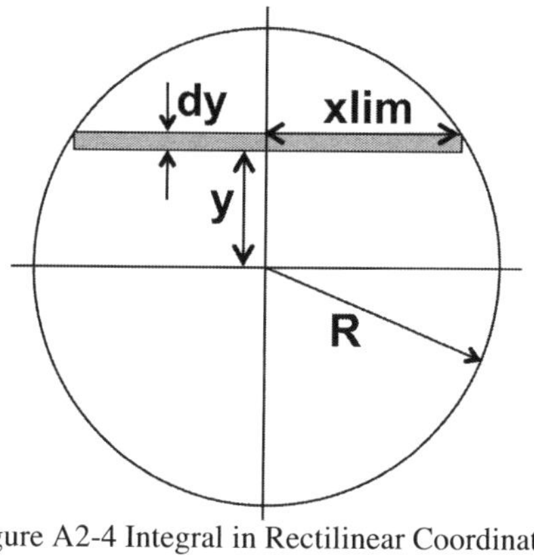

Figure A2-4 Integral in Rectilinear Coordinates

The integral is: $$Q = \int_{-R}^{+R} \int_{-\text{xlim}}^{+\text{xlim}} f \, dx \, dy$$

We integrate the prior equation over x first:

$$Q = f \int_{-R}^{+R} x \,|_{-xlim}^{+xlim} \, dy$$

$$Q = f \int_{-R}^{+R} 2 \sqrt{(R^2 - y^2)} \, dy$$

One can find this in integral tables, but let's calculate it directly by substituting:

$$\sin\theta = y / R$$

The differentials are: $dy \rightarrow R \cos\theta \, d\theta$

The limits become: $y = \pm R \rightarrow \theta = \pm \pi / 2$

$$Q = 2 f \int_{-\pi/2}^{+\pi/2} R \sqrt{(1 - \sin^2\theta)} \, (R \cos\theta) \, d\theta$$

$$Q = 2 f R^2 \int_{-\pi/2}^{+\pi/2} \cos^2\theta \, d\theta$$

$$Q = f R^2 \{ \theta + \sin\theta \cos\theta \} \,|_{-\pi/2}^{+\pi/2}$$

$$Q = f R^2 \{ + \pi / 2 + 0 - (- \pi / 2) - 0 \}$$

$$Q = f \pi R^2$$

You can confirm the $\cos^2\theta$ integral by taking the derivative of $\{ \theta + \sin\theta \cos\theta \}$.

The result Q is exactly what we should expect: the integral equals f multiplied by the area inside a circle of radius R.

Integration is a bit trickier in non-rectilinear coordinates.

Let's do the previous integral again, but now using polar coordinates r and θ. With r going from 0 to R, and θ going from 0 to 2π, the integral seems to be:

$$Q = \int_0^{2\pi} \int_0^{R} f \, dr \, d\theta$$

$$Q = f R \int_0^{2\pi} d\theta = 2 \pi f R \qquad \text{WRONG}$$

This calculation is wrong because dr dθ is not the area enclosed when the radius goes from r to r + dr and the angle goes from θ to θ + dθ. In fact (dr dθ) does not even have the units of area; its units are distance × radians. The problem is illustrated in Figure A2-5.

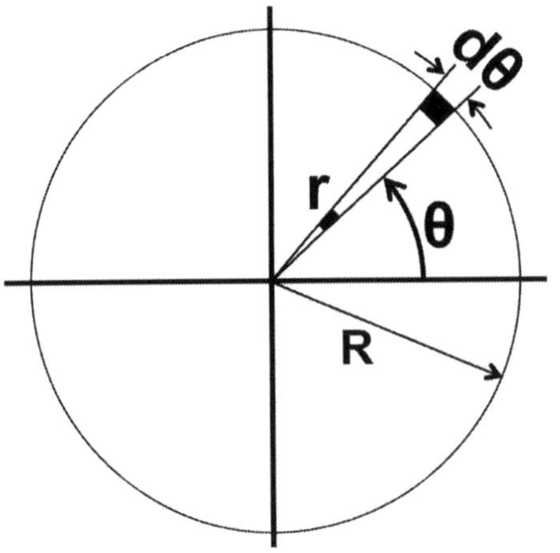

Figure A2-5 Integral in Polar Coordinates

Here, the two black enclosed regions have the same width dθ and the same length dr, but clearly have very different areas. The proper width is not dθ, but rather r dθ, and the incremental area is:

For 2-D polar coordinates: $r \, dr \, d\theta$

The prior expression is one example of a *volume element*, the measure of the space enclosed by infinitesimal changes in each coordinate axis. This name is commonly used regardless of the number of dimensions of that space. Every coordinate system has a volume element; for rectilinear coordinates, the volume element is the product of the coordinate differentials with no additional factors.

Let's redo the prior integral with the proper volume element, without the constant f:

$$Q = \int_0^{2\pi} \int_0^{R} r \, dr \, d\theta$$

$$Q = (R^2 / 2) \, (2\pi)$$

$$Q = \pi R^2$$

This is the correct result.

Here are the volume elements for all coordinate systems that physicists commonly use.

2-D polar:	$r \, dr \, d\theta$
2-D rectilinear:	$dx \, dy$
3-D rectilinear:	$dx \, dy \, dz$
3-D cylindrical:	$r \, dr \, d\phi \, dz$
3-D spherical:	$r^2 \sin\theta \, dr \, d\theta \, d\phi$
4-D spacetime:	$c \, dt \, dx \, dy \, dz$

We now know how to integrate in two other non-rectilinear coordinate systems: 3-D cylindrical and 3-D spherical, whose volumes are denoted C and S respectively.

$$C = \int_0^{Z} \int_0^{2\pi} \int_0^{R} r \, dr \, d\phi \, dz$$

$$C = (R^2 / 2) \, (2\pi) \, Z = \pi R^2 Z$$

$$S = \int_0^{2\pi} \int_0^{\pi} \int_0^{R} r^2 \sin\theta \, dr \, d\theta \, d\phi$$

$$S = (R^3 / 3) \, (2) \, (2\pi) = 4 \pi R^3 / 3$$

Q, C, and S are respectively: the area of a disk of radius R; the volume of a cylinder of radius R and length Z; and the volume of a ball of radius R. If we multiply each integrand by f, the integrals would become $< f >$ multiplied by volume for the disk, cylinder, or ball. This is exactly what integrals are supposed to do: sum an expression over a defined space, or equivalently, find an expression's average value multiplied by the volume of that space.

For any other coordinate system W, the volume element equals the determinant of the *Jacobian* matrix:

$$J_{kn} = \partial W_k / \partial X_n$$

Here W_k are the coordinates of W, and X_n are the rectilinear coordinates.

§A2.5 Derivatives of Common Functions

Differentiation is a *linear* operation. For any functions F and G, constants a and b, and variable q:

$$da/dq = 0$$

$$d(a F + b G)/dq = a\, dF/dq + b\, dG/dq$$

$$d(F G)/dq = d(G F)/dq = G\, dF/dq + F\, dG/dq$$

$$d(F / G)/dq = (1 / G)\, dF/dq - (F / G^2)\, dG/dq$$

$$d(F^n)/dq = n\, F^{n-1}\, dF/dq$$

For variables x and t:

$$d\, x^n / dt = n\, x^{n-1}\, dx/dt$$

$$d \sin(x) / dt = + \cos(x)\, dx/dt$$

$$d \cos(x) / dt = - \sin(x)\, dx/dt$$

$$d \tan(x) / dt = \cos^{-2}(x)\, dx/dt$$

$$d\, e^x / dt = e^x\, dx/dt$$

$$d \ln(x) / dt = (1 / x)\, dx/dt$$

$$d \sinh(x) / dt = \cosh(x)\, dx/dt$$

$$d \cosh(x) / dt = \sinh(x)\, dx/dt$$

$$d \tanh(x) / dt = \cosh^{-2}(x)\, dx/dt$$

For $r = \sqrt{(x^2 + a^2)}$, with constant a:

$$d\, r /dx = x / r$$

$$d\, (r)^{-1} /dx = - x / r^3$$

$$d\, (r)^{-2} /dx = - 2\, x / r^4$$

$$d\, (r)^{-3} /dx = - 3\, x / r^5$$

$$d\, (x / r) /dx = a^2 / r^3$$

Relativistic parameter γ:

$$\gamma = 1 / \sqrt{(1 - v^2 / c^2)}$$

$$d\gamma / dt = (- 1 / 2)\, \gamma^3\, (- 2\, v / c^2)\, dv/dt$$

$$d\gamma / dt = \gamma^3\, v\, dv/dt\, / c^2$$

§A2.6 Integrals of Common Functions

Some common integrals are listed below. Here, a and b are constants, x is the independent variable, C is the arbitrary integration constant, and F and G are functions.

$$\int (F + G)\ dx = (\int F\ dx) + (\int G\ dx) + C$$

$$\int (F\,G)\ dx \text{ IS NOT } (\int F\ dx)\ (\int G\ dx)$$

$$\int a\ F\ dx = a \int F\ dx + C$$

For b not = – 1:

$$\int a\ x^b\ dx = a\, x^{b+1} / (b + 1) + C$$

ln is the natural logarithm:

$$\int a\ x^{-1}\ dx = a\ \ln(x) + C$$

$$\int a\ e^{bx}\ dx = a\, e^{bx} / b + C$$

$$\int \sin(a\,x + b)\ dx = -\cos(a\,x + b) / a + C$$

$$\int \cos(a\,x + b)\ dx = +\sin(a\,x + b) / a + C$$

For $r = \sqrt{(x^2 + a^2)}$:

$$\int (1 / r)\ dx = \ln(x + r) + C$$

$$\int (1 / r^3)\ dx = x / (a^2\, r) + C$$

$$\int (x / r)\ dx = r + C$$

$$\int (x / r^3)\ dx = -1 / r + C$$

$$\int \sinh(x)\ dx = \cosh(x) + C$$

$$\int \cosh(x)\ dx = \sinh(x) + C$$

$$\int \tanh(x)\ dx = \ln\{ \cosh(x) \} + C$$

Appendix 3
Space Travel

Einstein's theory of special relativity has dramatic impacts on our hopes for future space travel, both favorable and unfavorable.

With c as a universal speed limit, we can never send a rocket to another galaxy and have it return in our lifetime. Even at nearly the speed of light, round trips lasting less than 100 Earth-years could only reach about 1000 stars, less than a billionth of our galaxy. But time dilation allows astronauts to reach more distant destinations within *their* lifetimes. The bad news is that they will never be able to go fast enough to return before everyone they knew on Earth has long since died.

With current technology, exploring beyond our solar system is completely unrealistic. The nearest star to the Sun, Proxima Centauri, is 4.2 light-years away. Since one light-year is the distance that light travels in one year, nearly 6 trillion miles, Proxima is 25 trillion miles away. A round trip to Proxima in the fastest rocket made so far would take over 100,000 years — just imagine how many times someone might ask: "are we there yet?" But why would we go there? We already know there is no suitable planet orbiting Proxima.

I estimate habitable planets are so rare that the nearest one might be 300 to 3000 light-years away, making a round trip to the next oasis a multi-million year odyssey, with current rockets. The cost of space travel is also a staggering challenge — one Space Shuttle launch cost $1.3 billion, and it only delivered a 20-ton payload to low Earth orbit. That is $65 million per ton — more expensive than gold or platinum.

§A3.1 What is Theoretically Possible?

Traveling to the stars requires vastly superior rocket technology. Antimatter fuel provides the ultimate propulsion. Antimatter annihilation converts 100% of the fuel's mass into energy, which is more than a billion times more efficient than chemical fuels. Also, the annihilation of an antielectron with an electron produces two photons. Since these travel at light speed, they provide the maximum momentum transfer per unit of energy.

If we could overcome every daunting engineering challenge, what is theoretically possible? What is the very best that nature allows?

Special relativity limits a rocket's speed. As we have discussed, objects appear to become heavier as their speed increases. Einstein's equations show that masses become infinite at the speed of light. Hence it would require an infinite amount of energy to accelerate a rocket, or anything else with mass, up to the speed of light. Since the entire observable universe does not contain an infinite amount of energy, traveling at light speed or faster is impossible.

This is well confirmed experimentally. In my own experience, the fastest massive particles we ever produced had velocities of 0.999,999,999,7 c and masses 40,000 times more than identical stationary particles. (That was at the Stanford Linear Accelerator Center, which is no longer doing particle physics.) While 40,000 is a bit short of infinity, the mass increase curve is exactly what Einstein predicted. There is no indication this curve will not continue rising into the wild black yonder. Even if there were a deviation from Einstein's theory at higher velocities, the measurements we already have pose insurmountable limitations: no spaceship will ever attain the velocities already achieved in particle accelerators.

Where special relativity helps space travel is in extending the lives of astronauts. Einstein showed that time runs slower in moving frames — the *time dilation* effect discussed in Chapter 2. For example, time runs ten times slower in a spaceship traveling at 0.995 c. At that speed, astronauts would age ten times slower, and we on Earth would observe them living ten times longer, perhaps 800 years.

Unlike Methuselah, the astronauts would perceive that they had a normal lifespan of 80 years. They would have no more heartbeats, no more thoughts of joy or sorrow, than their Earth-bound peers.

While their spaceship blasted through space at 668 million mph, The crew would seem to live a normal life, but would do so ten times slower. Thus, they would travel much farther within their lifespan, and still be alive seven centuries after their siblings on Earth had died.

Time dilation provides another critical benefit: reducing life support requirements. At 0.995 c, astronauts would breathe ten times slower, consume ten times less oxygen, eat ten times less food, and generate ten times less waste. All these greatly reduce the burden of keeping the crew alive during very long voyages.

A good flight plan starts with the acceleration we are used to on Earth's surface: 1 g (9.8 meters, or 32 feet, per second per second). This provides a comfortable environment with normal "gravity" during the long flight, while reducing flight time. For destinations within 400 light-years of Earth, the spaceship continues accelerating at 1 g up to the halfway point. It then decelerates at 1 g, and comes to a stop at the destination. The return trip would have the same acceleration and deceleration plan. All accelerations are measured in the spaceship's reference frame.

After accelerating at 1 g for six years on the spaceship's clock, and 201 years on Earth clocks, the spaceship would be 200 light-years from Earth, its relativistic γ factor would be 201, and its velocity would be 0.999,99 c (call that "Five 9's").

I do not recommend any greater velocity. To destinations farther than 400 light-years, the rocket should coast during the middle part of its journey. For example, on a voyage to a star 1000 light-years away, the spaceship should accelerate for the first 200 light-years, coast at maximum velocity for 600 light-years, and decelerate for the last 200 light-years.

The reasons for limiting the spaceship's maximum velocity are fuel consumption, space debris, and the ever-present cosmic microwave background radiation ("CMB"). At a relativistic γ factor of 201, space rocks, and even diffuse gas, strike the ship's nose with immense force. Steering to avoid debris would require intolerable accelerations and fuel expenditures. Even with an ideal deflector system, the momentum transferred by debris and gas pose a drag force that increases fuel consumption. Additionally, CMB photons are ubiquitous throughout the cosmos, outnumbering the particles of matter by a billion to one. We on Earth see these photons at an equivalent temperature of 2.72K (– 270 ºC, or – 454 °F). But at Five 9's, spaceships would face a withering headwind of blueshifted CMB photons at an apparent temperature of 525 °F (+274 ºC). With few ways to dump excess heat in space, there is a limit to the velocity that a spaceship and its crew can survive. My crystal ball says that limit is not above Five 9's.

With my flight plan, the table in Section §A3.2 offers a selection of one-way and round-trip space cruises people might someday enjoy.

As the table shows, you might cruise to a destination 20 light-years away, and then return to Earth. During this voyage, you would age only 12 years, while 43 years pass on Earth. If your less adventurous twin stayed home, you would be 31 years younger than your twin upon arrival and would remain 31 years younger for the rest of your lives. The trip's maximum velocity is 0.996 c, and 230,000 tons of fuel are consumed per ton of payload.

On a one-way trip to a destination 400 light-years away, 402 years pass on Earth while you age 12 years. The maximum velocity is Five 9's, and 160,000 tons of fuel are consumed per ton of payload.

For journeys beyond 1000 light-years, use the table's lowest two rows. On a one-way trip to a destination 3000 light-years away, use the tabulated numbers and x = 3000 – 1000 = 2000. This journey takes 3002 Earth years while the crew ages 25 years. The maximum velocity and fuel consumption per ton of payload are the same as the prior example because the spaceship coasts for the additional 2600 light-years, but the crew would need more supplies, hence more payload, to survive the longer voyage.

Recall that we are assuming the most efficient fuel possible: half matter and half antimatter. In 2007, scientists estimated the cost of producing antielectrons at $1 trillion per pound. The U.S. Congress may not fund such trips any time soon.

§A3.2 Potential Space Cruises

			<---- One-Way Trip ---->			<----- Round-Trip ----->		
Destination Distance (lt.-yrs.)	Max. Speed v / c	Max. γ Factor	Crew yrs.	Earth yrs.	Fuel / Payload	Crew yrs.	Earth yrs.	Fuel / Payload
1	0.74472	1.5	1.9	2.2	5.8	3.8	4.5	46
2	0.86629	2.0	2.7	3.5	13.3	5.3	7.0	200
4	0.94273	3.0	3.5	5.7	33	7.1	11	1150
10	0.98602	6.0	5.0	12	142	9.9	24	20k
20	0.99587	11	6.2	22	482	12	43	230k
40	0.99886	21	7.5	42	1771	15	84	3.1M
100	0.99981	51	9.3	102	10k	19	204	110M
200	0.99995	101	10.6	202	41k	21	404	1.7B
400	0.99999	201	12	402	160k	24	804	26B
1000	0.99999	201	15	1002	160k	30	2004	26B
+x	0.99999	201	+x/201	+x	160k	+x/100	+2x	26B

Notes for this Table:

1. The left column lists the one-way destination distance in light-years.
2. For destinations up to 400 light-years, the flight plan is to accelerate at 1 g to the midpoint, then decelerate at 1 g, stopping at the destination. For round trips, the return leg has the same acceleration plan. Beyond 400 light-years, the spaceship coasts when it is more than 200 light-years from either origin or destination.
3. Table assumes propulsion by electron-positron annihilation with 100% fuel efficiency, a billion-fold improvement over current technology. Fuel usage is shown in tons of fuel per ton of payload, with k denoting thousand, M million, and B billion.
4. Payload is everything on the launch pad except fuel — including crew, supplies, equipment, and the spaceship itself.
5. Planning longer trips is easy using the bottom two rows. For a round trip to a destination 1500 light-years away (x = 1500 – 1000 = 500), the crew ages 35 years (30 + x / 100) while 3004 years (2004 + 2 x) pass on Earth. Maximum speed, γ factor and fuel-to-payload ratio do not change, because the spaceship coasts throughout the added distance x.
6. Round trips assume all fuel for both legs is loaded on Earth — no refueling during the voyage. If refueling is available at the destination, the total fuel-to-payload ratio is just twice the stated ratio of a one-way trip.

§A3.3 Relativistic Space Travel Equations

For those wishing to plan their own journeys, with perhaps different accelerations or maximum velocities, here are the relevant equations from general relativity (special relativity does not address accelerations).

These equations assume constant acceleration u for spaceship time t, Earth time T, and travel distance D. They provide the data for the first half of a one-way trip without coasting. The same numbers apply to the second half of a one-way trip that has constant deceleration u.

Here, time is measured in years, velocities in fractions of the speed of light, distances in light-years, and accelerations in light-years/year/year. The acceleration of gravity on Earth's surface, 1 g, happens to be 1.0366 light-years/year/year, quite close to 1. Thus u is approximately a multiple of 1 g.

Sinh, cosh, and tanh are the hyperbolic sine, cosine, and tangent functions defined below.

$$T = \sinh(u\,t) \;/\; u$$

$$D = [\,\cosh(u\,t) - 1\,] \;/\; u$$

$$\text{maximum velocity} = \tanh(u\,t)$$

$$\text{fuel / payload} = e^{u\,t} - 1$$

For a full one-way trip, with constant acceleration u to the midpoint, followed by constant deceleration u, and stopping at the destination:

$$\text{Total Earth time} = 2\,T$$

$$\text{Total Crew time} = 2\,t$$

$$\text{Total distance} = 2\,D$$

$$\text{fuel / payload} = e^{2\,u\,t} - 1$$

For a round trip comprised of two one-way trips as above, without coasting or refueling:

$$\text{Total Earth time} = 4\,T$$

$$\text{Total Crew time} = 4\,t$$

$$\text{Total distance} = 2\,D \text{ each way}$$

$$\text{fuel / payload} = e^{4\,u\,t} - 1$$

When coasting (no fuel consumption) for C light-years at velocity v:

$$\text{Add to Earth time} = C \;/\; v$$

$$\text{Add to Crew time} = C\,\sqrt{(1 - v^2)} \;/\; v$$

$$\sinh(y) = (\,e^{y} - e^{-y}\,) \;/\; 2$$

$$\cosh(y) = (\,e^{y} + e^{-y}\,) \;/\; 2$$

$$\tanh(y) = \sinh(y) \;/\; \cosh(y)$$

Appendix 4
World Travel: Fast & Free

An amusing property of gravity allows for the possibility of rapid travel and transportation of freight between any two points A and B on Earth's surface. The transit would take about 42 minutes, and would not consume any energy or generate any pollution. Anyone could go from Los Angeles to London in 42 minutes. Cargo could be shipped from Shanghai to Chicago in 42 minutes, all without burning any fuel.

There is one catch: we need a tunnel straight through the Earth linking A and B.

Yet, with enough tunnels, we could freely and rapidly travel the globe and trade goods. The transit time through any tunnel would always be 42 minutes, whether the tunnel's ends are close together or on opposite sides of our planet.

Figure A4-1 shows a cross section of Earth, with R being Earth's radius, and various distances identified as a, b, r, and x. Let's assume a person (or freight container) drops into the tunnel at A with zero initial velocity. As the person drops, coming closer to Earth's center, their decreasing gravitational potential energy converts into kinetic energy of motion. They continue accelerating until they reach the tunnel's midpoint, after which they start moving away from Earth's center. From there on, they decelerate and their gravitational potential energy increases. They reach B at the instant their velocity equals zero once again. If they immediately step away from the tunnel, they will have arrived at B in 42 minutes. If they do nothing, they will fall back into the tunnel and return to A in another 42 minutes.

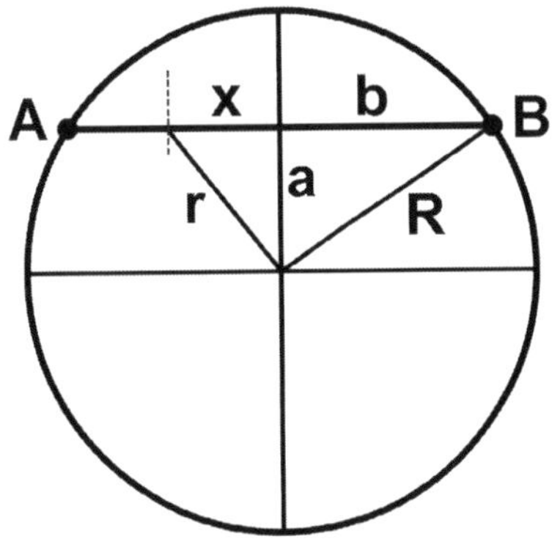

Figure A4-1 Tunnel Through Earth

Why does the trip take the same time regardless of the distance between A and B? The farther apart are A and B, the closer the A-to-B tunnel comes to Earth's center, and the higher the travelers' velocities become.

Why does the trip consume no energy? The traveler initially has a large gravitational potential energy, like a child on a swing at its maximum height. As with the swing, this potential energy is converted into kinetic energy of motion, and back to potential energy at the opposite side.

Here is the proof for an idealized situation.

The right triangle, whose sides are a, b, and R yields:

$$R^2 = a^2 + b^2$$

The right triangle whose sides are a, x, and r yields:

$$r^2 = a^2 + x^2$$

Subtracting these yields:

$$R^2 - r^2 = b^2 - x^2$$

We assume no friction with the tunnel walls, no issues with the molten core, and that A and B are at the same distance R from Earth's center — at the same elevation.

For ease of calculation, we also assume Earth's density is uniform. Density variations would alter the transit time slightly, but the overall scheme works for any reasonable density distribution.

The force of gravity for a traveler (or freight container) mass m at a location in the tunnel that is distance r from Earth's center is:

$$F(r) = G\ m\ (M\ r^3 / R^3) / r^2$$

$$F(r) = G\ m\ M\ r / R^3$$

Let's shorten the equations slightly by defining:

$$\mu^2 = G\ M / R^3$$

The factor in parentheses is Earth's total mass M times the fraction of M that is closer to Earth's center than r — for a uniform density that fraction is r^3 / R^3. Next, let's calculate Earth's gravitational potential energy U(r) at a distance r from its center. Since force equals minus the gradient of potential energy, U(r) equals minus the integral of F(ϱ) from Earth center, ϱ = 0, to ϱ = r.

$$U(r) = -\int_0^r F(\varrho)\ d\varrho$$

$$U(r) = -\int_0^r \mu^2\ m\ \varrho\ d\varrho$$

$$U(r) = C - \mu^2\ m\ r^2 / 2$$

Here, C is an arbitrary integration constant that sets the zero-point of potential energy. Let's elect to define U to be zero at Earth's surface. But as always, only potential energy differences are physically meaningful, so the choice of C has no bearing on our results. Our choice of C yields:

$$U(r) = -\mu^2\ m\ (R^2 - r^2) / 2$$

We next equate the traveler's kinetic energy to the decrease in gravitational potential energy from R (the surface) to some lesser radius r.

$$m\ v^2 / 2 = U(R) - U(r)$$

$$m\ v^2 / 2 = 0 + \mu^2\ m\ (R^2 - r^2) / 2$$

$$v^2 = \mu^2\ (R^2 - r^2)$$

Note that the mass m of the traveler (or freight container) cancels out; our calculations are valid for any transiting mass m.

Recall from the prior page that $R^2 - r^2 = b^2 - x^2$, where b is half the tunnel length, and x is the distance from the tunnel's midpoint (see Figure A3-1). The full journey is from x = – b to x = + b. With this substitution, we have:

$$v^2 = \mu^2\ (b^2 - x^2)$$

$$v = dx/dt = \mu\ \sqrt{(b^2 - x^2)}$$

$$dt = dx / \mu\ \sqrt{(b^2 - x^2)}$$

Total transit time t is: $$t = \int_{-b}^{+b} dx / \mu\ b\ \sqrt{(1 - x^2 / b^2)}$$

We can solve this integral by making the substitution:

$$\sin\theta = x / b$$

The full journey is now from θ = – π / 2 to θ = + π / 2. Note that θ is a mathematical convenience; it is not a physical angle related to any part of Figure A3-1.

The transit time integral becomes:

$$dx = b\ d(\sin\theta) = b\ \cos\theta\ d\theta$$

$$t = \int_{-\pi/2}^{+\pi/2} b\ \cos\theta\ d\theta\ /\ \mu\ b\ \sqrt{(1 - \sin\theta^2)}$$

$$t = \int_{-\pi/2}^{+\pi/2} d\theta\ /\ \mu$$

$$t = \theta\ \Big|_{-\pi/2}^{+\pi/2}\ /\ \mu$$

$$t = \pi\ /\ \mu$$

$$t = \pi\ /\ \sqrt{(G\ M / R^3)}$$

For Earth:

$$G\ M = 3.986 \times 10^5\ km^3 / sec^2$$

$$R = 6371\ km$$

$$\mu = \sqrt{(G\ M / R^3)} = 0.00124 / sec$$

$$t = 42 \text{ minutes, } 10.4 \text{ seconds}$$

Appendix 5
Geodesics & Derivatives in Curved Spacetime

This Appendix explores the mathematics underlying the definitions and capabilities of the curvature tensors used in general relativity. Since most physical theories describe how important quantities *change* over time or across space, derivatives and integrals are essential tools for physicists.

But how can we calculate the change of a function f with respect to x when the x-axis itself is stretching, contracting, and twisting in ever-changing spacetime? It seems like analyzing the flight of a paper airplane in a hurricane.

Yet, mathematicians have provided the tools to make this possible.

§A5.1 Basis Vectors & Their Inverses

In 3-D, we select three unit vectors that point in three orthogonal directions. We then define vectors as the sum of the three products, each product being a unit vector $\boldsymbol{e}_j$ multiplied by a coefficient v_j. These three coefficients v_j are called the vector's components along each axis. We write this:

$$\boldsymbol{v} = v_j \, \boldsymbol{e}_j \quad \text{summed over } j = 1 \text{ to } 3$$

In 4-D spacetime, 4-vectors are defined similarly, as the sum of four products of basis vectors $\boldsymbol{e}_\mu$ multiplied by coefficients v^μ. We write this:

$$\boldsymbol{v} = v^\mu \, \boldsymbol{e}_\mu \quad \text{summed over } \mu = 0 \text{ to } 3$$

Please note an aspect of our notation that can easily lead to confusion: e_μ does not denote one component of a basis vector in the μ-direction, but rather, it denotes the entire μ basis vector — μ is an axis designator, not a component index. In the Cartesian coordinates we know and love, the j basis vector always points in the j-direction, so this was never an issue. But, in the Wonderland of the dynamic, curved, spacetime continuum, basis vectors and their inverse operators can have components in each and every direction, and those components can change over time and / or across space.

As discussed in Section §20.3, a rank n tensor can be formed by the uncontracted tensor product of n 4-vectors, for example:

$$W_{\alpha}{}^{\beta}{}_{\gamma} = A_\alpha \, B^\beta \, G_\gamma$$

As we see here, tensors have an additional complexity: some indices are subscripted, while some are superscripted.

Basis vectors with subscripted designators are mathematically equivalent to directional partial derivatives, while those with superscripted designators are the inverse operators of their subscripted cousins (for more on these inverse operators see Section §A5.7). We write this:

$$e^\alpha \bullet e_\beta = \delta^\alpha{}_\beta$$

The left side of this equation denotes the tensor contraction (" • ") of e^α and e_β. The equation says the contraction of inverse operator e^α with basis vector e_α of the same axis equals 1, while the contraction of e^α with any other basis vector equals 0 — the basis vectors comprise an orthonormal set.

We can write a rank 3 tensor S as:

$$S = S^\alpha{}_{\beta\delta} \, e_\alpha \, e^\beta \, e^\delta$$

The collection of coefficients, $S^\alpha{}_{\beta\delta}$ in this case, are the components of the tensor S.

Recall that a tensor contraction is similar to a 3-vector dot product, but in more dimensions. Between two 4-vectors A^α and B_α, we normally write this: $A^\alpha B_\alpha$ with an implied sum over all values of α. I will write the contraction of a basis vector and its inverse operator differently (" • ") in hopes of minimizing the confusion of which label is a component index and which is an axis designator.

§A5.2 Derivatives of Base Vectors & Inverses

Connection coefficients describe how coordinate axes change throughout spacetime. Their primary definition is:

$$e_{\alpha;\beta} = + \; e_\mu \; \Gamma^\mu{}_{\alpha\beta}$$

Here, " ;β " denotes the covariant derivative in the β direction, e_α is the α basis vector, and connection coefficient $\Gamma^\mu{}_{\alpha\beta}$ is the μ-component of change of basis vector α as we move in the β-direction. By summing over μ, we combine all changes in all directions to basis vector α.

The above equation provides the covariant derivative of a basis vector with subscripted designator α. We also need to know how the covariant derivative acts on an inverse operator with a superscripted designator.

Let's now derive this, using the fact that partial derivatives and sums commute — these operations can be performed in any order we wish. The same is true for multiple contractions — we get the same final result regardless of order of the contractions.

To begin, we take the covariant derivative in the μ-direction of the contraction of e^α with e_β. Since that contraction produces a scalar, its derivative is identically zero.

$$(e^\alpha \bullet e_\beta)_{;\mu} = \delta^\alpha{}_{\beta;\mu} = 0$$

$$e^\alpha{}_{;\mu} \bullet e_\beta + e^\alpha \bullet e_{\beta;\mu} = 0$$

By definition:

$$e^\alpha{}_{;\mu} \bullet e_\beta + e^\alpha \bullet e_\gamma \; \Gamma^\gamma{}_{\beta\mu} = 0$$

$$e^\alpha{}_{;\mu} \bullet e_\beta = - \; \delta^\alpha{}_\gamma \; \Gamma^\gamma{}_{\beta\mu}$$

$$e^\alpha{}_{;\mu} \bullet e_\beta = - \; \Gamma^\alpha{}_{\beta\mu}$$

Multiply all by e^β:

$$e^\alpha{}_{;\mu} \bullet e_\beta \; e^\beta = - \; e^\beta \; \Gamma^\alpha{}_{\beta\mu}$$

Contract ( ) on β:

$$e^\alpha{}_{;\mu} (e_\beta \bullet e^\beta) = - \; e^\beta \; \Gamma^\alpha{}_{\beta\mu}$$

$$e^\alpha{}_{;\mu} = - \; e^\beta \; \Gamma^\alpha{}_{\beta\mu}$$

Next, let's apply the covariant derivative to a tensor S with one superscripted and one subscripted index. As above, we write S as a sum of components multiplied by basis vectors and their inverse operators.

$$S = S^\alpha{}_\beta \; e_\alpha \; e^\beta$$

$$S_{;\mu} = S^\alpha{}_{\beta;\mu} \; e_\alpha \; e^\beta + S^\alpha{}_\beta \; e_{\alpha;\mu} \; e^\beta + S^\alpha{}_\beta \; e_\alpha \; e^\beta{}_{;\mu}$$

Since the component $S^\alpha{}_\beta$ is a scalar, its covariant derivative equals its partial derivative.

$$S_{;\mu} = S^\alpha{}_{\beta,\mu} \; e_\alpha \; e^\beta + S^\alpha{}_\beta \; (e_\gamma \; \Gamma^\gamma{}_{\alpha\mu}) \, e^\beta + S^\alpha{}_\beta \; e_\alpha (- \; e^\nu \; \Gamma^\beta{}_{\nu\mu})$$

Now, let's swap summation indices in the last two terms so each term ends up with $e_\alpha \, e^\beta$.

$$S_{;\mu} = S^\alpha{}_{\beta,\mu} \; e_\alpha \; e^\beta + S^\gamma{}_\beta \; e_\alpha \; \Gamma^\alpha{}_{\gamma\mu} \; e^\beta - S^\alpha{}_\gamma \; e_\alpha \; e^\beta \; \Gamma^\gamma{}_{\beta\mu}$$

$$S^\alpha{}_{\beta;\mu} = S^\alpha{}_{\beta,\mu} + S^\gamma{}_\beta \; \Gamma^\alpha{}_{\gamma\mu} - S^\alpha{}_\gamma \; \Gamma^\gamma{}_{\beta\mu}$$

The last equation, in our usual notation without basis vectors, matches that in Section §22.2.

§A5.3 Parallel Transport & Covariant Derivative

In analyzing phenomena in any geometry, we need to know how things change from one location to another, and from one moment in time to another.

In the flat spacetime of Newton, or of special relativity, the components of a vector might change for two reasons: due to physics, or due to a change of coordinate axes. Examples of the latter are rotations and boosts that we the observers may perform for our own mathematical convenience. Examples of the former are interactions among physical entities that exert forces upon one another. Physicists may use the latter to understand the former, but only the former have physical reality. In flat spacetime, we know how to separate coordinate changes from the real physics.

In the curved spacetime of general relativity, the components of a vector might change for three reasons: due to physics, due to a change of coordinate axes, or due to spacetime curvature. Even in the absence of our performing coordinate transformations, even in a system without any external or internal forces, even in a system moving along a geodesic, the components of vectors can change. Geodesics are the "straightest" paths through curved spacetime, but they need not be straight. We need to learn how to separate changes due to curved spacetime from the other types of changes.

To cope with this complexity, we employ the *parallel transport* method — a means of moving tensors from one event to another without changing their "true direction". For example, we can parallel transport a particle's 4-velocity v^μ from event A to event B, call that $v^\mu(A\rightarrow B)$, and then compare the transported velocity with the particle's actual velocity at B, call that $v^\mu(B)$. If $v^\mu(A\rightarrow B) = v^\mu(B)$, the particle's true velocity has not changed, even if its components have. If $v^\mu(B) - v^\mu(A\rightarrow B)$ is nonzero, the particle's true velocity really did change.

How does parallel transport work?

Let's build a *Schild's ladder*. Our task is to parallel transport a vector ***Aa*** along a curve AE, as illustrated in Figure A5-1. Along AE, we define an *affine* parameter λ that measures distance along AE. The affine parameter can be any measurement technique that yields equal distances along AE for equal increments of λ. For example, λ could be proper time as measured by a perfect clock traversing AE.

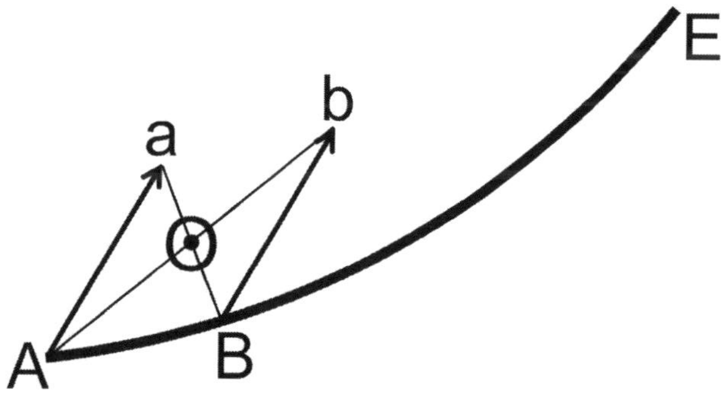

Figure A5-1 Schild's Ladder Construction

The steps in building Schild's ladder are:

Step 1: For a tiny increment of the affine parameter $d\lambda$, move distance $d\lambda$ from event A to event B.

Step 2: Draw a geodesic line from B to a, and define an affine parameter ξ along ***Ba***. If the length of ***Ba*** is $\Delta\xi$, the midpoint O is the point that is $\Delta\xi/2$ from both ***B*** and ***a*** along ***Ba***. Midpoint O is at the black dot in Figure A5-1.

Step 3: Draw a geodesic line from A to O, and define an affine parameter ζ along ***AO***. If the length of ***AO*** is $\Delta\zeta$, extend this geodesic to $2\,\Delta\zeta$, and call that event b. In the limit that $d\lambda$ and the length of ***Aa*** go to zero, the vector ***Bb*** is the vector ***Aa*** parallel transported from event A to event B.

We can indefinitely repeat Steps 1 through 3, all the way to the end of curve AE. The vector ***Ee*** is the vector ***Aa*** parallel transported from event A to event E, as shown in Figure A5-2.

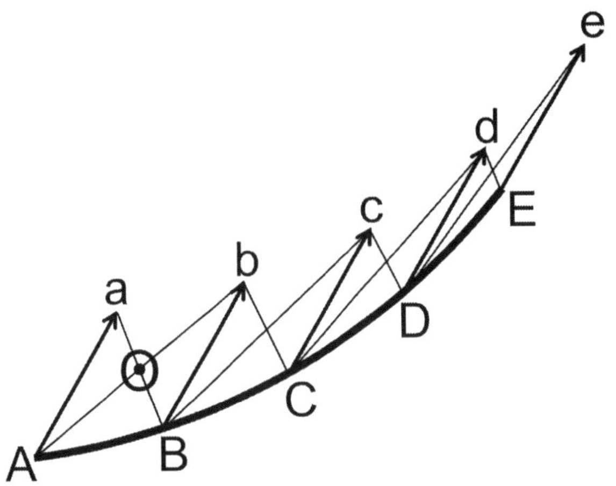

Figure A5-2 Parallel Transport of Aa to Ee

Let's define $\boldsymbol{u}$ to be the tangent vector field of curve AE; at each event along AE, u is given by:

$$\boldsymbol{u} = \partial/\partial\lambda$$

With what we just developed, we can now define an effective covariant derivative. For any vector $\boldsymbol{v}$, defined at a spacetime event A, and an affine parameter λ that defines a tangent vector $\boldsymbol{u}$ (see last equation), the covariant derivative of $\boldsymbol{v}$ at event A is defined by:

$$\boldsymbol{v}(A);u = \text{limit as } d\lambda \rightarrow 0 \ [\ \{\ \boldsymbol{v}(\lambda_A + d\lambda) - \boldsymbol{v}^*(\lambda_A)\ \}\ /\ d\lambda\ ]$$

Here, λ_A is the value of the affine parameter at event A, and the asterisk on $\boldsymbol{v}^*(\lambda_A)$ denotes the vector $\boldsymbol{v}(\lambda_A)$ parallel transported to $(\lambda_A + d\lambda)$ from λ_A.

This definition is easily generalized to encompass tensors of any rank.

A fundamental property of all geodesics is that they parallel transport their tangent vectors. We write this:

Along any geodesic: $u_{;u} = 0$

We can understand why this is true by imagining a gyroscope moving along a geodesic curve — a free-falling, zero-gravity reference frame. Let's assume the gyroscope is not subject to any forces, and that its rotational axis is parallel to the geodesic's tangent vector u.

In curved spacetime, a geodesic curve is the "straightest" possible curve, but it need not be straight. Geodesics may stretch or turn — changing the directions of its coordinate axes — but faithfully following the geometry of spacetime, as does the axis of the gyroscope, which will always remain parallel to the geodesics tangent vector u. Hence, its derivative along u is zero.

Let's expand the prior equation according to the rules we derived in Section §A5.2, with the derivative operator u acting on a vector x^α:

$$u^\alpha = \partial x^\alpha/\partial\lambda$$

$$0 = u_{;u} = (\partial x^\alpha/\partial\lambda)_{;u}$$

$$0 = (\partial x^\alpha/\partial\lambda)_{,u} + (\partial x^\beta/\partial\lambda)\ (\partial x^\mu/\partial\lambda)\ \Gamma^\alpha{}_{\beta\mu}$$

$$0 = \partial^2 x^\alpha/\partial\lambda^2 + \Gamma^\alpha{}_{\beta\mu}\ (\partial x^\beta/\partial\lambda)\ (\partial x^\mu/\partial\lambda)$$

This is the geodesic equation quoted in Section §24.1, and derived in another way in Section §A5.5.

This equation proves that the connection coefficient $\Gamma^\alpha{}_{\beta\mu}$ is symmetric in its last two indices.

Einstein said: "The world line of a free-falling test body is independent of its composition or structure." This means all free-falling bodies that start at a specified place, at a specified time, with a specified initial velocity, will follow the same path (world line) through spacetime — for any set of specified initial conditions, the geodesic is unique.

A test body must be electrically neutral to eliminate electromagnetic forces. It must also be small enough that it does not appreciably alter its environment or have its structure distorted by non-uniform force fields.

§A5.4 Derivation of Connection Coefficients

Now, let's derive the equation for $\Gamma_{\alpha\beta\gamma}$ in terms of the metric components in a coordinate basis. We take covariant derivatives of the metric, which is always zero, with three different sets of indices (our brilliant choice of indices results from knowing the answer).

$$g_{\alpha\beta;\gamma} = 0 = g_{\alpha\beta,\gamma} - g_{\mu\beta}\,\Gamma^{\mu}{}_{\alpha\gamma} - g_{\alpha\mu}\,\Gamma^{\mu}{}_{\beta\gamma}$$
$$g_{\alpha\gamma;\beta} = 0 = g_{\alpha\gamma,\beta} - g_{\mu\gamma}\,\Gamma^{\mu}{}_{\alpha\beta} - g_{\alpha\mu}\,\Gamma^{\mu}{}_{\gamma\beta}$$
$$g_{\beta\gamma;\alpha} = 0 = g_{\beta\gamma,\alpha} - g_{\mu\gamma}\,\Gamma^{\mu}{}_{\beta\alpha} - g_{\beta\mu}\,\Gamma^{\mu}{}_{\gamma\alpha}$$

Rearranging the prior 3 equations:

$$+\,\Gamma_{\beta\alpha\gamma} + \Gamma_{\alpha\beta\gamma} = g_{\alpha\beta,\gamma}$$
$$+\,\Gamma_{\gamma\alpha\beta} + \Gamma_{\alpha\gamma\beta} = g_{\alpha\gamma,\beta}$$
$$+\,\Gamma_{\gamma\beta\alpha} + \Gamma_{\beta\gamma\alpha} = g_{\beta\gamma,\alpha}$$

Let's now add first two equations and subtract the last.

$$\Gamma_{\beta\alpha\gamma} + \Gamma_{\alpha\beta\gamma} + \Gamma_{\gamma\alpha\beta} + \Gamma_{\alpha\gamma\beta} - \Gamma_{\gamma\beta\alpha} - \Gamma_{\beta\gamma\alpha}$$
$$= g_{\alpha\beta,\gamma} + g_{\alpha\gamma,\beta} - g_{\beta\gamma,\alpha}$$

We proved in the last section that the connection coefficient is symmetric in its last two indices. On the left side above, therefore, the first and sixth terms cancel one another, as do the third and fifth. Also, the second and fourth terms are equal. The result is:

$$2\,\Gamma_{\alpha\beta\gamma} = g_{\alpha\beta,\gamma} + g_{\alpha\gamma,\beta} - g_{\beta\gamma,\alpha}$$
$$\Gamma_{\alpha\beta\gamma} = (\,g_{\alpha\beta,\gamma} + g_{\alpha\gamma,\beta} - g_{\beta\gamma,\alpha}\,)\,/\,2$$

This proves the connection coefficient equation quoted in Section §24.3.

§A5.5 Derivation of Geodesic Equation

Employing variational calculus, we derive here the geodesic equation, which defines the paths that minimize the interval ds^2 ("shortest distance"), and maximize the proper time τ ($c^2\,d\tau^2 = -\,ds^2$), between any two points A and B.

Let's characterize a geodesic path with an *affine* parameter u that measures the distance traveled along the path. Each component of position x^α is expressed as a function of u: $x^\alpha = x^\alpha(u)$.

This derivation demonstrates the power of *calculus of variations*: we will vary the path from A to B, and identify which path maximizes proper time.

With c = 1:

$$d\tau^2 = -\ ds^2$$

$$d\tau^2 = -\ g_{\alpha\beta}\ dx^\alpha\ dx^\beta$$

$$d\tau/du = \sqrt{(-\ g_{\alpha\beta}\ dx^\alpha/du\ \ dx^\beta/du\ )}$$

$$\tau = \int (d\tau/du)\ du = \int du \sqrt{(-\ g_{\alpha\beta}\ dx^\alpha/du\ \ dx^\beta/du\ )}$$

Let the path variations have the form:

$$x^\alpha(u) \rightarrow x^\alpha(u) + \delta x^\alpha(u)$$

Variation δ is a form of differentiation; it obeys the chain rule, and commutes with other derivatives:

$$\delta(dx^\alpha/du) = d(\delta x^\alpha)/du$$

$$\delta g_{\alpha\beta} = (\partial g_{\alpha\beta}/\partial x^\mu)\ \delta x^\mu = g_{\alpha\beta,\mu}\ \delta x^\mu$$

To find the path that minimizes proper time from A to B, begin by setting the variation of the above integral to zero. For brevity, define $v^\alpha = dx^\alpha/du$.

$$0 = \delta\tau = \delta \int du \sqrt{(-\ g_{\alpha\beta}\ v^\alpha\ v^\beta\ )}$$

$$0 = \int du\ \ \delta \{\ (-\ g_{\alpha\beta}\ v^\alpha\ v^\beta\ )^{+1/2}\ \}$$

$$0 = \int du\ \ (1\ /\ 2)\ (-\ g_{\alpha\beta}\ v^\alpha\ v^\beta\ )^{-1/2}\ \{\ \delta\ (-\ g_{\alpha\beta}\ v^\alpha\ v^\beta\ )\ \}$$

Now substitute $du/d\tau = 1\ /\ \sqrt{(-\ g_{\alpha\beta}\ v^\alpha\ v^\beta\ )}$, and expand the variational derivative in { }'s.

$$0 = \int du\ \ (du/d\tau)\ \{\ -\ (\delta g_{\alpha\beta})\ v^\alpha\ v^\beta$$

$$-\ g_{\alpha\beta}\ (\delta v^\alpha)\ v^\beta\ -\ g_{\alpha\beta}\ v^\alpha\ (\delta v^\beta\ )\ \}$$

The second and third terms in { }'s are identical on the exchange of α and β, allowing us to combine these into one term.

$$0 = \int du\ (du/d\tau)\ \{\ g_{\alpha\beta,\mu}\ \delta x^\mu\ v^\alpha\ v^\beta + 2\ g_{\alpha\beta}\ v^\beta\ (\delta v^\alpha)\ \}$$

$$0 = \int du\ (du/d\tau)\ \{\ g_{\alpha\beta,\mu}\ \delta x^\mu\ v^\alpha\ v^\beta + 2\ g_{\alpha\beta}\ v^\beta\ d(\delta x^\alpha)/du\ \}$$

Next, we integrate the last term by parts. Recall how this works:

$$\int_A^B z\ dw = \int_A^B d(z\ w)\ -\ \int_A^B w\ dz$$

$$\int_A^B z\ dw = (\ z\ w\ |_A^B\ )\ -\ \int_A^B w\ dz$$

Here, $(z\ w\ |_A^B)$ the value of (z w) at event B minus the value of (z w) at event A.

To employ integration by parts, let:

$$z = 2\ (du/d\tau)\ g_{\alpha\beta}\ v^\beta$$

$$dw = d(\delta x^\alpha)/du$$

Integrating the last term of our main equation by parts yields:

$$\int_A^B du\ [\ 2\ (du/d\tau)\ g_{\alpha\beta}\ v^\beta\ ]\ d(\delta x^\alpha)/du$$

$$= [\ 2\ (du/d\tau)\ g_{\alpha\beta}\ v^\beta\ \delta x^\alpha\ ]\ |_A^B$$

$$-\ \int du\ \ \delta x^\alpha\ \ d[\ 2\ (du/d\tau)\ g_{\alpha\beta}\ v^\beta\ ]/du$$

Let's focus on the middle line. Recall that all paths being considered here must start at event A and end at event B — the path variation δx^α can have any value *between* the limits, but must be zero *at* both limits. Hence, middle line vanishes.

Putting the lower line of the prior equation into the variational integral yields:

$$0 = \int du \{ (du/d\tau)\ g_{\alpha\beta,\mu}\ \delta x^\mu\ v^\alpha\ v^\beta$$
$$-\ \delta x^\alpha\ d[\ 2\ (du/d\tau)\ (g_{\alpha\beta}\ v^\beta\ )\ ]/du \}$$

We next replace the summation index α with μ in the last term. We can choose any label we wish for a repeated index. By choosing μ, we match the variation δx^μ in both terms. We can also cancel the du's in the last term.

$$0 = \int du \{ (du/d\tau)\ g_{\alpha\beta,\mu}\ \delta x^\mu\ v^\alpha\ v^\beta$$
$$-\ 2\ \delta x^\mu\ (d/d\tau)\ (g_{\mu\beta}\ v^\beta\ ) \}$$
$$0 = \int du\ \delta x^\mu\ \{ (du/d\tau)\ g_{\alpha\beta,\mu}\ v^\alpha\ v^\beta\ -\ 2\ (d/d\tau)\ (g_{\mu\beta}\ v^\beta\ ) \}$$

Since δx^μ is arbitrary, the integral must be zero for any δx^μ, which can be true only if the expression in { }'s is zero everywhere.

$$0 = (du/d\tau)\ g_{\alpha\beta,\mu}\ v^\alpha\ v^\beta\ -\ 2\ (d/d\tau)\ (g_{\mu\beta}\ v^\beta\ )$$

Next, let's expand the proper time derivative in the last term.

$$0 = (du/d\tau)\ g_{\alpha\beta,\mu}\ v^\alpha\ v^\beta - 2\ [\ g_{\mu\beta,\alpha}\ v^\beta\ dx^\alpha/d\tau\ +\ g_{\mu\beta}\ dv^\beta/d\tau]$$

Since we have eliminated the integral, let's multiply everything by $du/d\tau$ to switch our time basis from the affine parameter u to proper time τ, and define 4-velocities as $w^\alpha = dx^\alpha/d\tau$.

$$0 = (du/d\tau)^2\ g_{\alpha\beta,\mu}\ dx^\alpha/du\ \ dx^\beta/du$$
$$-\ 2\ (du/d\tau)\ [\ g_{\mu\beta,\alpha}\ dx^\beta/du\ \ dx^\alpha/d\tau\ +\ g_{\mu\beta}\ d^2x^\beta/du\, d\tau]$$

$$0 = g_{\alpha\beta,\mu}\ w^\alpha\ w^\beta\ -\ 2\ [g_{\mu\beta,\alpha}\ w^\beta\ w^\alpha\ +\ g_{\mu\beta}\ dw^\beta/d\tau\ ]$$

Because $w^\alpha\ w^\beta$ is symmetric on indices α and β, we can split the middle term into two terms, and swap summation indices in the second of those terms.

$$0 = g_{\alpha\beta,\mu}\, w^\alpha\, w^\beta\ -\ g_{\mu\beta,\alpha}\, w^\alpha\, w^\beta\ -\ g_{\mu\alpha,\beta}\, w^\alpha\, w^\beta\ -\ 2\ g_{\mu\beta}\ dw^\beta/d\tau$$
$$0 = [-\ g_{\mu\beta,\alpha}\ -\ g_{\mu\alpha,\beta}\ +\ g_{\alpha\beta,\mu}\,]\ w^\alpha\ w^\beta\ /\ 2\ -\ g_{\mu\beta}\ dw^\beta/d\tau$$

Now raise index μ by multiplying everything by $-g^{\varepsilon\mu}$, contracting on μ, and recall $g^{\varepsilon\mu}\ g_{\mu\beta} = \delta^\varepsilon{}_\beta = 1$ if $\varepsilon = \beta$ and $= 0$ otherwise.

$$0 = g^{\varepsilon\mu}\ [\ g_{\mu\beta,\alpha}\ +\ g_{\mu\alpha,\beta}\ -\ g_{\alpha\beta\mu}\,]\ w^\alpha\ w^\beta\ /\ 2\ +\ g^{\varepsilon\mu}\ g_{\mu\beta}\ dw^\beta/d\tau$$
$$0 = \{\ g^{\varepsilon\mu}\ [\ g_{\mu\beta,\alpha} + g_{\mu\alpha,\beta} - g_{\alpha\beta,\mu}\,]\ /\ 2\ \}\ w^\alpha\ w^\beta\ +\ dw^\varepsilon/d\tau$$

The quantity in { }'s is *connection coefficient* $\Gamma^\varepsilon{}_{\alpha\beta}$, as defined in Section §22.3. Our result is the geodesic equation:

$$0 = \Gamma^\varepsilon{}_{\alpha\beta}\ (dx^\alpha/d\tau)\ (dx^\beta/d\tau)\ +\ d^2x^\varepsilon/d\tau^2$$

This equation provides the acceleration of a *free* body, with Γ describing the twists and turns of the curved spacetime through which the body passes. In this context, *free* means absent all forces. (Recall Einstein said gravity is not a "force".)

Sometimes, it is more convenient to use a different basis for measuring velocities and accelerations. For example, multiplying everything by $(d\tau^2/dt^2)$ changes the time base from τ to t, yielding:

$$0 = \Gamma^\varepsilon{}_{\alpha\beta}\ (dx^\alpha/dt)\ (dx^\beta/dt)\ +\ d^2x^\varepsilon/dt^2$$

§A5.6 Reimann Tensor & Geodesic Deviation

Gravity produces four effects: acceleration, time dilation, tidal forces, and gravity waves. We explore the first three in Chapter 14, and gravity waves in Chapter 34. As we discovered in Chapter 13, Einstein's equivalence principle states accelerations due to uniform gravitational fields are indistinguishable from those due to mechanical accelerations. In Chapter 26, we found that in uniform gravity, spacetime is flat, as it is with zero gravity.

Only non-uniform fields provide definitive signatures of gravity's presence: gravitational time dilation, tidal forces, and gravity waves. Gravity waves are incredibly hard to detect. Gravitational time dilation can be detected by comparing ultra-high precision clocks in different locations. Tidal forces offer another, perhaps easier, opportunity to detect a unique signature of gravity: diverging geodesics.

Figure A5-3 shows two neighboring geodesics. We imagine a family of geodesics, one for each value of a variable ξ. We call the lower curve the fiducial geodesic with $\xi = n$, and the upper curve the test geodesic with $\xi = n + \Delta n$ (n is not an integer). Along each geodesic, we define an affine parameter ζ.

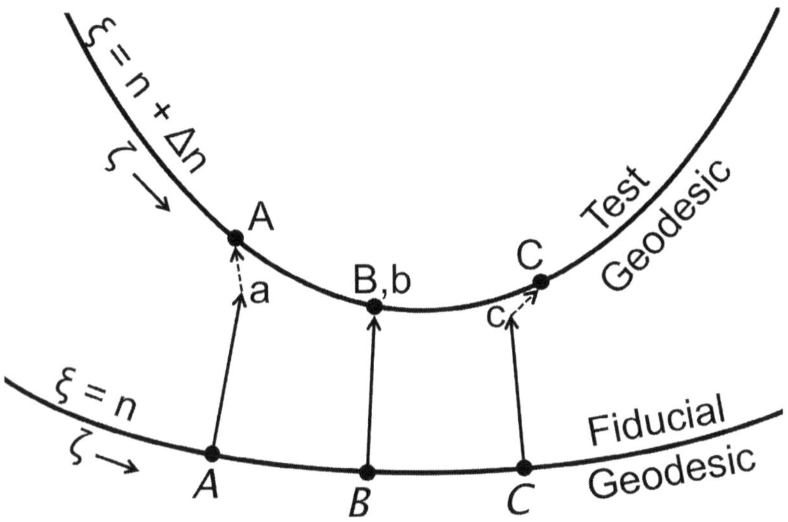

Figure A5-3 Tidal Effect on Two Geodesics

We define these events:

Along Fiducial Geodesic	Along Test Geodesic	In Between
A at $\zeta = \lambda - \Delta\lambda$	A at $\zeta = \lambda - \Delta\lambda$	a at $A + (\partial/\partial\xi)_A\ \Delta n$
B at $\zeta = \lambda$	B at $\zeta = \lambda$	b at $B + (\partial/\partial\xi)_B\ \Delta n$
C at $\zeta = \lambda + \Delta\lambda$	C at $\zeta = \lambda + \Delta\lambda$	c at $C + (\partial/\partial\xi)_C\ \Delta n$

Here, $(\partial/\partial\xi)_Q$ is a unit vector that is normal to the fiducial geodesic at event Q. We have chosen λ and Δn such that event B and event b are coincident.

If spacetime were flat, A and a would be also be coincident, as would C and c. It is the deviations — the two dotted arrows in Figure A5-3 — that reveal the tidal effect of spacetime curvature.

These deviations are clearly proportional to both Δn and $\Delta\lambda$. Let's factor out those Δ's to isolate a scale-independent expression for the spacetime curvature effect.

We define 4-vectors δ_A and δ_C:

$$\delta_A = (A - a) \ / \ \Delta n \ (-\Delta\lambda)$$
$$\delta_C = (C - c) \ / \ \Delta n \ (+\Delta\lambda)$$

The *relative acceleration* 4-vector Ψ of the test geodesic relative to the fiducial geodesic equals the the rate of change — the covariant derivative with respect to λ — of the deviations at C and A, specifically:

Relative acceleration: $$\Psi = (\delta_C - \delta_A) \,/\, 2\,\Delta\lambda$$

The deviations δ_A and δ_C are each first order derivatives in λ, hence the relative acceleration Ψ is the second order covariant derivative with respect to λ — the second order derivative of $\partial/\partial\xi$, the unit vector normal to the fiducial geodesic. We therefore define two vector fields u and n, and write:

$$u = \partial/\partial\lambda$$

$$n = \partial/\partial\xi$$

Relative acceleration: $$\Psi = n_{;\,u\,u}$$

Let's evaluate this expression. The first step is recognizing that, like all partial derivatives, u and n commute:

$$u\ n = \partial/\partial\lambda\ \partial/\partial\xi = \partial/\partial\xi\ \partial/d\lambda = n\ u$$

Hence:

$$(n_{;\,u})_{;\,u} = (u_{;\,n})_{;\,u}$$

We discovered in Section §A5.3 that geodesics parallel transport their tangent vectors, which we write:

$$u_{;\,u} = 0$$

Now take the covariant derivative of this equation with respect to n.

$$(u_{;\,u})_{;\,n} = 0$$

Next subtract this from the prior $n_{;u\,u}$ equation, yielding:

$$(n_{;\,u})_{;\,u} = (u_{;\,n})_{;\,u} - (u_{;\,u})_{;\,n}$$

$$(n_{;\,u})_{;\,u} + (u_{;\,u})_{;\,n} - (u_{;\,n})_{;\,u} = 0$$

The first term in this equation is the relative acceleration, while the last two terms describe the tidal forces of non-uniform gravity — how geodesics are distorted by non-uniform gravity.

We capture the essence of the gravitational tidal effect by defining the Riemann tensor:

$$R^{\alpha}{}_{\beta\gamma\varepsilon} = e^{\alpha} \bullet [\,(e_{\beta;\,\varepsilon})_{;\,\gamma} - (e_{\beta;\,\gamma})_{;\,\varepsilon}\,]$$

Let's expand the first covariant derivative.

$$(e_{\beta;\,\varepsilon})_{;\,\gamma} = (e_{\mu}\,\Gamma^{\mu}{}_{\beta\delta})_{;\,\gamma}$$

$$(e_{\beta;\,\varepsilon})_{;\,\gamma} = e_{\mu}\,(\Gamma^{\mu}{}_{\beta\varepsilon})_{;\,\gamma} + e_{\mu;\gamma}\,\Gamma^{\mu}{}_{\beta\varepsilon}$$

In the first term, since $\Gamma^{\mu}{}_{\beta\delta}$ is a scalar, the covariant derivative reduces to the partial derivative. Now let's expand the second term of the last equation, using $e_{\mu;\gamma} = e_{\sigma}\,\Gamma^{\sigma}{}_{\mu\gamma}$.

$$(e_{\beta;\,\varepsilon})_{;\,\gamma} = e_{\mu}\,\Gamma^{\mu}{}_{\beta\varepsilon,\,\gamma} + \Gamma^{\mu}{}_{\beta\varepsilon}\,e_{\sigma}\,\Gamma^{\sigma}{}_{\mu\gamma}$$

Now multiply by e^{α}: $$e^{\alpha} \bullet (e_{\beta;\,\varepsilon})_{;\,\gamma} = e^{\alpha} \bullet e_{\mu}\,\Gamma^{\mu}{}_{\beta\varepsilon,\,\gamma} + e^{\alpha} \bullet e_{\sigma}\,\Gamma^{\sigma}{}_{\mu\gamma}\ \Gamma^{\mu}{}_{\beta\varepsilon}$$

Use $e^{\alpha} \bullet e_{\nu} = \delta^{\alpha}{}_{\nu}$: $$e^{\alpha} \bullet (e_{\beta;\,\varepsilon})_{;\,\gamma} = \Gamma^{\alpha}{}_{\beta\varepsilon,\,\gamma} + \Gamma^{\alpha}{}_{\mu\gamma}\ \Gamma^{\mu}{}_{\beta\varepsilon}$$

Similarly for the second covariant derivative in the equation for $R^{\alpha}{}_{\beta\gamma\varepsilon}$:

$$e^{\alpha} \bullet (e_{\beta;\,\gamma})_{;\,\varepsilon} = \Gamma^{\alpha}{}_{\beta\gamma,\varepsilon} + \Gamma^{\alpha}{}_{\mu\varepsilon}\ \Gamma^{\mu}{}_{\beta\gamma}$$

Subtracting the prior two equations yields the component form of the Reimann tensor in a coordinate basis:

$$R^{\alpha}{}_{\beta\gamma\varepsilon} = \Gamma^{\alpha}{}_{\beta\varepsilon,\gamma} - \Gamma^{\alpha}{}_{\beta\gamma,\varepsilon} + \Gamma^{\alpha}{}_{\mu\gamma}\,\Gamma^{\mu}{}_{\beta\varepsilon} - \Gamma^{\alpha}{}_{\mu\varepsilon}\,\Gamma^{\mu}{}_{\beta\gamma}$$

§A5.7 Basis Vector Inverse Operators

In this Appendix and in Appendix 6, I have simplified the notation somewhat by denoting the inverse operator of basis vector e_{α} as e^{α}. It is more mathematically precise, and more common in the literature, to define this operator as a *1-form*, written ω^{α}, defined by:

$$\omega^{\alpha}\ e_{\beta} = \delta^{\alpha}{}_{\beta}$$

I have never seen any advantage to this additional complication to the tensor calculus formalism, but evidently many others do. I include this note so you will not be overly surprised to read about *forms* in other texts.

Appendix 6
Coordinate & Orthonormal Curvature Tensors

This Appendix provides calculations of curvature tensors in both coordinate bases and orthonormal, non-coordinate bases.

To facilitate the learning experience, every intermediate step is shown in full detail. Hopefully, you will not find this too tedious. After seeing the calculations for a few components, you might try to derive the rest yourself. Practice and precision are essential to mastering tensor calculus.

In a **coordinate** basis, the metric equation (the invariant interval) is:

$$ds^2 = g_{\alpha\beta}\ dx^\alpha\ dx^\beta$$

We can in general diagonalize the metric $g_{\alpha\beta}$, which may require redefining our coordinate system. This greatly simplifies subsequent calculations, reducing the above equation from 16 to 4 terms, becoming this sum over α:

$$ds^2 = g_{\alpha\alpha}\ dx^\alpha\ dx^\alpha$$

One can always multiply any basis vector by a constant and still have a coordinate basis. But, when one multiplies a basis vector by a function of one or more spacetime coordinates, the result is a non-coordinate basis.

The most effective approach is an *orthonormal* basis, where the basis vectors absorb the curvature characteristics, as follows:

For each α no sum on α: $\quad dX^\alpha = dx^\alpha\ \sqrt{(\,g_{\alpha\alpha}\,)}$

Normal sum on α: $\quad ds^2 = dX_\alpha\ dX^\alpha$

This greatly simplifies the metric, reducing it to that of flat spacetime, but it complicates the coordinates. In most cases, the trade-offs favor an orthonormal basis. Whichever basis one chooses, the principal final result — the Einstein tensor $G^\alpha{}_\alpha$ — will be the same.

I will begin with the most general case, and later show the reduction in complexity that results from special symmetries and metric forms.

I find, perhaps surprisingly, that factoring complex expressions makes these calculation more difficult — it is better to write AB + AC rather than A(B + C). When A, B and C become long and complicated, it is easier to manage them, and easier to avoid errors, if each term is fully expanded.

§A6.1 $\Gamma_{\alpha\beta\delta}$ in Coordinate Basis, Any Geometry or Metric

In a coordinate basis, the axes behave in the normal Cartesian manner that we know so well in classical 3-D, zero-curvature space. The general equation for connection coefficients in a coordinate basis, for any metric in any geometry, is:

$$\Gamma_{\alpha\beta\delta} = (g_{\alpha\beta,\delta} + g_{\alpha\delta,\beta} - g_{\beta\delta,\alpha})\,/\,2$$

In a coordinate basis, Γ is always symmetric in its last two indices, reducing the number of independent components from $4^3 = 64$ to $4\times(4+6) = 40$.

For specificity, let's label the coordinates: t, r, θ, and ϕ, since most problems are best analyzed in spherical coordinates. The 40 independent individual components of Γ are then:

$$\Gamma_{ttt} = (g_{tt,t} + g_{tt,t} - g_{tt,t}) / 2$$
$$\Gamma_{ttr} = (g_{tt,r} + g_{tr,t} - g_{tr,t}) / 2$$
$$\Gamma_{tt\theta} = (g_{tt,\theta} + g_{t\theta,t} - g_{t\theta,t}) / 2$$
$$\Gamma_{tt\phi} = (g_{tt,\phi} + g_{t\phi,t} - g_{t\phi,t}) / 2$$
$$\Gamma_{trr} = (g_{tr,r} + g_{tr,r} - g_{rr,t}) / 2$$
$$\Gamma_{tr\theta} = (g_{tr,\theta} + g_{t\theta,r} - g_{r\theta,t}) / 2$$
$$\Gamma_{tr\phi} = (g_{tr,\phi} + g_{t\phi,r} - g_{r\phi,t}) / 2$$
$$\Gamma_{t\theta\theta} = (g_{t\theta,\theta} + g_{t\theta,\theta} - g_{\theta\theta,t}) / 2$$
$$\Gamma_{t\theta\phi} = (g_{t\theta,\phi} + g_{t\phi,\theta} - g_{\theta\phi,t}) / 2$$
$$\Gamma_{t\phi\phi} = (g_{t\phi,\phi} + g_{t\phi,\phi} - g_{\phi\phi,t}) / 2$$

$$\Gamma_{rtt} = (g_{rt,t} + g_{rt,t} - g_{tt,r}) / 2$$
$$\Gamma_{rtr} = (g_{rt,r} + g_{rr,t} - g_{tr,r}) / 2$$
$$\Gamma_{rt\theta} = (g_{rt,\theta} + g_{r\theta,t} - g_{t\theta,r}) / 2$$
$$\Gamma_{rt\phi} = (g_{rt,\phi} + g_{r\phi,t} - g_{t\phi,r}) / 2$$
$$\Gamma_{rrr} = (g_{rr,r} + g_{rr,r} - g_{rr,r}) / 2$$
$$\Gamma_{rr\theta} = (g_{rr,\theta} + g_{r\theta,r} - g_{r\theta,r}) / 2$$
$$\Gamma_{rr\phi} = (g_{rr,\phi} + g_{r\phi,r} - g_{r\phi,r}) / 2$$
$$\Gamma_{r\theta\theta} = (g_{r\theta,\theta} + g_{r\theta,\theta} - g_{\theta\theta,r}) / 2$$
$$\Gamma_{r\theta\phi} = (g_{r\theta,\phi} + g_{r\phi,\theta} - g_{\theta\phi,r}) / 2$$
$$\Gamma_{r\phi\phi} = (g_{r\phi,\phi} + g_{r\phi,\phi} - g_{\phi\phi,r}) / 2$$

$$\Gamma_{\theta tt} = (g_{\theta t,t} + g_{\theta t,t} - g_{tt,\theta}) / 2$$
$$\Gamma_{\theta tr} = (g_{\theta t,r} + g_{\theta r,t} - g_{tr,\theta}) / 2$$
$$\Gamma_{\theta t\theta} = (g_{\theta t,\theta} + g_{\theta\theta,t} - g_{t\theta,\theta}) / 2$$
$$\Gamma_{\theta t\phi} = (g_{\theta t,\phi} + g_{\theta\phi,t} - g_{t\phi,\theta}) / 2$$
$$\Gamma_{\theta rr} = (g_{\theta r,r} + g_{\theta r,r} - g_{rr,\theta}) / 2$$
$$\Gamma_{\theta r\theta} = (g_{\theta r,\theta} + g_{\theta\theta,r} - g_{r\theta,\theta}) / 2$$
$$\Gamma_{\theta r\phi} = (g_{\theta r,\phi} + g_{\theta\phi,r} - g_{r\phi,\theta}) / 2$$
$$\Gamma_{\theta\theta\theta} = (g_{\theta\theta,\theta} + g_{\theta\theta,\theta} - g_{\theta\theta,\theta}) / 2$$
$$\Gamma_{\theta\theta\phi} = (g_{\theta\theta,\phi} + g_{\theta\phi,\theta} - g_{\theta\phi,\theta}) / 2$$
$$\Gamma_{\theta\phi\phi} = (g_{\theta\phi,\phi} + g_{\theta\phi,\phi} - g_{\phi\phi,\theta}) / 2$$

$$\Gamma_{\phi tt} = (g_{\phi t,t} + g_{\phi t,t} - g_{tt,\phi}) / 2$$
$$\Gamma_{\phi tr} = (g_{\phi t,r} + g_{\phi r,t} - g_{tr,\phi}) / 2$$
$$\Gamma_{\phi t\theta} = (g_{\phi t,\theta} + g_{\phi\theta,t} - g_{t\theta,\phi}) / 2$$
$$\Gamma_{\phi t\phi} = (g_{\phi t,\phi} + g_{\phi\phi,t} - g_{t\phi,\phi}) / 2$$
$$\Gamma_{\phi rr} = (g_{\phi r,r} + g_{\phi r,r} - g_{rr,\phi}) / 2$$
$$\Gamma_{\phi r\theta} = (g_{\phi r,\theta} + g_{\phi\theta,r} - g_{r\theta,\phi}) / 2$$
$$\Gamma_{\phi r\phi} = (g_{\phi r,\phi} + g_{\phi\phi,r} - g_{r\phi,\phi}) / 2$$
$$\Gamma_{\phi\theta\theta} = (g_{\phi\theta,\theta} + g_{\phi\theta,\theta} - g_{\theta\theta,\phi}) / 2$$
$$\Gamma_{\phi\theta\phi} = (g_{\phi\theta,\phi} + g_{\phi\phi,\theta} - g_{\theta\phi,\phi}) / 2$$
$$\Gamma_{\phi\phi\phi} = (g_{\phi\phi,\phi} + g_{\phi\phi,\phi} - g_{\phi\phi,\phi}) / 2$$

§A6.2 $R^{\alpha}{}_{\beta\delta e}$ in Coordinate Basis, Any Geometry or Metric

The general equation for the Reimann tensor in a coordinate basis, for any metric in any geometry, is:

$$R^{\alpha}{}_{\beta\delta e} = \Gamma^{\alpha}{}_{\beta e,\delta} - \Gamma^{\alpha}{}_{\beta\delta,e} + \Gamma^{\alpha}{}_{\mu\delta}\,\Gamma^{\mu}{}_{\beta e} - \Gamma^{\alpha}{}_{\mu e}\,\Gamma^{\mu}{}_{\beta\delta}$$

In any basis, R is always antisymmetric in its first index pair, antisymmetric in its last index pair, and symmetric in the exchange of the first pair with the last pair. All this reduces the number of independent components that we need to calculate from $4^4 = 256$ to 21.

Again, let's label the coordinates: t, r, θ, and ϕ. The 21 independent components of R are then:

$$R^{t}{}_{rtr} = \Gamma^{t}{}_{rr,t} - \Gamma^{t}{}_{rt,r} + \Gamma^{t}{}_{\mu t}\,\Gamma^{\mu}{}_{rr} - \Gamma^{t}{}_{\sigma r}\,\Gamma^{\sigma}{}_{rt}$$
$$R^{t}{}_{rt\theta} = \Gamma^{t}{}_{r\theta,t} - \Gamma^{t}{}_{rt,\theta} + \Gamma^{t}{}_{\mu t}\,\Gamma^{\mu}{}_{r\theta} - \Gamma^{t}{}_{\sigma\theta}\,\Gamma^{\sigma}{}_{rt}$$
$$R^{t}{}_{rt\phi} = \Gamma^{t}{}_{r\phi,t} - \Gamma^{t}{}_{rt,\phi} + \Gamma^{t}{}_{\mu t}\,\Gamma^{\mu}{}_{r\phi} - \Gamma^{t}{}_{\sigma\phi}\,\Gamma^{\sigma}{}_{rt}$$
$$R^{t}{}_{rr\theta} = \Gamma^{t}{}_{r\theta,r} - \Gamma^{t}{}_{rr,\theta} + \Gamma^{t}{}_{\mu r}\,\Gamma^{\mu}{}_{r\theta} - \Gamma^{t}{}_{\sigma\theta}\,\Gamma^{\sigma}{}_{rr}$$
$$R^{t}{}_{rr\phi} = \Gamma^{t}{}_{r\phi,r} - \Gamma^{t}{}_{rr,\phi} + \Gamma^{t}{}_{\mu r}\,\Gamma^{\mu}{}_{r\phi} - \Gamma^{t}{}_{\sigma\phi}\,\Gamma^{\sigma}{}_{rr}$$
$$R^{t}{}_{r\theta\phi} = \Gamma^{t}{}_{r\phi,\theta} - \Gamma^{t}{}_{r\theta,\phi} + \Gamma^{t}{}_{\mu\theta}\,\Gamma^{\mu}{}_{r\phi} - \Gamma^{t}{}_{\sigma\phi}\,\Gamma^{\sigma}{}_{r\theta}$$

$$R^{t}{}_{\theta t\theta} = \Gamma^{t}{}_{\theta\theta,t} - \Gamma^{t}{}_{\theta t,\theta} + \Gamma^{t}{}_{\mu t}\,\Gamma^{\mu}{}_{\theta\theta} - \Gamma^{t}{}_{\sigma\theta}\,\Gamma^{\sigma}{}_{\theta t}$$
$$R^{t}{}_{\theta t\phi} = \Gamma^{t}{}_{\theta\phi,t} - \Gamma^{t}{}_{\theta t,\phi} + \Gamma^{t}{}_{\mu t}\,\Gamma^{\mu}{}_{\theta\phi} - \Gamma^{t}{}_{\sigma\phi}\,\Gamma^{\sigma}{}_{\theta t}$$
$$R^{t}{}_{\theta r\theta} = \Gamma^{t}{}_{\theta\theta,r} - \Gamma^{t}{}_{\theta r,\theta} + \Gamma^{t}{}_{\mu r}\,\Gamma^{\mu}{}_{\theta\theta} - \Gamma^{t}{}_{\sigma\theta}\,\Gamma^{\sigma}{}_{\theta r}$$
$$R^{t}{}_{\theta r\phi} = \Gamma^{t}{}_{\theta\phi,r} - \Gamma^{t}{}_{\theta r,\phi} + \Gamma^{t}{}_{\mu r}\,\Gamma^{\mu}{}_{\theta\phi} - \Gamma^{t}{}_{\sigma\phi}\,\Gamma^{\sigma}{}_{\theta r}$$
$$R^{t}{}_{\theta\theta\phi} = \Gamma^{t}{}_{\theta\phi,\theta} - \Gamma^{t}{}_{\theta\theta,\phi} + \Gamma^{t}{}_{\mu\theta}\,\Gamma^{\mu}{}_{\theta\phi} - \Gamma^{t}{}_{\sigma\phi}\,\Gamma^{\sigma}{}_{\theta\theta}$$

$$R^t{}_{\phi t\phi} = \Gamma^t{}_{\phi\phi,t} - \Gamma^t{}_{\phi t,\phi} + \Gamma^t{}_{\mu t}\,\Gamma^\mu{}_{\phi\phi} - \Gamma^t{}_{\sigma\phi}\,\Gamma^\sigma{}_{\phi t}$$
$$R^t{}_{\phi r\theta} = \Gamma^t{}_{\phi\theta,r} - \Gamma^t{}_{\phi r,\theta} + \Gamma^t{}_{\mu r}\,\Gamma^\mu{}_{\phi\theta} - \Gamma^t{}_{\sigma\theta}\,\Gamma^\sigma{}_{\phi r}$$
$$R^t{}_{\phi r\phi} = \Gamma^t{}_{\phi\phi,r} - \Gamma^t{}_{\phi r,\phi} + \Gamma^t{}_{\mu r}\,\Gamma^\mu{}_{\phi\phi} - \Gamma^t{}_{\sigma\phi}\,\Gamma^{\mu\sigma}{}_{\phi r}$$
$$R^t{}_{\phi\theta\phi} = \Gamma^t{}_{\phi\phi,\theta} - \Gamma^t{}_{\phi\theta,\phi} + \Gamma^t{}_{\mu\theta}\,\Gamma^\mu{}_{\phi\phi} - \Gamma^t{}_{\sigma\phi}\,\Gamma^\sigma{}_{\phi\theta}$$

$$R^r{}_{\theta r\theta} = \Gamma^r{}_{\theta\theta,r} - \Gamma^r{}_{\theta r,\theta} + \Gamma^r{}_{\mu r}\,\Gamma^\mu{}_{\theta\theta} - \Gamma^r{}_{\sigma r}\,\Gamma^\sigma{}_{\theta\theta}$$
$$R^r{}_{\theta r\phi} = \Gamma^r{}_{\theta\phi,r} - \Gamma^r{}_{\theta r,\phi} + \Gamma^r{}_{\mu r}\,\Gamma^\mu{}_{\theta\phi} - \Gamma^r{}_{\sigma\phi}\,\Gamma^\sigma{}_{\theta r}$$
$$R^r{}_{\theta\theta\phi} = \Gamma^r{}_{\theta\phi,\theta} - \Gamma^r{}_{\theta\theta,\phi} + \Gamma^r{}_{\mu\theta}\,\Gamma^\mu{}_{\theta\phi} - \Gamma^r{}_{\sigma\phi}\,\Gamma^\sigma{}_{\theta\theta}$$

$$R^r{}_{\phi r\phi} = \Gamma^r{}_{\phi\phi,r} - \Gamma^r{}_{\phi r,\phi} + \Gamma^r{}_{\mu r}\,\Gamma^\mu{}_{\phi\phi} - \Gamma^r{}_{\sigma\phi}\,\Gamma^\sigma{}_{\phi r}$$
$$R^r{}_{\phi\theta\phi} = \Gamma^r{}_{\phi\phi,\theta} - \Gamma^r{}_{\phi\theta,\phi} + \Gamma^r{}_{\mu\theta}\,\Gamma^\mu{}_{\phi\phi} - \Gamma^r{}_{\sigma\phi}\,\Gamma^\sigma{}_{\phi\theta}$$

$$R^\theta{}_{\phi\theta\phi} = \Gamma^\theta{}_{\phi\phi,\theta} - \Gamma^\theta{}_{\phi\theta,\phi} + \Gamma^\theta{}_{\mu\theta}\,\Gamma^\mu{}_{\phi\phi} - \Gamma^\theta{}_{\sigma\phi}\,\Gamma^\sigma{}_{\phi\theta}$$

§A6.3 $\Gamma_{\alpha\beta\delta}$ in Coordinate Basis, in Spherical Symmetry

Considerable simplification is possible if the geometry and metric are spherically symmetric. Many cosomological situations are spherically symmetric, or at least approximately so.

In a spherically symmetric geometry, the metric is diagonal and has the form:

$$ds^2 = g_{tt}\,dt^2 + g_{rr}\,dr^2 + g_{\theta\theta}\,d\theta^2 + g_{\phi\phi}\,d\phi^2$$

Metric components with different indices, and their derivatives, are always zero. This eliminates 12 independent Γ components, leaving $4\times7 = 28$ Γ components.

In addition, spherical symmetry precludes curvature in the θ or ϕ directions, reducing those metric components to their flat spacetime values:

$$ds^2 = g_{tt}\,dt^2 + g_{rr}\,dr^2 + r^2\,d\theta^2 + r^2\sin^2\theta\,d\phi^2$$

No components can be a function of ϕ, and only $g_{\phi\phi}$ can be a function of θ. Spherical symmetry does not prevent r from being a function of time. The remaining 16 independent components of Γ are:

$$\Gamma_{ttt} = +\,g_{tt,t}\,/\,2$$
$$\Gamma_{ttr} = +\,g_{tt,r}\,/\,2$$
$$\Gamma_{trr} = -\,g_{rr,t}\,/\,2$$
$$\Gamma_{t\theta\theta} = -\,g_{\theta\theta,t}\,/\,2$$
$$\Gamma_{t\phi\phi} = -\,g_{\phi\phi,t}\,/\,2$$
$$\Gamma_{rtt} = -\,g_{tt,r}\,/\,2$$
$$\Gamma_{rtr} = +\,g_{rr,t}\,/\,2$$
$$\Gamma_{rrr} = +\,g_{rr,r}\,/\,2$$

$$\Gamma_{r\theta\theta} = -\ g_{\theta\theta,r}\ /\ 2$$
$$\Gamma_{r\phi\phi} = -\ g_{\phi\phi,r}\ /\ 2$$

$$\Gamma_{\theta t\theta} = +\ g_{\theta\theta,t}\ /\ 2$$
$$\Gamma_{\theta r\theta} = +\ g_{\theta\theta,r}\ /\ 2$$
$$\Gamma_{\theta\phi\phi} = -\ g_{\phi\phi,\theta}\ /\ 2$$

$$\Gamma_{\phi t\phi} = +\ g_{\phi\phi,t}\ /\ 2$$
$$\Gamma_{\phi r\phi} = +\ g_{\phi\phi,r}\ /\ 2$$
$$\Gamma_{\phi\theta\phi} = +\ g_{\phi\phi,\theta}\ /\ 2$$

§A6.4 $R^{\alpha}{}_{\beta\delta e}$ in Coordinate Basis, in Spherical Symmetry

In a spherically symmetric geometry, the metric is diagonal and has the form:

$$ds^2 = g_{tt}\ dt^2 + g_{rr}\ dr^2 + g_{\theta\theta}\ d\theta^2 + g_{\phi\phi}\ d\phi^2$$

Metric components with different indices, and their derivatives, are always zero. This simplifies, but does not reduce the number of R components to be calculated. These 21 components are:

Sum μ = t, r and σ = t, r: $R^t{}_{rtr} = \Gamma^t{}_{rr,t} - \Gamma^t{}_{rt,r} + \Gamma^t{}_{\mu t}\,\Gamma^\mu{}_{rr} - \Gamma^t{}_{\sigma r}\,\Gamma^\sigma{}_{rt}$

Sum μ = none; σ = none: $R^t{}_{rt\theta} = 0 - 0 + 0 - 0$

Sum μ = none; σ = none: $R^t{}_{rt\phi} = 0 - 0 + 0 - 0$

Sum μ = none; σ = none: $R^t{}_{rr\theta} = 0 - 0 + 0 - 0$

Sum μ = none; σ = none: $R^t{}_{rr\phi} = 0 - 0 + 0 - 0$

Sum μ = none; σ = none: $R^t{}_{r\theta\phi} = 0 - 0 + 0 - 0$

Sum μ = t, r; σ = θ: $R^t{}_{\theta t\theta} = \Gamma^t{}_{\theta\theta,t} - 0 + \Gamma^t{}_{\mu t}\,\Gamma^\mu{}_{\theta\theta} - \Gamma^t{}_{\sigma\theta}\,\Gamma^\sigma{}_{\theta t}$

Sum μ = none; σ = none: $R^t{}_{\theta t\phi} = 0 - 0 + 0 - 0$

Sum μ = t, r; σ = θ: $R^t{}_{\theta r\theta} = \Gamma^t{}_{\theta\theta,r} - 0 + \Gamma^t{}_{\mu r}\,\Gamma^\mu{}_{\theta\theta} - \Gamma^t{}_{\sigma\theta}\,\Gamma^\sigma{}_{\theta r}$

Sum μ = none; σ = none: $R^t{}_{\theta r\phi} = 0 - 0 + 0 - 0$

Sum μ = none; σ = none: $R^t{}_{\theta\theta\phi} = 0 - 0 + 0 - 0$

Sum μ = t, r; σ = ϕ: $R^t{}_{\phi t\phi} = \Gamma^t{}_{\phi\phi,t} - 0 + \Gamma^t{}_{\mu t}\,\Gamma^\mu{}_{\phi\phi} - \Gamma^t{}_{\sigma\phi}\,\Gamma^\sigma{}_{\phi t}$

Sum μ = none; σ = none: $R^t{}_{\phi r\theta} = 0 - 0 + 0 - 0$

Sum μ = t, r; σ = ϕ: $R^t{}_{\phi r\phi} = \Gamma^t{}_{\phi\phi,r} - 0 + \Gamma^t{}_{\mu r}\,\Gamma^\mu{}_{\phi\phi} - \Gamma^t{}_{\sigma\phi}\,\Gamma^\sigma{}_{\phi r}$

Sum μ = none; σ = none: $R^t{}_{\phi\theta\phi} = \Gamma^t{}_{\phi\phi,\theta} - 0 + 0 - 0$

Sum μ = t, r; σ = t, r: $R^r{}_{\theta r\theta} = \Gamma^r{}_{\theta\theta,r} - 0 + \Gamma^r{}_{\mu r}\Gamma^\mu{}_{\theta\theta} - \Gamma^r{}_{\sigma\theta}\Gamma^\sigma{}_{\theta r}$

Sum μ = none; σ = none: $R^r{}_{\theta r\phi} = 0 - 0 + 0 - 0$

Sum μ = none; σ = none: $R^r{}_{\theta\theta\phi} = 0 - 0 + 0 - 0$

Sum μ = t, r; σ = ϕ: $R^r{}_{\phi r\phi} = \Gamma^r{}_{\phi\phi,r} - 0 + \Gamma^r{}_{\mu r}\Gamma^\mu{}_{\phi\phi} - \Gamma^r{}_{\sigma\phi}\Gamma^\sigma{}_{\phi r}$

Sum μ = θ; σ = ϕ: $R^r{}_{\phi\theta\phi} = \Gamma^r{}_{\phi\phi,\theta} - 0 + \Gamma^r{}_{\mu\theta}\Gamma^\mu{}_{\phi\phi} - \Gamma^r{}_{\sigma\phi}\Gamma^\sigma{}_{\phi\theta}$

Sum μ = t, r; σ = ϕ: $R^\theta{}_{\phi\theta\phi} = \Gamma^\theta{}_{\phi\phi,\theta} - 0 + \Gamma^\theta{}_{\mu\theta}\Gamma^\mu{}_{\phi\phi} - \Gamma^\theta{}_{\sigma\phi}\Gamma^\sigma{}_{\phi\theta}$

The 10 potentially nonzero components of R are:

Sum μ = t, r and σ = t, r:

$$R^t{}_{rtr} = \Gamma^t{}_{rr,t} - \Gamma^t{}_{rt,r} + \Gamma^t{}_{\mu t}\Gamma^\mu{}_{rr} - \Gamma^t{}_{\sigma r}\Gamma^\sigma{}_{rt}$$

$$\begin{aligned} R^t{}_{rtr} = {} & (-g^{tt}g_{rr,t}),t/2 - (+g^{tt}g_{tt,r}),r/2 \\ & + (+g^{tt}g_{tt,t})(-g^{tt}g_{rr,t})/4 + (+g^{tt}g_{tt,r})(+g^{rr}g_{rr,r})/4 \\ & - (+g^{tt}g_{tt,r})(+g^{tt}g_{tt,r})/4 - (-g^{tt}g_{rr,t})(-g^{rr}g_{rr,t})/4 \end{aligned}$$

Sum μ = t, r; σ = θ:

$$R^t{}_{\theta t\theta} = \Gamma^t{}_{\theta\theta,t} - 0 + \Gamma^t{}_{\mu t}\Gamma^\mu{}_{\theta\theta} - \Gamma^t{}_{\sigma\theta}\Gamma^\sigma{}_{\theta t}$$

$$\begin{aligned} R^t{}_{\theta t\theta} = {} & (-g^{tt}g_{\theta\theta,t}),t/2 \\ & + (+g^{tt}g_{tt,t})(-g^{tt}g_{\theta\theta,t})/4 + (+g^{tt}g_{tt,r})(-g^{rr}g_{\theta\theta,r})/4 \\ & - (-g^{tt}g_{\theta\theta,t})(+g^{\theta\theta}g_{\theta\theta,t})/4 \end{aligned}$$

Sum μ = t, r; σ = θ:

$$R^t{}_{\theta r\theta} = \Gamma^t{}_{\theta\theta,r} - 0 + \Gamma^t{}_{\mu r}\Gamma^\mu{}_{\theta\theta} - \Gamma^t{}_{\sigma\theta}\Gamma^\sigma{}_{\theta r}$$

$$\begin{aligned} R^t{}_{\theta r\theta} = {} & (-g^{tt}g_{\theta\theta,t}),r/2 \\ & + (+g^{tt}g_{tt,r})(-g^{tt}g_{\theta\theta,t})/4 + (-g^{tt}g_{rr,t})(-g^{rr}g_{\theta\theta,r})/4 \\ & - (-g^{tt}g_{\theta\theta,t})(+g^{\theta\theta}g_{\theta\theta,r})/4 \end{aligned}$$

Sum μ = t, r; σ = ϕ:

$$R^t{}_{\phi t\phi} = \Gamma^t{}_{\phi\phi,t} - 0 + \Gamma^t{}_{\mu t}\Gamma^\mu{}_{\phi\phi} - \Gamma^t{}_{\sigma\phi}\Gamma^\sigma{}_{\phi t}$$

$$\begin{aligned} R^t{}_{\phi t\phi} = {} & (-g^{tt}g_{\phi\phi,t}),t/2 + (+g^{tt}g_{tt,t})(-g^{tt}g_{\phi\phi,t})/4 \\ & + (+g^{tt}g_{tt,r})(-g^{rr}g_{\phi\phi,r})/4 - (-g^{tt}g_{\phi\phi,t})(+g^{\phi\phi}g_{\phi\phi,t})/4 \end{aligned}$$

Sum μ = t, r; σ = ϕ:

$$R^t{}_{\phi r\phi} = \Gamma^t{}_{\phi\phi,r} - 0 + \Gamma^t{}_{\mu r}\Gamma^\mu{}_{\phi\phi} - \Gamma^t{}_{\sigma\phi}\Gamma^\sigma{}_{\phi r}$$

$$\begin{aligned} R^t{}_{\phi r\phi} = {} & (-g^{tt}g_{\phi\phi,t}),r/2 \\ & + (+g^{tt}g_{tt,r})(-g^{tt}g_{\phi\phi,t})/4 + (-g^{tt}g_{rr,t})(-g^{rr}g_{\phi\phi,r})/4 \\ & - (-g^{tt}g_{\phi\phi,t})(+g^{\phi\phi}g_{\phi\phi,r})/4 \end{aligned}$$

Sum μ = none; σ = none:

$$R^{t}{}_{\phi\theta\phi} = \Gamma^{t}{}_{\phi\phi,\theta} - 0 + 0 - 0$$

$$R^{t}{}_{\phi\theta\phi} = (-g^{tt}\, g_{\phi\phi,t}),\theta / 2$$

Sum μ = t, r; $\sigma = \theta$:

$$R^{r}{}_{\theta r\theta} = \Gamma^{r}{}_{\theta\theta,r} - 0 + \Gamma^{r}{}_{\mu r}\, \Gamma^{\mu}{}_{\theta\theta} - \Gamma^{r}{}_{\sigma\theta}\, \Gamma^{\sigma}{}_{\theta r}$$

$$\begin{aligned} R^{r}{}_{\theta r\theta} &= (-g^{rr}\, g_{\theta\theta,r}),r / 2 \\ &\quad + (+g^{rr}\, g_{rr,t})(-g^{tt}\, g_{\theta\theta,t}) / 4 + (+g^{rr}\, g_{rr,r})(-g^{rr}\, g_{\theta\theta,r}) / 4 \\ &\quad - (+g^{rr}\, g_{\theta\theta,r})(-g^{\theta\theta}\, g_{\theta\theta,r}) / 4 \end{aligned}$$

Sum μ = t, r; $\sigma = \phi$:

$$R^{r}{}_{\phi r\phi} = \Gamma^{r}{}_{\phi\phi,r} - 0 + \Gamma^{r}{}_{\mu r}\, \Gamma^{\mu}{}_{\phi\phi} - \Gamma^{r}{}_{\sigma\phi}\, \Gamma^{\sigma}{}_{\phi r}$$

$$\begin{aligned} R^{r}{}_{\phi r\phi} &= (-g^{rr}\, g_{\phi\phi,r}),r / 2 \\ &\quad + (+g^{rr}\, g_{rr,t})(-g^{tt}\, g_{\phi\phi,t}) / 4 + (+g^{rr}\, g_{rr,r})(-g^{rr}\, g_{\phi\phi,r}) / 4 \\ &\quad - (-g^{rr}\, g_{\phi\phi,r})(+g^{\phi\phi}\, g_{\phi\phi,r}) / 4 \end{aligned}$$

Sum $\mu = \theta$; $\sigma = \phi$:

$$R^{r}{}_{\phi\theta\phi} = \Gamma^{r}{}_{\phi\phi,\theta} - 0 + \Gamma^{r}{}_{\mu\theta}\, \Gamma^{\mu}{}_{\phi\phi} - \Gamma^{r}{}_{\sigma\phi}\, \Gamma^{\sigma}{}_{\phi\theta}$$

$$\begin{aligned} R^{r}{}_{\phi\theta\phi} &= (-g^{rr}\, g_{\phi\phi,r}),\theta / 2 \\ &\quad + (-g^{rr}\, g_{\theta\theta,r})(-g^{\theta\theta}\, g_{\phi\phi,\theta}) / 4 \\ &\quad - (-g^{rr}\, g_{\phi\phi,r})(+g^{\phi\phi}\, g_{\phi\phi,\theta}) / 4 \end{aligned}$$

Sum μ = t, r; $\sigma = \phi$:

$$R^{\theta}{}_{\phi\theta\phi} = \Gamma^{\theta}{}_{\phi\phi,\theta} - 0 + \Gamma^{\theta}{}_{\mu\theta}\, \Gamma^{\mu}{}_{\phi\phi} - \Gamma^{\theta}{}_{\sigma\phi}\, \Gamma^{\sigma}{}_{\phi\theta}$$

$$\begin{aligned} R^{\theta}{}_{\phi\theta\phi} &= (-g^{\theta\theta}\, g_{\phi\phi,\theta}),\theta / 2 \\ &\quad + (+g^{\theta\theta}\, g_{\theta\theta,t})(-g^{tt}\, g_{\phi\phi,t}) / 4 \\ &\quad + (+g^{\theta\theta}\, g_{\theta\theta,r})(-g^{rr}\, g_{\phi\phi,r}) / 4 \\ &\quad - (-g^{\theta\theta}\, g_{\phi\phi,\theta})(+g^{\phi\phi}\, g_{\phi\phi,\theta}) / 4 \end{aligned}$$

Did I mention that calculating curvature tensors can be quite tedious?

§A6.5 $R^{\alpha}{}_{\beta\delta e}$ in Coordinate Basis, Schwarzschild Geometry

Let's explore the curvature tensors in an even simpler geometry. The results of Section §A6.4 are for a spherically symmetric geometry that could vary over time. Let's now eliminate that degree of freedom and examine a static geometry, where all time derivatives are zero. We will employ a coordinate basis.

We write the interval as:

$$ds^2 = -\Lambda^2 dt^2 + \Sigma^2 dr^2 + r^2 d\theta^2 + r^2 \sin^2\theta \, d\phi^2$$

Here, Λ and Σ are functions of r only.

Using the results of Section §A6.3, and setting all time derivatives to zero, 9 independent connection coefficients remain to be calculated:

$$\Gamma_{ttr} = +g_{tt,r} / 2 = -\Lambda \Lambda,r$$

$$\Gamma_{rtt} = -g_{tt,r} / 2 = +\Lambda \Lambda,r$$

$$\Gamma_{rrr} = +g_{rr,r} / 2 = +\Sigma \Sigma,r$$

$$\Gamma_{r\theta\theta} = -g_{\theta\theta,r} / 2 = -r$$

$$\Gamma_{r\phi\phi} = -g_{\phi\phi,r} / 2 = -r \sin^2\theta$$

$$\Gamma_{\theta r\theta} = +g_{\theta\theta,r} / 2 = +r$$

$$\Gamma_{\theta\phi\phi} = -g_{\phi\phi,\theta} / 2 = -r^2 \sin\theta \cos\theta$$

$$\Gamma_{\phi r\phi} = +g_{\phi\phi,r} / 2 = +r \sin^2\theta$$

$$\Gamma_{\phi\theta\phi} = +g_{\phi\phi,\theta} / 2 = +r^2 \sin\theta \cos\theta$$

Using the results of Section §A6.4, and setting all time derivatives to zero, these Riemann components remain to be calculated:

$$R^t{}_{rtr} = -(+g^{tt} g_{tt,r}),r / 2 + (+g^{tt} g_{tt,r})(+g^{rr} g_{rr,r}) / 4$$
$$- (+g^{tt} g_{tt,r})(+g^{tt} g_{tt,r}) / 4$$

$$R^t{}_{rtr} = +(-\Lambda,r / \Lambda),r$$
$$-(-\Lambda \Lambda,r)(\Sigma \Sigma,r) / \Lambda^2 \Sigma^2$$
$$-(-\Lambda \Lambda,r)(-\Lambda \Lambda,r) / \Lambda^4$$

$$R^t{}_{rtr} = -\Lambda,rr / \Lambda - (-\Lambda,r \Lambda,r) / \Lambda^2$$
$$-(-\Lambda \Lambda,r)(\Sigma \Sigma,r) / \Lambda^2 \Sigma^2$$
$$-(-\Lambda \Lambda,r)(-\Lambda \Lambda,r) / \Lambda^4$$

$$R^t{}_{rtr} = -\Lambda,rr / \Lambda + \Lambda,r \Sigma,r / \Lambda \Sigma$$

$$R^t{}_{\theta t\theta} = (+g^{tt} g_{tt,r})(-g^{rr} g_{\theta\theta,r}) / 4$$

$$R^t{}_{\theta t\theta} = (+\Lambda \Lambda,r)(-r) / \Lambda^2 \Sigma^2$$

$$R^t{}_{\theta t\theta} = -r \Lambda,r / \Lambda \Sigma^2$$

$$R^{t}{}_{\phi t\phi} = (+\, g^{tt}\, g_{tt,r})\,(-\, g^{rr}\, g_{\phi\phi,r}) / 4$$
$$R^{t}{}_{\phi t\phi} = -\, r \sin^2\theta\ \Lambda{,}r\ /\ \Lambda\, \Sigma^2$$

$$R^{r}{}_{\theta r\theta} = (-\, g^{rr}\, g_{\theta\theta,r}){,}r / 2 + (+\, g^{rr}\, g_{rr,r})\,(-\, g^{rr}\, g_{\theta\theta,r}) / 4$$
$$-\ (+\, g^{rr}\, g_{\theta\theta,r})\,(-\, g^{\theta\theta}\, g_{\theta\theta,r}) / 4$$
$$R^{r}{}_{\theta r\theta} = (-\, r / \Sigma^2){,}r\ -\ (\Sigma\, \Sigma{,}r)\,(r) / \Sigma^4 + (r)\,(r) / r^2\, \Sigma^2$$
$$R^{r}{}_{\theta r\theta} = -\, 1 / \Sigma^2 + 2\, r\, \Sigma{,}r / \Sigma^3\ -\ r\, \Sigma{,}r / \Sigma^3\ +\ 1 / \Sigma^2$$
$$R^{r}{}_{\theta r\theta} = +\ r\, \Sigma{,}r / \Sigma^3$$

$$R^{r}{}_{\phi r\phi} = (-\, g^{rr}\, g_{\phi\phi,r}){,}r / 2 + (+\, g^{rr}\, g_{rr,r})\,(-\, g^{rr}\, g_{\phi\phi,r}) / 4$$
$$-\ (-\, g^{rr}\, g_{\phi\phi,r})\,(+\, g^{\phi\phi}\, g_{\phi\phi,r}) / 4$$
$$R^{r}{}_{\phi r\phi} = (-\, r \sin^2\theta / \Sigma^2){,}r\ -\ (\Sigma\, \Sigma{,}r)\,(r \sin^2\theta) / \Sigma^4$$
$$+\ (r \sin^2\theta)\,(r \sin^2\theta) / r^2 \sin^2\theta\, \Sigma^2$$
$$R^{r}{}_{\phi r\phi} = -\, \sin^2\theta / \Sigma^2 + 2r \sin^2\theta\, \Sigma{,}r / \Sigma^3$$
$$-\ r \sin^2\theta\, \Sigma{,}r\ / \Sigma^3 + \sin^2\theta\, \Sigma^2$$
$$R^{r}{}_{\phi r\phi} = +\, r \sin^2\theta\, \Sigma{,}r / \Sigma^3$$

$$R^{r}{}_{\phi\theta\phi} = (-\, g^{rr}\, g_{\phi\phi,r}){,}\theta / 2 + (-\, g^{rr}\, g_{\theta\theta,r})\,(-\, g^{\theta\theta}\, g_{\phi\phi,\theta}) / 4$$
$$-\ (-\, g^{rr}\, g_{\phi\phi,r})\,(+\, g^{\phi\phi}\, g_{\phi\phi,\theta}) / 4$$
$$R^{r}{}_{\phi\theta\phi} = (-\, r \sin^2\theta){,}\theta / \Sigma^2 + (r)\,(r^2 \sin\theta \cos\theta) / r^2\, \Sigma^2$$
$$+\ (r \sin^2\theta)\,(r^2 \sin\theta \cos\theta) / r^2 \sin^2\theta\, \Sigma^2$$
$$R^{r}{}_{\phi\theta\phi} = -\, 2\, r \sin\theta \cos\theta / \Sigma^2 + 2\, r \sin\theta \cos\theta\ / \Sigma^2$$
$$R^{r}{}_{\phi\theta\phi} = 0$$

$$R^{\theta}{}_{\phi\theta\phi} = (-\, g^{\theta\theta}\, g_{\phi\phi,\theta}){,}\theta / 2 + (+\, g^{\theta\theta}\, g_{\theta\theta,r})\,(-\, g^{rr}\, g_{\phi\phi,r}) / 4$$
$$-\ (-\, g^{\theta\theta}\, g_{\phi\phi,\theta})\,(+\, g^{\phi\phi}\, g_{\phi\phi,\theta}) / 4$$
$$R^{\theta}{}_{\phi\theta\phi} = -\, (\sin\theta \cos\theta){,}\theta\ -\ (r)\,(r \sin^2\theta) / r^2\, \Sigma^2$$
$$+\ (r^2 \sin\theta \cos\theta)\,(r^2 \sin\theta \cos\theta) / r^4 \sin^2\theta$$
$$R^{\theta}{}_{\phi\theta\phi} = -\, \cos^2\theta + \sin^2\theta\ -\ \sin^2\theta / \Sigma^2 + (\cos\theta)\,(\cos\theta)$$
$$R^{\theta}{}_{\phi\theta\phi} = +\ \sin^2\theta\, (1 - 1 / \Sigma^2)$$

From these calculations, we discover that there are only 6 independent, nonzero Riemann components in the Schwarzschild metric. In all of these components, the first index pair exactly matches the second index pair. Thus, the only nonzero components are those on the diagonal

The effect on the components of the Reimann tensor from various symmetries in the spacetime geometry is graphically shown in Figure A6-1.

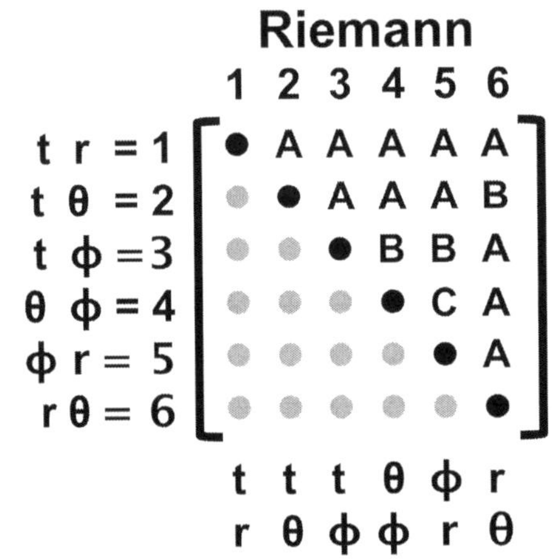

Figure A6-1 Some Riemann Tensor Symmetries

In this figure, the letter A denotes a component that is nonzero in general, but is zero in a spherically symmetric geometry. The letter B denotes a component that is nonzero in general, but is zero in a static, spherically symmetric geometry. The letter C denotes a component that is nonzero in general, but is zero in the Schwarzschild geometry.

The black dots denote components that are in general nonzero, even in the static, spherically symmetric, Schwarzschild geometry. The gray dots denote below-diagonal components that equal their above-diagonal counterparts.

§A6.6 $\Gamma_{\alpha\beta\delta}$ in Orthonormal Basis, Any Geometry or Metric

In a general, non-coordinate basis, the connection coefficients (see Section §22.3) are given by:

$$\Gamma_{\alpha\beta\mu} = (g_{\alpha\beta,\mu} + g_{\alpha\mu,\beta} - g_{\beta\mu,\alpha}) / 2$$
$$+ \{ c_{\alpha\beta\mu} + c_{\alpha\mu\beta} - c_{\beta\mu\alpha} \} / 2$$

Here, the c's are called commutation coefficients, and are determined by the commutator $[e_\beta , e_\mu]$ for each pair of basis vectors e_β and e_μ.

The commutators are calculated by treating each basis vector as a directional partial derivative, as shown by:

$$e_\alpha \, c_{\beta\mu}{}^{\alpha} = [\partial_\beta , \partial_\mu] = \partial_\beta \, \partial_\mu - \partial_\mu \, \partial_\beta$$

Clearly, the $c_{\beta\mu}{}^{\alpha}$ are antisymmetric in their first two indices β and μ.

While adding the c's seems to add more complex math, we can gain substantial simplification by choosing basis vectors that make the metric orthonormal.

For a comparison of the basis alternatives, consider a spherical coordinate basis, with a diagonalized metric, whose invariant interval has the form:

$$ds^2 = -\Lambda^2\, dt^2 + \Sigma^2\, dr^2 + r^2\, d\theta^2 + r^2 \sin^2\theta\, d\phi^2$$

We can describe the same geometry in an orthonormal non-coordinate basis as:

$$ds^2 = -d\sigma^2 + d\varrho^2 + d\Theta^2 + d\Phi^2$$

The new basis vectors are:

$$d\sigma = \Lambda\, dt$$

$$d\varrho = \Sigma\, dr$$

$$d\Theta = r\, d\theta$$

$$d\Phi = r\, \sin\theta\, d\phi$$

In the orthonormal metric, $g_{\alpha\beta} = 0$ unless $\alpha = \beta$, and $g_{\alpha\alpha} = +1$ or -1. Raising or lowering a σ index inverts that term's polarity, while raising or lowering any other index produces no change. Additionally, any derivative of any $g_{\alpha\beta}$ is zero. The connection coefficients in an orthonormal basis are:

$$\Gamma_{\alpha\beta\mu} = \{ c_{\alpha\beta\mu} + c_{\alpha\mu\beta} - c_{\beta\mu\alpha} \} / 2$$

Swaping μ and β:

$$\Gamma_{\alpha\mu\beta} = \{ c_{\alpha\mu\beta} + c_{\alpha\beta\mu} - c_{\mu\beta\alpha} \} / 2$$

Since c is antisymmetric in its first two indices, the right hand sides of two prior equations are not equal: the first two terms in each equation are equal, but the last terms are of opposite polarity. The operation of swapping μ and β is therefore neither symmetric nor antisymmetric.

Hence, unlike Γ in a coordinate basis, Γ is NOT symmetric in its last two indices in non-coordinate bases. The number of independent Γ's is $4\times6 = 24$ — for each of the 4 possible values of the last index, there are 4×4 combinations of the first two indices. But since those indices are antisymmetric, only 6 combinations are nonzero and independent; 4 others are zero, and 6 other are redundant with the opposite polarity.

The orthonormal connection coefficients are:

$$\Gamma_{\sigma\varrho\sigma} = \{ c_{\sigma\varrho\sigma} + c_{\sigma\sigma\varrho} - c_{\varrho\sigma\sigma} \} / 2 = c_{\sigma\varrho\sigma}$$

$$\Gamma_{\sigma\Theta\sigma} = \{ c_{\sigma\Theta\sigma} + c_{\sigma\sigma\Theta} - c_{\Theta\sigma\sigma} \} / 2 = c_{\sigma\Theta\sigma}$$

$$\Gamma_{\sigma\Phi\sigma} = \{ c_{\sigma\Phi\sigma} + c_{\sigma\sigma\Phi} - c_{\Phi\sigma\sigma} \} / 2 = c_{\sigma\Phi\sigma}$$

$$\Gamma_{\varrho\Theta\sigma} = \{ c_{\varrho\Theta\sigma} + c_{\varrho\sigma\Theta} - c_{\Theta\sigma\varrho} \} / 2$$

$$\Gamma_{\varrho\Phi\sigma} = \{ c_{\varrho\Phi\sigma} + c_{\varrho\sigma\Phi} - c_{\Phi\sigma\varrho} \} / 2$$

$$\Gamma_{\Theta\Phi\sigma} = \{ c_{\Theta\Phi\sigma} + c_{\Theta\sigma\Phi} - c_{\Phi\sigma\Theta} \} / 2$$

$$\Gamma_{\sigma\varrho\varrho} = \{ c_{\sigma\varrho\varrho} + c_{\sigma\varrho\varrho} - c_{\varrho\varrho\sigma} \} / 2 = c_{\sigma\varrho\varrho}$$

$$\Gamma_{\sigma\Theta\varrho} = \{ c_{\sigma\Theta\varrho} + c_{\sigma\varrho\Theta} - c_{\Theta\varrho\sigma} \} / 2$$

$$\Gamma_{\sigma\Phi\varrho} = \{ c_{\sigma\Phi\varrho} + c_{\sigma\varrho\Phi} - c_{\Phi\varrho\sigma} \} / 2$$

$$\Gamma_{\varrho\Theta\varrho} = \{ c_{\varrho\Theta\varrho} + c_{\varrho\varrho\Theta} - c_{\Theta\varrho\varrho} \} / 2 = c_{\varrho\Theta\varrho}$$

$$\Gamma_{\varrho\Phi\varrho} = \{ c_{\varrho\Phi\varrho} + c_{\varrho\varrho\Phi} - c_{\Phi\varrho\varrho} \} / 2 = c_{\varrho\Phi\varrho}$$

$$\Gamma_{\Theta\Phi\varrho} = \{ c_{\Theta\Phi\varrho} + c_{\Theta\varrho\Phi} - c_{\Phi\varrho\Theta} \} / 2$$

$$\Gamma_{\sigma\varrho\Theta} = \{ c_{\sigma\varrho\Theta} + c_{\sigma\Theta\varrho} - c_{\varrho\Theta\sigma} \} / 2$$
$$\Gamma_{\sigma\Theta\Theta} = \{ c_{\sigma\Theta\Theta} + c_{\sigma\Theta\Theta} - c_{\Theta\Theta\sigma} \} / 2 = c_{\sigma\Theta\Theta}$$
$$\Gamma_{\sigma\Phi\Theta} = \{ c_{\sigma\Phi\Theta} + c_{\sigma\Theta\Phi} - c_{\Phi\Theta\sigma} \} / 2$$
$$\Gamma_{\varrho\Theta\Theta} = \{ c_{\varrho\Theta\Theta} + c_{\varrho\Theta\Theta} - c_{\Theta\Theta\varrho} \} / 2 = c_{\varrho\Theta\Theta}$$
$$\Gamma_{\varrho\Phi\Theta} = \{ c_{\varrho\Phi\Theta} + c_{\varrho\Theta\Phi} - c_{\Phi\Theta\varrho} \} / 2$$
$$\Gamma_{\Theta\Phi\Theta} = \{ c_{\Theta\Phi\Theta} + c_{\Theta\Theta\Phi} - c_{\Phi\Theta\Theta} \} / 2 = c_{\Theta\Phi\Theta}$$

$$\Gamma_{\sigma\varrho\Phi} = \{ c_{\sigma\varrho\Phi} + c_{\sigma\Phi\varrho} - c_{\Phi\sigma} \} / 2$$
$$\Gamma_{\sigma\Theta\Phi} = \{ c_{\sigma\Theta\Phi} + c_{\sigma\Phi\Theta} - c_{\Theta\Phi\sigma} \} / 2$$
$$\Gamma_{\sigma\Phi\Phi} = \{ c_{\sigma\Phi\Phi} + c_{\sigma\Phi\Phi} - c_{\Phi\Phi\sigma} \} / 2 = c_{\sigma\Phi\Phi}$$
$$\Gamma_{\varrho\Theta\Phi} = \{ c_{\varrho\Theta\Phi} + c_{\varrho\Phi\Theta} - c_{\Theta\Phi\varrho} \} / 2$$
$$\Gamma_{\varrho\Phi\Phi} = \{ c_{\varrho\Phi\Phi} + c_{\varrho\Phi\Phi} - c_{\Phi\Phi\varrho} \} / 2 = c_{\varrho\Phi\Phi}$$
$$\Gamma_{\Theta\Phi\Phi} = \{ c_{\Theta\Phi\Phi} + c_{\Theta\Phi\Phi} - c_{\Phi\Phi\Theta} \} / 2 = c_{\Theta\Phi\Phi}$$

§A6.7 $R^{\alpha}{}_{\beta}$ in Any Basis or Geometry

After calculating the Riemann tensor components $R^{\alpha}{}_{\beta\delta\epsilon}$ in any basis or geometry, we can then calculate the Ricci tensor $R^{\alpha}{}_{\beta}$. The general equation is:

$$R^{\alpha}{}_{\beta} = R^{\alpha\mu}{}_{\beta\mu}$$

Let's assume a diagonal metric and label the coordinates: t, r, θ, and ϕ. The Ricci components are:

$$R^{t}{}_{t} = R^{t\mu}{}_{t\mu} = R^{\mu t}{}_{\mu t}$$
$$R^{t}{}_{t} = R^{tr}{}_{tr} + R^{t\theta}{}_{t\theta} + R^{t\phi}{}_{t\phi}$$
$$R^{t}{}_{t} = g^{rr} R^{t}{}_{rtr} + g^{\theta\theta} R^{t}{}_{\theta t\theta} + g^{\phi\phi} R^{t}{}_{\phi t\phi}$$
$$R^{r}{}_{r} = R^{rt}{}_{rt} + R^{r\theta}{}_{r\theta} + R^{r\phi}{}_{r\phi}$$
$$R^{r}{}_{r} = g^{rr} R^{t}{}_{rtr} + g^{\theta\theta} R^{r}{}_{\theta r\theta} + g^{\phi\phi} R^{r}{}_{\phi r\phi}$$

$$R^{\theta}{}_{\theta} = R^{\theta t}{}_{\theta t} + R^{\theta r}{}_{\theta r} + R^{\theta\phi}{}_{\theta\phi}$$
$$R^{\theta}{}_{\theta} = g^{\theta\theta} R^{t}{}_{\theta t\theta} + g^{\theta\theta} R^{r}{}_{\theta r\theta} + g^{\phi\phi} R^{\theta}{}_{\phi\theta\phi}$$

$$R^{\phi}{}_{\phi} = R^{t\phi}{}_{t\phi} + R^{r\phi}{}_{r\phi} + R^{\theta\phi}{}_{\theta\phi}$$
$$R^{\phi}{}_{\phi} = g^{\phi\phi} R^{t}{}_{\phi t\phi} + g^{\phi\phi} R^{r}{}_{\phi r\phi} + g^{\phi\phi} R^{\theta}{}_{\phi\theta\phi}$$

$$R^{t}{}_{r} = R^{tr}{}_{rr} + R^{t\theta}{}_{r\theta} + R^{t\phi}{}_{r\phi}$$
$$R^{t}{}_{r} = 0 + g^{\theta\theta} R^{t}{}_{\theta r\theta} + g^{\phi\phi} R^{t}{}_{\phi r\phi}$$

$$R^t{}_\theta = R^{tr}{}_{\theta r} + R^{t\theta}{}_{\theta\theta} + R^{t\phi}{}_{\theta\phi}$$
$$R^t{}_\theta = g^{rr}\,R^t{}_{rr\theta} + 0 + g^{\phi\phi}\,R^t{}_{\phi\theta\phi}$$

$$R^t{}_\phi = R^{tr}{}_{\phi r} + R^{t\theta}{}_{\phi\theta} + R^{t\phi}{}_{\phi\phi}$$
$$R^t{}_\phi = g^{rr}\,R^t{}_{r\phi r} + g^{\theta\theta}\,R^t{}_{\theta\phi\theta} + 0$$

$$R^r{}_\theta = R^{tr}{}_{t\theta} + R^{rr}{}_{r\theta} + R^{r\phi}{}_{\theta\phi}$$
$$R^r{}_\theta = g^{rr}\,R^t{}_{rt\theta} + 0 + g^{\phi\phi}\,R^r{}_{\phi\theta\phi}$$

$$R^r{}_\phi = R^{tr}{}_{t\phi} + R^{rr}{}_{r\phi} - R^{r\theta}{}_{\theta\phi}$$
$$R^r{}_\phi = g^{rr}\,R^t{}_{rt\phi} + 0 - g^{\theta\theta}\,R^r{}_{\theta\theta\phi}$$

$$R^\theta{}_\phi = R^{t\theta}{}_{t\phi} + R^{r\theta}{}_{r\phi} + R^{\theta\phi}{}_{\phi\phi}$$
$$R^\theta{}_\phi = g^{\theta\theta}\,R^t{}_{\theta t\phi} + g^{\theta\theta}\,R^r{}_{\theta r\phi} + 0$$

In a spherically symmetric geometry, the following Ricci components are always zero:

$$R^t{}_\phi = R^r{}_\phi = R^\theta{}_\phi = 0$$

In a static, spherically symmetric geometry, the following Ricci components are always zero:

$$R^t{}_r = R^t{}_\theta = R^t{}_\phi = R^r{}_\phi = R^\theta{}_\phi = 0$$

In the Schwarzschild geometry, all off-diagonal Ricci components are always zero:

$$R^t{}_r = R^t{}_\theta = R^t{}_\phi = R^r{}_\theta = R^r{}_\phi = R^\theta{}_\phi = 0$$

We also need the invariant curvature scalar R — the *trace* of the Ricci tensor — defined by:

$$R = R^\mu{}_\mu$$
$$R = R^t{}_t + R^r{}_r + R^\theta{}_\theta + R^\phi{}_\phi$$
$$\begin{aligned} R = {} & g^{rr}\,R^t{}_{rtr} + g^{\theta\theta}\,R^t{}_{\theta t\theta} + g^{\phi\phi}\,R^t{}_{\phi t\phi} \\ & + g^{rr}\,R^t{}_{rtr} + g^{\theta\theta}\,R^r{}_{\theta r\theta} + g^{\phi\phi}\,R^r{}_{\phi r\phi} \\ & + g^{\theta\theta}\,R^t{}_{\theta t\theta} + g^{\theta\theta}\,R^r{}_{\theta r\theta} + g^{\phi\phi}\,R^\theta{}_{\phi\theta\phi} \\ & + g^{\phi\phi}\,R^t{}_{\phi t\phi} + g^{\phi\phi}\,R^r{}_{\phi r\phi} + g^{\phi\phi}\,R^\theta{}_{\phi\theta\phi} \end{aligned}$$

We actually need R / 2.

$$\begin{aligned} R/2 = {} & g^{rr}\,R^t{}_{rtr} + g^{\theta\theta}\,R^t{}_{\theta t\theta} + g^{\phi\phi}\,R^t{}_{\phi t\phi} \\ & + g^{\theta\theta}\,R^r{}_{\theta r\theta} + g^{\phi\phi}\,R^r{}_{\phi r\phi} + g^{\phi\phi}\,R^\theta{}_{\phi\theta\phi} \end{aligned}$$

§A6.8 $G^{\alpha}{}_{\beta}$ in Any Basis or Geometry

Having calculated the Ricci tensor components $R^{\alpha}{}_{\beta}$ in any basis or geometry, we can finally calculate the Einstein tensor $G^{\alpha}{}_{\beta}$. The general equation is:

$$G^{\alpha}{}_{\beta} = R^{\alpha}{}_{\beta} - R\, g^{\alpha}{}_{\beta} / 2$$

Recall that $g^{\alpha}{}_{\beta}$ is equivalent to the Kronecker delta; it equals 1 if $\alpha = \beta$, and $g^{\alpha}{}_{\beta}$ equals zero if α and β are different.

Let's assume a diagonal metric and label the coordinates: t, r, θ, and ϕ. The diagonal components of the Einstein tensor are:

$$G^{t}{}_{t} = R^{t}{}_{t} - R / 2$$

$$\begin{aligned} G^{t}{}_{t} = {} & g^{rr} R^{t}{}_{rtr} + g^{\theta\theta} R^{t}{}_{\theta t\theta} + g^{\phi\phi} R^{t}{}_{\phi t\phi} \\ & - g^{rr} R^{t}{}_{rtr} - g^{\theta\theta} R^{t}{}_{\theta t\theta} - g^{\phi\phi} R^{t}{}_{\phi t\phi} \\ & - g^{\theta\theta} R^{r}{}_{\theta r\theta} - g^{\phi\phi} R^{r}{}_{\phi r\phi} - g^{\phi\phi} R^{\theta}{}_{\phi\theta\phi} \end{aligned}$$

$$G^{t}{}_{t} = - g^{\theta\theta} R^{r}{}_{\theta r\theta} - g^{\phi\phi} R^{r}{}_{\phi r\phi} - g^{\phi\phi} R^{\theta}{}_{\phi\theta\phi}$$

$$G^{r}{}_{r} = R^{r}{}_{r} - R / 2$$

$$\begin{aligned} G^{r}{}_{r} = {} & g^{rr} R^{t}{}_{rtr} + g^{\theta\theta} R^{r}{}_{\theta r\theta} + g^{\phi\phi} R^{r}{}_{\phi r\phi} \\ & - g^{rr} R^{t}{}_{rtr} - g^{\theta\theta} R^{t}{}_{\theta t\theta} - g^{\phi\phi} R^{t}{}_{\phi t\phi} \\ & - g^{\theta\theta} R^{r}{}_{\theta r\theta} - g^{\phi\phi} R^{r}{}_{\phi r\phi} - g^{\phi\phi} R^{\theta}{}_{\phi\theta\phi} \end{aligned}$$

$$G^{r}{}_{r} = - g^{\theta\theta} R^{t}{}_{\theta t\theta} - g^{\phi\phi} R^{t}{}_{\phi t\phi} - g^{\phi\phi} R^{\theta}{}_{\phi\theta\phi}$$

$$G^{\theta}{}_{\theta} = R^{\theta}{}_{\theta} - R / 2$$

$$\begin{aligned} G^{\theta}{}_{\theta} = {} & g^{\theta\theta} R^{t}{}_{\theta t\theta} + g^{\theta\theta} R^{r}{}_{\theta r\theta} + g^{\phi\phi} R^{\theta}{}_{\phi\theta\phi} \\ & - g^{rr} R^{t}{}_{rtr} - g^{\theta\theta} R^{t}{}_{\theta t\theta} - g^{\phi\phi} R^{t}{}_{\phi t\phi} \\ & - g^{\theta\theta} R^{r}{}_{\theta r\theta} - g^{\phi\phi} R^{r}{}_{\phi r\phi} - g^{\phi\phi} R^{\theta}{}_{\phi\theta\phi} \end{aligned}$$

$$G^{\theta}{}_{\theta} = - g^{rr} R^{t}{}_{rtr} - g^{\phi\phi} R^{t}{}_{\phi t\phi} - g^{\phi\phi} R^{r}{}_{\phi r\phi}$$

$$G^{\phi}{}_{\phi} = R^{\phi}{}_{\phi} - R / 2$$

$$\begin{aligned} G^{\phi}{}_{\phi} = {} & g^{\phi\phi} R^{t}{}_{\phi t\phi} + g^{\phi\phi} R^{r}{}_{\phi r\phi} + g^{\phi\phi} R^{\theta}{}_{\phi\theta\phi} \\ & - g^{rr} R^{t}{}_{rtr} - g^{\theta\theta} R^{t}{}_{\theta t\theta} - g^{\phi\phi} R^{t}{}_{\phi t\phi} \\ & - g^{\theta\theta} R^{r}{}_{\theta r\theta} - g^{\phi\phi} R^{r}{}_{\phi r\phi} - g^{\phi\phi} R^{\theta}{}_{\phi\theta\phi} \end{aligned}$$

$$G^{\phi}{}_{\phi} = - g^{rr} R^{t}{}_{rtr} - g^{\theta\theta} R^{t}{}_{\theta t\theta} - g^{\theta\theta} R^{r}{}_{\theta r\theta}$$

The off-diagonal Einstein tensor components are simpler, but are not commonly needed.

$$G^{\alpha}{}_{\beta} = R^{\alpha}{}_{\beta}$$

$$G^{t}{}_{r} = + g^{\theta\theta} R^{t}{}_{\theta r\theta} + g^{\phi\phi} R^{t}{}_{\phi r\phi}$$

$$G^t{}_\theta = + g^{\phi\phi} R^t{}_{\phi\theta\phi}$$

$$G^t{}_\phi = g^{rr} R^t{}_{r\phi r} + g^{\theta\theta} R^t{}_{\theta\phi\theta}$$

$$G^r{}_\theta = + g^{\phi\phi} R^r{}_{\phi\theta\phi}$$

$$G^r{}_\phi = g^{rr} R^t{}_{rt\phi} - g^{\theta\theta} R^r{}_{\theta\theta\phi}$$

$$G^\theta{}_\phi = g^{\theta\theta} R^t{}_{\theta t\phi} + g^{\theta\theta} R^r{}_{\theta r\phi}$$

In a spherically symmetric geometry, the following Einstein components are always zero:

$$G^t{}_\phi = G^r{}_\phi = G^\theta{}_\phi = 0$$

In a static, spherically symmetric geometry, the following Einstein components are always zero:

$$G^t{}_r = G^t{}_\theta = G^t{}_\phi = G^r{}_\phi = G^\theta{}_\phi = 0$$

In the Schwarzschild geometry, all off-diagonal Einstein components are always zero:

$$G^t{}_r = G^t{}_\theta = G^t{}_\phi = G^r{}_\theta = G^r{}_\phi = G^\theta{}_\phi = 0$$

The symmetries of the Ricci and Einstein tensors are graphically shown in Figure A6-2.

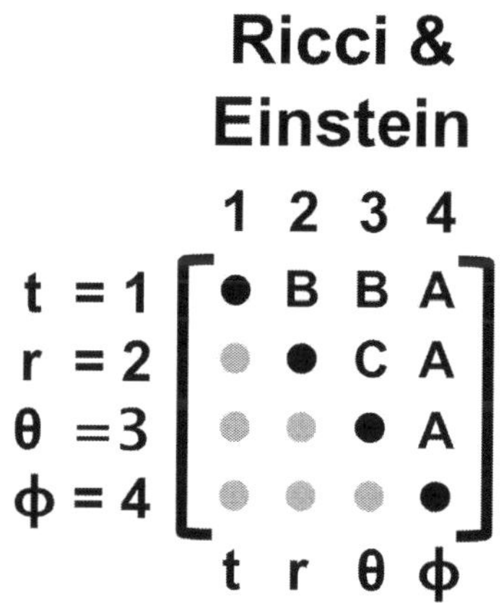

Figure A6-2 Some Ricci & Einstein Tensor Symmetries

In this figure, the letter A denotes a component that is nonzero in general, but is zero in a spherically symmetric geometry. The letter B denotes a component that is nonzero in general, but is zero in a static, spherically symmetric geometry. The letter C denotes a component that is nonzero in general, but is zero in the Schwarzschild geometry.

The black dots denote components that are in general nonzero, even in the static, spherically symmetric, Schwarzschild geometry. The gray dots denote below-diagonal components that equal their above-diagonal counterparts.

§A6.9 Coordinate – Orthonormal Basis Comparison

The axes of a coordinate basis measure 4-D spacetime displacements as they would be if the space had zero curvature everywhere. Conversely, the axes of an orthonormal basis measure 4-D spacetime displacements as they are with curvature. For an illustration, see Figure 22-1 at the end of Section §22.5.

In a coordinate basis, curvature is described by the metric components $g_{\alpha\beta}$. In an orthonormal basis, curvature is described by the basis vectors. These two types of bases are mathematically equivalent; we are free to choose whichever seems most suitable in any specific situation.

We can relate certain curvature tensors calculated in an orthonormal basis with the same tensor calculated in a coordinate basis. For example, consider the 4-index contraction of the Riemann tensor with basis vectors in an orthonormal basis.

$$b = R^{\sigma}{}_{\varrho\Theta\Phi}\; e_{\sigma}\; e^{\varrho}\; e^{\Theta}\; e^{\Phi}$$

In this equation, b is an invariant scalar, because it is the contraction on all indices of the product of 5 tensors. This means b has the same value in any reference frame or coordinate system.

Let's derive equations for the four basis vectors in the last equation, using the spherical orthonormal basis of Section §A6.5.

Coordinate basis interval: $ds^2 = -\Lambda^2\, dt^2 + \Sigma^2\, dr^2 + r^2\, d\theta^2 + r^2 \sin^2\theta\, d\phi^2$

Orthonormal basis interval: $ds^2 = -d\sigma^2 + d\varrho^2 + d\Theta^2 + d\Phi^2$

Orthonormal basis vectors:

$$d\sigma = \Lambda\; dt$$

$$d\varrho = \Sigma\; dr$$

$$d\Theta = r\; d\theta$$

$$d\Phi = r\; \sin\theta\; d\phi$$

The key point is the orthonormal basis vectors are scaled (multiplied by) functions that vary throughout spacetime.

Let's next relate these basis vectors, with subscripted designators, to partial derivatives along actual coordinate axes:

$$e_{\sigma} = \partial/\partial\sigma = \partial/\partial(\Lambda\, dt) = (1 / \Lambda)\, \partial/\partial t$$

$$e_{\varrho} = \partial/\partial\varrho = \partial/\partial(\Sigma\, dr) = (1 / \Sigma)\, \partial/\partial r$$

$$e_{\Theta} = \partial/\partial\Theta = \partial/\partial(r\, d\theta) = (1 / r)\, \partial/\partial\theta$$

$$e_{\Phi} = \partial/\partial\Phi = \partial/\partial(r \sin\theta\, d\phi) = (1 / r \sin\theta)\, \partial/\partial\phi$$

Note that the designators specify basis vectors' coordinate axis, not one of its components; it may have components in any direction (see Section §A5.1). We also need the inverse operators with superscripted designators, which are:

$$e^{\sigma} = \Lambda\; e^{t}$$

$$e^{\varrho} = \Sigma\; e^{r}$$

$$e^{\Theta} = r\; e^{\theta}$$

$$e^{\Phi} = r \sin\theta\; e^{\phi}$$

Here, each operator e^{α} is the inverse operator of the directional derivative ∂_{α}, so that the tensor contraction $e^{\alpha} \bullet \partial_{\alpha} = 1$, and $e^{\alpha} \bullet \partial_{\beta} = 0$ if α is not equal to β. (See Section §A5.7 for more on these inverse operators.)

With these expressions, the equation for scalar b in an orthonormal basis becomes:

$$b = R^{\sigma}{}_{\varrho\Theta\Phi}\ e_{\sigma}\ e^{\varrho}\ e^{\Theta}\ e^{\Phi}$$

$$b = R^{\sigma}{}_{\varrho\Theta\Phi}\ [\,(1/\Lambda)\ \partial/\partial t\,]\ [\,\Sigma\ /\ \partial/\partial r\,]$$
$$\times\ [\,r\ /\ \partial/\theta\,]\ [\,r\sin\theta\ /\ \partial/\partial\phi\,]$$

$$b = R^{\sigma}{}_{\varrho\Theta\Phi}\ (r^2 \sin\theta\ \Sigma\ \partial/\partial t\,)\,/\,(\,\Lambda\ \partial/\partial r\ \partial/\theta\ \partial/\partial\phi\,)$$

In a coordinate basis, basis vectors are much simpler — they are pure coordinate derivatives without scaling functions that vary through 4-D spacetime. In a spherical coordinate system, these basis vectors, and their corresponding directional derivatives are:

$$e_t = \partial/\partial t$$
$$e_r = \partial/\partial r$$
$$e_\theta = \partial/\partial\theta$$
$$e_\phi = \partial/\partial\phi$$
$$e^{\alpha} \bullet e_{\beta} = \delta^{\alpha}{}_{\beta}$$

Since it is an invariant scalar, b must have the same value in a coordinate basis, hence:

$$b = R^{t}{}_{r\theta\phi}\ e_t\ e^r\ e^\theta\ e^\phi$$
$$b = R^{t}{}_{r\theta\phi}\ (\partial/\partial t\,)\,/\,(\,\partial/\partial r\ \partial/\theta\ \partial/\partial\phi\,)$$

Equating the two equations for b, we obtain:

$$R^{t}{}_{r\theta\phi} = R^{\sigma}{}_{\varrho\Theta\Phi}\ (r^2 \sin\theta\ \Sigma\ )\,/\,\Lambda$$

This example, with four different indices, illustrates how each axis contributes; the same logic clearly applies to any combination of indices.

One important consequence of this correspondence of tensors in different bases, is that, for any specific geometry, the number of nonzero, independent components of the Riemann tensor is the same in all bases. The same is true for any other tensor.

We found in Section §A6.5 that, in the Schwarzschild metric, Riemann has 6 nonzero, independent components in a coordinate basis. It must therefore also have 6 nonzero, independent components, though with different values, in an orthonormal basis.

Note that we cannot make a similar comparison between a connection coefficient $\Gamma^{\alpha}{}_{\beta\gamma}$ calculated in a coordinate basis and in an orthonormal basis. This is because Γ is not a tensor, it does transform properly from one coordinate system to another.

Let's take this correspondence and calculate the Riemann components for the Schwarzschild metric in an orthonormal basis. In each pair of equations below, the upper one is the Riemann component in a coordinate basis, while the lower one is the same component in the orthonormal basis described above.

$$R^{t}{}_{rtr} = -\,\Lambda_{,rr}\,/\,\Lambda\ +\ \Lambda_{,r}\ \Sigma_{,r}\ /\,\Lambda\,\Sigma$$

Divide by $g_{rr} = \Sigma^2$:

$$R^{\sigma}{}_{\varrho\sigma\varrho} = -\,\Lambda_{,rr}\,/\,\Lambda\,\Sigma^2\ +\ \Lambda_{,r}\ \Sigma_{,r}\ /\,\Lambda\,\Sigma^3$$

$$R^{t}{}_{\theta t\theta} = -r\,\Lambda,r\ /\ \Lambda\,\Sigma^2$$

Divide by $g_{\theta\theta} = r^2$: $$R^{\sigma}{}_{\Theta\sigma\Theta} = -\Lambda,r\ /\ r\,\Lambda\,\Sigma^2$$

$$R^{t}{}_{\phi t\phi} = -r\sin^2\theta\,\Lambda,r\ /\ \Lambda\,\Sigma^2$$

Divide by $g_{\phi\phi} = r^2\sin^2\theta$: $$R^{\sigma}{}_{\Phi\sigma\Phi} = -\Lambda,r\ /\ r\,\Lambda\,\Sigma^2$$

$$R^{r}{}_{\theta r\theta} = +\ r\,\Sigma,r\ /\ \Sigma^3$$

Divide by $g_{\theta\theta} = r^2$: $$R^{\varrho}{}_{\Theta\varrho\Theta} = +\ \Sigma,r\ /\ r\,\Sigma^3$$

$$R^{r}{}_{\phi r\phi} = +r\sin^2\theta\,\Sigma,r\ /\ \Sigma^3$$

Divide by $g_{\phi\phi} = r^2\sin^2\theta$: $$R^{\varrho}{}_{\Phi\varrho\Phi} = +\Sigma,r\ /\ r\,\Sigma^3$$

$$R^{\theta}{}_{\phi\theta\phi} = +\ \sin^2\theta\,(\,1 - 1\ /\ \Sigma^2\,)$$

Divide by $g_{\phi\phi} = r^2\sin^2\theta$: $$R^{\Theta}{}_{\Phi\Theta\Phi} = +\ (\,1 - 1\ /\ \Sigma^2\,)\ /\ r^2$$

$$R^{r}{}_{\phi\theta\phi} = +r\sin\theta\cos\theta\ /\ \Sigma^2$$

Multiply by $\Sigma\ /\ r^3\sin^2\theta$: $$R^{\varrho}{}_{\Phi\Theta\Phi} = +\cot\theta\ /\ r^2\,\Sigma^2$$

For $R^{\varrho}{}_{\Phi\Theta\Phi}$ we must multiply by $\sqrt{(g_{rr})} = \Sigma$, and divide by $g_{\phi\phi}\sqrt{(g_{\theta\theta})} = (r^2\sin^2\theta)\,(r)$.

We will find in Section §A7.1 that our predicted values above match the detailed calculations in an orthonormal basis.

Appendix 7
Derivation of Schwarzschild Metric

Karl Schwarzschild derived the first exact solution of Einstein's field equations: $G = 8\pi T$.

This Appendix presents a complete derivation of the Schwarzschild metric for spacetime outside a massive, symmetric body, assuming nothing changes over time.

Evaluating the curvature tensors requires a lot of complex differentiation and algebra. Those who wish to see only the conclusion can skip to Section §A7.2.

For those who wish to learn how to do the math, this is an enlightening example. See how I calculate a few components, and then try to do the others by yourself. Section §A7.7 provides more practice.

In general, solving the field equations is complicated by the fact that G depends on T, T depends on G, and at the outset, we often know neither. Our approach is: (1) find the most general form of the metric, which of necessity incorporates some unknown functions; (2) express G in terms of those functions; and (3) use the field equations, and any knowledge of T, to calculate the previously unknown functions.

§A7.1 Metric in Orthonormal Basis

In a spherical coordinate basis, centered on a static, spherically symmetric mass M, the metric can always be diagonalized, making the most general form of the static, spherically symmetric interval:

$$ds^2 = -\Lambda^2 \, dt^2 + \Sigma^2 \, dr^2 + r^2 \, d\theta^2 + r^2 \sin^2\theta \, d\phi^2$$

Here, Λ and Σ are two unknown functions, but due to the static, spherical symmetry, they can only be functions of the radial coordinate r. At very large r, far from the gravitating mass, spacetime curvature approaches zero, and therefore Λ and Σ must both approach 1, matching the metric of flat spacetime.

Examination shows that r is indeed the proper radial coordinate. Each great circle (θ going from 0 to 2π for any ϕ, or ϕ going from 0 to 2π for $\sin\theta = 1$) has a circumference of 2π r, which is all we can expect in curved spacetime. Due to spherical symmetry, the circumferences of great circles are not distorted by curvature, but they are no longer equal to their diameters multiplied by π, because curvature stretches the diameters.

The mathematics of this derivation is somewhat simpler in an orthonormal non-coordinate basis, defined by:

$$d\sigma = \Lambda \; dt$$

$$d\varrho = \Sigma \; dr$$

$$d\Theta = r \; d\theta$$

$$d\Phi = r \; \sin\theta \; d\phi$$

This reduces the interval to:

$$ds^2 = -d\sigma^2 + d\varrho^2 + d\Theta^2 + d\phi^2$$

The first step is computing the coordinate commutators $c_{\alpha\beta}{}^{\mu}$:

$$[\partial_\alpha, \partial_\beta] = \partial_\alpha \partial_\beta - \partial_\beta \partial_\alpha = \partial_\mu \; c_{\alpha\beta}{}^{\mu}$$

This may be easier to follow if we consider the operators ∂_σ and ∂_ϱ acting on some function A.

$$[\partial_\sigma, \partial_\varrho]\ A = \partial_\sigma\ \partial_\varrho\ A - \partial_\varrho\ \partial_\sigma\ A$$

$$[\partial_\sigma, \partial_\varrho]\ A = (\partial/\partial\sigma)\ (\partial/\partial\varrho)\ A - (\partial/\partial\varrho)\ (\partial/\partial\sigma)\ A$$

$$[\partial_\sigma, \partial_\varrho]\ A = (\partial/\Lambda\ \partial t)\ (\partial/\Sigma\ \partial r)\ A - (\partial/\Sigma\ \partial r)\ (\partial/\Lambda\ \partial t)\ A$$

$$[\partial_\sigma, \partial_\varrho]\ A = (1\ /\ \Lambda\ \Sigma)\ (\partial/\partial t)\ (\partial/\partial r)\ A$$
$$- (1\ /\ \Lambda\ \Sigma)\ (\partial/\partial r)\ (\partial/\partial t)\ A$$
$$- A\ (1\ /\ \Sigma)\ (\partial/\partial t)\ (\partial/\partial r)\ (1\ /\ \Lambda)$$

We can henceforth drop A: $[\partial_\sigma, \partial_\varrho] = -\ (1\ /\ \Sigma)\ (-\ \Lambda{,}r\ /\ \Lambda^2)\ (\partial/\partial t)$

$$[\partial_\sigma, \partial_\varrho] = +\ (\Lambda{,}r\ /\ \Sigma\ \Lambda)\ (\partial/\Lambda\ \partial t) = +\ (\Lambda{,}r\ /\ \Sigma\ \Lambda)\ \partial_\sigma$$

This implies: $c_{\sigma\varrho}{}^{\sigma} = +\ (\Lambda{,}r\ /\ \Sigma\ \Lambda)$

And: $c_{\sigma\varrho}{}^{\varrho} = c_{\sigma\varrho}{}^{\Theta} = c_{\sigma\varrho}{}^{\Phi} = 0$

Since Λ and Σ are not functions of θ or ϕ, and nothing is a function of time:

$$[\partial_\sigma, \partial_\Theta] = [\partial_\sigma, \partial_\Phi] = 0$$

$$[\partial_\varrho, \partial_\Theta] = (\partial/\Sigma\ \partial r)\ (\partial/r\ \partial\theta) - (\partial/r\ \partial\theta)\ (\partial/\Sigma\ \partial r)$$

$$[\partial_\varrho, \partial_\Theta] = (-1\ /\ \Sigma\ r^2)\ (\partial/\partial\theta)$$
$$+\ (1\ /\ \Sigma\ r)\ (\partial/\partial r\ \ \partial/\partial\theta - \partial/\partial\theta\ \ \partial/\partial r)$$

$$[\partial_\varrho, \partial_\Theta] = (-1\ /\ \Sigma\ r)\ (\partial/r\ \partial\theta) + 0 = -\ (1\ /\ \Sigma\ r)\ \partial_\Theta$$

$$[\partial_\varrho, \partial_\Phi] = (\partial/\Sigma\ \partial r)\ (\partial/r\ \sin\theta\ \partial\phi) - (\partial/r\ \sin\theta\ \partial\phi)\ (\partial/\Sigma\ \partial r)$$

$$[\partial_\varrho, \partial_\Phi] = (-1\ /\ \Sigma\ r^2)\ (\partial/\sin\theta\ \partial\phi)$$

$$[\partial_\varrho, \partial_\Phi] = (-1\ /\ \Sigma\ r)\ (\partial/r\ \sin\theta\ \partial\phi) = -\ (1\ /\ \Sigma\ r)\ \partial_\Phi$$

$$[\partial_\Theta, \partial_\Phi] = (\partial/r\ \partial\theta)\ (\partial/r\ \sin\theta\ \partial\phi) - (\partial/r\ \sin\theta\ \partial\phi)\ (\partial/r\ \partial\theta)$$

$$[\partial_\Theta, \partial_\Phi] = (1\ /\ r)\ (-\cos\theta\ /\ \sin^2\theta)\ (\partial/r\ \partial\phi)$$

$$[\partial_\Theta, \partial_\Phi] = (-\cot\theta\ /\ r)\ (\partial/r\ \sin\theta\ \partial\phi) = -\ (\cot\theta\ /\ r)\ \partial_\Phi$$

The nonzero c's are listed below. Since $g_{\sigma\sigma} = -1$ and $g_{\varrho\varrho} = g_{\Theta\Theta} = g_{\Phi\Phi} = +1$, the c's change sign when a σ index is raised or lowered, but do not change when other indices are raised or lowered. All c's change sign when the first two indices are swapped. In this case, all nonzero c's have two equal indices.

$$c_{\sigma\varrho}{}^{\sigma} = c_{\varrho\sigma\sigma} = c^{\sigma}{}_{\varrho\sigma} = +\ \Lambda{,}r\ /\ \Sigma\ \Lambda$$

$$c_{\varrho\sigma}{}^{\sigma} = c_{\sigma\varrho\sigma} = -\ \Lambda{,}r\ /\ \Sigma\ \Lambda$$

$$c_{\varrho\Theta\Theta} = c_{\varrho\Phi\Phi} = -\ 1\ /\ \Sigma\ r$$

$$c_{\Phi\varrho\Phi} = c_{\Theta\varrho\Theta} = +\ 1\ /\ \Sigma\ r$$

$$c_{\Phi\Theta\Phi} = -c_{\Theta\Phi\Phi} = +\ \cot\theta\ /\ r$$

Next are the connection coefficients, which are easier in an orthonormal basis, where all derivatives of the metric components are zero.

General orthonormal rule: $\Gamma_{\alpha\beta\gamma} = (c_{\alpha\beta\gamma} + c_{\alpha\gamma\beta} - c_{\beta\gamma\alpha}) / 2$

Since $c_{\alpha\alpha\beta} = 0$: $\Gamma_{\alpha\alpha\beta} = (c_{\alpha\alpha\beta} + c_{\alpha\beta\alpha} - c_{\alpha\beta\alpha}) / 2 = 0$

Since $c_{\alpha\beta\alpha} = - c_{\beta\alpha\alpha}$: $\Gamma_{\alpha\beta\alpha} = (c_{\alpha\beta\alpha} + c_{\alpha\alpha\beta} - c_{\beta\alpha\alpha}) / 2 = c_{\alpha\beta\alpha}$

$$\Gamma_{\beta\alpha\alpha} = (c_{\beta\alpha\alpha} + c_{\beta\alpha\alpha} - c_{\alpha\alpha\beta}) / 2 = c_{\beta\alpha\alpha} = - \Gamma_{\alpha\beta\alpha}$$

From above, we see that $\Gamma_{\alpha\alpha\beta} = 0$ whenever the first pair of indices are the same. Also, $\Gamma_{\alpha\beta\gamma} = 0$ if all indices are different, because all $c_{\alpha\beta\gamma} = 0$ if all indices are different. This means we can choose the first index pair in 6 different ways (without them being the same), and can then choose the last index in two ways (without all three being different). Thus, there are at most 12 nonzero, independent components.

$$\Gamma^{\sigma}{}_{\varrho\sigma} = - \Gamma_{\sigma\varrho\sigma} = \Gamma_{\varrho\sigma\sigma} = \Gamma^{\varrho}{}_{\sigma\sigma} = c^{\varrho}{}_{\sigma\sigma} = + \Lambda{,}r / \Sigma\, \Lambda$$
$$\Gamma^{\sigma}{}_{\varrho\varrho} = + \Gamma^{\varrho}{}_{\sigma\varrho} = c^{\varrho}{}_{\sigma\varrho} = 0$$
$$\Gamma^{\sigma}{}_{\Theta\sigma} = \Gamma^{\Theta}{}_{\sigma\sigma} = c^{\sigma}{}_{\Theta\sigma} = 0$$
$$\Gamma^{\sigma}{}_{\Theta\Theta} = \Gamma^{\Theta}{}_{\sigma\Theta} = c^{\sigma}{}_{\Theta\Theta} = 0$$
$$\Gamma^{\sigma}{}_{\Phi\sigma} = \Gamma^{\Phi}{}_{\sigma\sigma} = c^{\sigma}{}_{\Phi\sigma} = 0$$
$$\Gamma^{\sigma}{}_{\Phi\Phi} = \Gamma^{\Phi}{}_{\sigma\Phi} = c^{\sigma}{}_{\Phi\Phi} = 0$$
$$\Gamma^{\varrho}{}_{\Theta\varrho} = - \Gamma^{\Theta}{}_{\varrho\varrho} = c^{\varrho}{}_{\Theta\varrho} = 0$$
$$\Gamma^{\varrho}{}_{\Theta\Theta} = - \Gamma^{\Theta}{}_{\varrho\Theta} = c^{\varrho}{}_{\Theta\Theta} = - 1 / \Sigma\, r$$
$$\Gamma^{\varrho}{}_{\Phi\varrho} = - \Gamma^{\Phi}{}_{\varrho\varrho} = c^{\varrho}{}_{\Phi\varrho} = 0$$
$$\Gamma^{\varrho}{}_{\Phi\Phi} = - \Gamma^{\Phi}{}_{\varrho\Phi} = c^{\varrho}{}_{\Phi\Phi} = - 1 / \Sigma\, r$$
$$\Gamma^{\Theta}{}_{\Phi\Theta} = - \Gamma^{\Phi}{}_{\Theta\Theta} = c^{\Theta}{}_{\Phi\Theta} = 0$$
$$\Gamma^{\Theta}{}_{\Phi\Phi} = - \Gamma^{\Phi}{}_{\Theta\Phi} = - c^{\Phi}{}_{\Theta\Phi} = - \cot\theta / r$$

We see that only 4 independent components are nonzero.

Before we go further, let's calculate two tricky derivatives that we will need shortly.

$$\Gamma^{\sigma}{}_{\varrho\sigma,\varrho} = (\Lambda{,}r / \Sigma\, \Lambda){,}\varrho = \partial(\Lambda{,}r / \Sigma\, \Lambda)/\partial r \;\; (\partial r/\partial\varrho)$$
$$\Gamma^{\sigma}{}_{\varrho\sigma,\varrho} = \partial(\Lambda{,}r / \Sigma\, \Lambda)/\partial r \;\; (1 / \Sigma)$$
$$\Gamma^{\sigma}{}_{\varrho\sigma,\varrho} = \Lambda{,}rr / \Sigma^2\, \Lambda \; - \; \Lambda{,}r \;\; \Sigma{,}r / \Sigma^3\, \Lambda \; - \; \Lambda{,}r \;\; \Lambda{,}r / \Sigma^2\, \Lambda^2$$

$$\Gamma^{\varrho}{}_{\Theta\Theta,\varrho} = \Gamma^{\varrho}{}_{\Phi\Phi,\varrho} = (-1 / \Sigma\, r){,}\varrho$$
$$\Gamma^{\varrho}{}_{\Theta\Theta,\varrho} = \Gamma^{\varrho}{}_{\Phi\Phi,\varrho} = (-1 / \Sigma\, r){,}r \;\; (\partial r/\partial\varrho)$$
$$\Gamma^{\varrho}{}_{\Theta\Theta,\varrho} = \Gamma^{\varrho}{}_{\Phi\Phi,\varrho} = + \; \Sigma{,}r / r\, \Sigma^3 \; + \; 1 / r^2\, \Sigma^2$$

Next are the Riemann components.

The general rule for the Riemann components in an orthonormal basis is:

$$R^{\alpha}{}_{\beta\gamma e} = \Gamma^{\alpha}{}_{\beta e,\gamma} - \Gamma^{\alpha}{}_{\beta\gamma,e} + \Gamma^{\alpha}{}_{\mu\gamma}\,\Gamma^{\mu}{}_{\beta e} - \Gamma^{\alpha}{}_{\xi e}\,\Gamma^{\xi}{}_{\beta\gamma} - \Gamma^{\alpha}{}_{\beta\delta}\,c_{\gamma e}{}^{\delta}$$

In Section §A6.9, we discovered that for a Schwarzschild metric in any basis, Riemann has only 6 nonzero, independent components.

Let's start with:

$$R^{\sigma}{}_{\varrho\sigma\varrho} = \Gamma^{\sigma}{}_{\varrho\varrho,\sigma} - \Gamma^{\sigma}{}_{\varrho\sigma,\varrho} + \Gamma^{\sigma}{}_{\mu\sigma}\,\Gamma^{\mu}{}_{\varrho\varrho} - \Gamma^{\sigma}{}_{\xi\varrho}\,\Gamma^{\xi}{}_{\varrho\sigma} - \Gamma^{\sigma}{}_{\varrho\delta}\,c_{\sigma\varrho}{}^{\delta}$$

The free indices being summed over are μ, ξ, and δ; the other indices are component labels. In the last term, $c_{\sigma\varrho}{}^{\delta} = 0$ unless $\delta = \sigma$. The next to last term is zero, because $\Gamma^{\xi}{}_{\varrho\sigma} = 0$ unless $\xi = \sigma$, but that would be multiplied by $\Gamma^{\sigma}{}_{\sigma\varrho} = 0$. The middle term is also zero, because $\Gamma^{\sigma}{}_{\mu\sigma} = 0$ unless $\mu = \varrho$, but that would be multiplied by $\Gamma^{\varrho}{}_{\varrho\varrho} = 0$. Hence:

$$R^{\sigma}{}_{\varrho\sigma\varrho} = 0 - \Gamma^{\sigma}{}_{\varrho\sigma,\varrho} - \Gamma^{\sigma}{}_{\varrho\sigma}\,c_{\sigma\varrho}{}^{\sigma}$$

$$R^{\sigma}{}_{\varrho\sigma\varrho} = -\,\Lambda{,}rr / \Sigma^2\,\Lambda \;+\; \Lambda{,}r\;\Sigma{,}r / \Sigma^3\,\Lambda \;+\; \Lambda{,}r\;\Lambda{,}r / \Sigma^2\,\Lambda^2$$
$$-\;(\Lambda{,}r / \Sigma\,\Lambda)\,(\Lambda{,}r / \Sigma\,\Lambda)$$

$$R^{\sigma}{}_{\varrho\sigma\varrho} = -\,\Lambda{,}rr / \Sigma^2\,\Lambda \;+\; \Lambda{,}r\;\Sigma{,}r \;/\; \Sigma^3\,\Lambda$$

$$R^{\sigma}{}_{\Theta\sigma\Theta} = \Gamma^{\sigma}{}_{\Theta\Theta,\sigma} - \Gamma^{\sigma}{}_{\Theta\sigma,\Theta} + \Gamma^{\sigma}{}_{\mu\sigma}\,\Gamma^{\mu}{}_{\Theta\Theta} - \Gamma^{\sigma}{}_{\xi\Theta}\,\Gamma^{\xi}{}_{\Theta\sigma}$$
$$-\;\Gamma^{\sigma}{}_{\Theta\delta}\,c_{\Theta\sigma}{}^{\delta}$$

$$R^{\sigma}{}_{\Theta\sigma\Theta} = 0 - 0 + \Gamma^{\sigma}{}_{\varrho\sigma}\,\Gamma^{\varrho}{}_{\Theta\Theta} - 0 - 0$$

$$R^{\sigma}{}_{\Theta\sigma\Theta} = (\Lambda{,}r / \Sigma\,\Lambda)\,(-1 / \Sigma\,r)$$

$$R^{\sigma}{}_{\Theta\sigma\Theta} = -\,\Lambda{,}r / r\,\Sigma^2\,\Lambda$$

$$R^{\sigma}{}_{\Phi\sigma\Phi} = \Gamma^{\sigma}{}_{\Phi\Phi,\sigma} - \Gamma^{\sigma}{}_{\Phi\sigma,\Phi} + \Gamma^{\sigma}{}_{\mu\sigma}\,\Gamma^{\mu}{}_{\Phi\Phi} - \Gamma^{\sigma}{}_{\xi\Phi}\,\Gamma^{\xi}{}_{\Phi\sigma}$$
$$-\;\Gamma^{\sigma}{}_{\Phi\delta}\,c_{\Phi\sigma}{}^{\delta}$$

$$R^{\sigma}{}_{\Phi\sigma\Phi} = 0 - 0 + \Gamma^{\sigma}{}_{\varrho\sigma}\,\Gamma^{\varrho}{}_{\Phi\Phi} - 0 - 0$$

$$R^{\sigma}{}_{\Phi\sigma\Phi} = +\;(\Lambda{,}r / \Sigma\,\Lambda)\,(-1 / \Sigma\,r)$$

$$R^{\sigma}{}_{\Phi\sigma\Phi} = -\,\Lambda{,}r / r\,\Sigma^2\,\Lambda$$

$$R^{\varrho}{}_{\Theta\varrho\Theta} = \Gamma^{\varrho}{}_{\Theta\Theta,\varrho} - \Gamma^{\varrho}{}_{\Theta\varrho,\Theta} + \Gamma^{\varrho}{}_{\mu\varrho}\,\Gamma^{\mu}{}_{\Theta\Theta} - \Gamma^{\varrho}{}_{\xi\Theta}\,\Gamma^{\xi}{}_{\Theta\varrho}$$
$$-\;\Gamma^{\varrho}{}_{\Theta\delta}\,c_{\varrho\Theta}{}^{\delta}$$

$$R^{\varrho}{}_{\Theta\varrho\Theta} = \Gamma^{\varrho}{}_{\Theta\Theta,\varrho} - 0 + (0)\,\Gamma^{\mu}{}_{\Theta\Theta} - \Gamma^{\varrho}{}_{\Theta\Theta}\,(0) - \Gamma^{\varrho}{}_{\Theta\Theta}\,c_{\varrho\Theta}{}^{\Theta}$$

$$R^{\varrho}{}_{\Theta\varrho\Theta} = (-\,1 / \Sigma\,r){,}\varrho \;-\; (-1 / \Sigma\,r)\,(-\,1 / \Sigma\,r)$$

$$R^{\varrho}{}_{\Theta\varrho\Theta} = +\,\Sigma{,}r / r\,\Sigma^3 \;+\; 1 / r^2\,\Sigma^2 \;-\; (+1 / \Sigma^2\,r^2)$$

$$R^{\varrho}{}_{\Theta\varrho\Theta} = +\,\Sigma{,}r / r\,\Sigma^3$$

$$R^{\varrho}{}_{\Phi\varrho\Phi} = \Gamma^{\varrho}{}_{\Phi\Phi,\varrho} - \Gamma^{\varrho}{}_{\Phi\varrho,\Phi} + \Gamma^{\varrho}{}_{\mu\varrho}\,\Gamma^{\mu}{}_{\Phi\Phi} - \Gamma^{\varrho}{}_{\xi\Phi}\,\Gamma^{\xi}{}_{\Phi\varrho} - \Gamma^{\varrho}{}_{\Phi\delta}\,c_{\varrho\Phi}{}^{\delta}$$

$$R^{\varrho}{}_{\Phi\varrho\Phi} = \Gamma^{\varrho}{}_{\Phi\Phi,\varrho} - 0 + (0)\,\Gamma^{\mu}{}_{\Phi\Phi} - \Gamma^{\varrho}{}_{\Phi\Phi}\,(0) - \Gamma^{\varrho}{}_{\Phi\Phi}\,c_{\varrho\Phi}{}^{\Phi}$$

$$R^{\varrho}{}_{\Phi\varrho\Phi} = +\,\Sigma{,}r / r\,\Sigma^3 + 1 / r^2\,\Sigma^2 - (+1 / \Sigma^2\,r^2)$$

$$R^{\varrho}{}_{\Phi\varrho\Phi} = +\,\Sigma{,}r / r\,\Sigma^3$$

$$R^{\Theta}{}_{\Phi\Theta\Phi} = \Gamma^{\Theta}{}_{\Phi\Phi,\Theta} - \Gamma^{\Theta}{}_{\Phi\Theta,\Phi} + \Gamma^{\Theta}{}_{\mu\Theta}\,\Gamma^{\mu}{}_{\Phi\Phi} - \Gamma^{\Theta}{}_{\xi\Phi}\,\Gamma^{\xi}{}_{\Phi\Theta} - \Gamma^{\Theta}{}_{\Phi\delta}\,c_{\Theta\Phi}{}^{\delta}$$

$$R^{\Theta}{}_{\Phi\Theta\Phi} = \Gamma^{\Theta}{}_{\Phi\Phi,\Theta} - 0 + \Gamma^{\Theta}{}_{\varrho\Theta}\,\Gamma^{\varrho}{}_{\Phi\Phi} - \Gamma^{\Theta}{}_{\Phi\Phi}\,\Gamma^{\Phi}{}_{\Phi\Theta} - \Gamma^{\Theta}{}_{\Phi\Phi}\,c_{\Theta\Phi}{}^{\Phi}$$

$$R^{\Theta}{}_{\Phi\Theta\Phi} = (-\cot\theta / r){,}\Theta - 0 + (0)\,\Gamma^{\sigma}{}_{\Phi\Phi} + (1 / \Sigma\,r)\,(-1 / \Sigma\,r) - \Gamma^{\Theta}{}_{\Phi\Phi}\,(0) - (-\cot\theta / r)\,(-\cot\theta / r)$$

$$R^{\Theta}{}_{\Phi\Theta\Phi} = \partial(-\cot\theta / r){,}\partial\theta\,(\partial\theta/\partial\Theta) - 1 / r^2\,\Sigma^2 - (\cot\theta / r)^2$$

$$R^{\Theta}{}_{\Phi\Theta\Phi} = (1 / r\,\sin^2\theta)\,(1 / r) - \cos^2\theta / (r^2\sin^2\theta) - 1 / r^2\,\Sigma^2$$

$$R^{\Theta}{}_{\Phi\Theta\Phi} = 1 / r^2 - 1 / r^2\,\Sigma^2$$

The next step is calculating the 4 nonzero Ricci components in the Schwarzschild metric:

$$R^{\alpha}{}_{\alpha} = R^{\alpha\mu}{}_{\alpha\mu} = R^{\alpha\sigma}{}_{\alpha\sigma} + R^{\alpha\varrho}{}_{\alpha\varrho} + R^{\alpha\Theta}{}_{\alpha\Theta} + R^{\alpha\Phi}{}_{\alpha\Phi}$$

$$R^{\sigma}{}_{\sigma} = R^{\sigma\varrho}{}_{\sigma\varrho} + R^{\sigma\Theta}{}_{\sigma\Theta} + R^{\sigma\Phi}{}_{\sigma\Phi}$$

$$R^{\sigma}{}_{\sigma} = -\,\Lambda{,}rr / \Sigma^2\,\Lambda + \Lambda{,}r\ \Sigma{,}r / \Sigma^3\,\Lambda - 2\Lambda{,}r / r\,\Sigma^2\,\Lambda$$

$$R^{\varrho}{}_{\varrho} = R^{\sigma\varrho}{}_{\sigma\varrho} + R^{\varrho\Theta}{}_{\varrho\Theta} + R^{\varrho\Phi}{}_{\varrho\Phi}$$

$$R^{\varrho}{}_{\varrho} = -\,\Lambda{,}rr / \Sigma^2\,\Lambda + \Lambda{,}r\ \Sigma{,}r / \Sigma^3\,\Lambda + 2\Sigma{,}r / r\,\Sigma^3$$

$$R^{\Theta}{}_{\Theta} = R^{\sigma\Theta}{}_{\upsilon\Theta} + R^{\varrho\Theta}{}_{\varrho\Theta} + R^{\Theta\Phi}{}_{\Theta\Phi}$$

$$R^{\Theta}{}_{\Theta} = -\,\Lambda{,}r / r\,\Sigma^2\Lambda + r\,\Sigma{,}r / r^2\,\Sigma^3 + 1 / r^2 - 1 / r^2\,\Sigma^2$$

$$R^{\Phi}{}_{\Phi} = R^{\sigma\Phi}{}_{\sigma\Phi} + R^{\varrho\Phi}{}_{\varrho\Phi} + R^{\Theta\Phi}{}_{\Theta\Phi}$$

$$R^{\Phi}{}_{\Phi} = -\Lambda{,}r / r\,\Sigma^2\,\Lambda + r\,\Sigma{,}r / r^2\,\Sigma^3 + 1 / r^2 - 1 / r^2\,\Sigma^2$$

Now the curvature scalar:

$$R = R^{\sigma}{}_{\sigma} + R^{\varrho}{}_{\varrho} + R^{\Theta}{}_{\Theta} + R^{\Phi}{}_{\Phi}$$

$$\begin{aligned} R = & -\Lambda{,}rr / \Sigma^2 \Lambda + \Lambda{,}r\ \Sigma{,}r / \Sigma^3 \Lambda - 2\Lambda{,}r / r\,\Sigma^2 \Lambda \\ & -\Lambda{,}rr / \Sigma^2 \Lambda + \Lambda{,}r\ \Sigma{,}r / \Sigma^3 \Lambda + 2\Sigma{,}r / r\,\Sigma^3 \\ & -2\,\Lambda{,}r / r\,\Sigma^2\Lambda + 2\,r\,\Sigma{,}r / r^2\,\Sigma^3 + 2 / r^2 - 2 / r^2\,\Sigma^2 \end{aligned}$$

$$\begin{aligned} R / 2 = & -\Lambda{,}rr / \Sigma^2 \Lambda + \Lambda{,}r\ \Sigma{,}r / \Sigma^3 \Lambda - 2\Lambda{,}r / r\,\Sigma^2 \Lambda \\ & + 2\,\Sigma{,}r / r\,\Sigma^3 + 1 / r^2 - 1 / r^2\,\Sigma^2 \end{aligned}$$

Finally, the Einstein tensor:

$$G_{\alpha\beta} = R_{\alpha\beta} - g_{\alpha\beta}\ R / 2$$

Multiply by $g^{\alpha\mu}$ and sum on α to raise first index:

$$g^{\alpha\mu}\ G_{\alpha\beta} = g^{\alpha\mu}\ R_{\alpha\beta} - g^{\alpha\mu}\ g_{\alpha\beta}\ R / 2$$

$$G^{\mu}{}_{\beta} = R^{\mu}{}_{\beta} - g^{\alpha\mu}\ g_{\alpha\beta}\, R / 2$$

($g^{\alpha\mu}\ g_{\alpha\beta}$) is zero unless $\mu = \beta$.

The 4 nonzero $G^{\beta}{}_{\beta}$ are:

$$\begin{aligned} G^{\sigma}{}_{\sigma} = & -\Lambda{,}rr / \Sigma^2 \Lambda + \Lambda{,}r\ \Sigma{,}r / \Sigma^3 \Lambda - 2\Lambda{,}r / r\,\Sigma^2 \Lambda \\ & + \Lambda{,}rr / \Sigma^2 \Lambda - \Lambda{,}r\ \Sigma{,}r / \Sigma^3 \Lambda + 2\Lambda{,}r / r\,\Sigma^2 \Lambda \\ & - 2\,\Sigma{,}r / r\,\Sigma^3 - 1 / r^2 + 1 / r^2\,\Sigma^2 \end{aligned}$$

$$G^{\sigma}{}_{\sigma} = -2\,\Sigma{,}r / r\,\Sigma^3 - 1 / r^2 + 1 / r^2\,\Sigma^2$$

$$\begin{aligned} G^{\varrho}{}_{\varrho} = & -\Lambda{,}rr / \Sigma^2 \Lambda + \Lambda{,}r\ \Sigma{,}r / \Sigma^3 \Lambda + 2\Sigma{,}r / r\,\Sigma^3 \\ & + \Lambda{,}rr / \Sigma^2 \Lambda - \Lambda{,}r\ \Sigma{,}r / \Sigma^3 \Lambda + 2\Lambda{,}r / r\,\Sigma^2 \Lambda \\ & - 2\,\Sigma{,}r / r\,\Sigma^3 - 1 / r^2 + 1 / r^2\,\Sigma^2 \end{aligned}$$

$$G^{\varrho}{}_{\varrho} = +2\,\Lambda{,}r / r\,\Sigma^2\,\Lambda - 1 / r^2 + 1 / r^2\,\Sigma^2$$

$$\begin{aligned} G^{\Theta}{}_{\Theta} = & -\Lambda{,}r / r\,\Sigma^2\Lambda + r\,\Sigma{,}r / r^2\,\Sigma^3 + 1 / r^2 - 1 / r^2\,\Sigma^2 \\ & + \Lambda{,}rr / \Sigma^2 \Lambda - \Lambda{,}r\ \Sigma{,}r / \Sigma^3 \Lambda + 2\Lambda{,}r / r\,\Sigma^2 \Lambda \\ & - 2\,\Sigma{,}r / r\,\Sigma^3 - 1 / r^2 + 1 / r^2\,\Sigma^2 \end{aligned}$$

$$\begin{aligned} G^{\Theta}{}_{\Theta} = & +\Lambda{,}r / r\,\Sigma^2\Lambda + r\,\Sigma{,}r / r^2\,\Sigma^3 \\ & + \Lambda{,}rr / \Sigma^2 \Lambda - \Lambda{,}r\ \Sigma{,}r / \Sigma^3 \Lambda - 2\,\Sigma{,}r / r\,\Sigma^3 \end{aligned}$$

$$G^{\Phi}{}_{\Phi} = G^{\Theta}{}_{\Theta}$$

Now, we can solve for Σ, using the equation for $G^{\sigma}{}_{\sigma}$. We start by lowering the first index, which simply means multiplying everything by –1.

$$G_{\sigma\sigma} = g_{\sigma\mu}\ G^{\mu}{}_{\sigma} = -G^{\sigma}{}_{\sigma}$$

Evaluating $G_{\sigma\sigma}$ yields:

$$G_{\sigma\sigma} = +2\,\Sigma{,}r / r\,\Sigma^3 + 1/r^2 - 1/r^2\,\Sigma^2$$

$$G_{\sigma\sigma} = (1/r^2)\ (2\,r\,\Sigma{,}r/\Sigma^3 + 1 - 1/\Sigma^2)$$

$$G_{\sigma\sigma} = (1/r^2)\ \partial\,[r\,(1 - 1/\Sigma^2)]\,/\partial r$$

You can verify the last step by doing the partial derivative — solving differential equations is part art and part persistence.

§A7.2 Combining Metric & Field Equations

The final step employs Einstein's field equations of general relativity: $G_{\alpha\beta} = 8\pi\,T_{\alpha\beta}$.

The energy-momentum tensor T for a stationary massive body in orthonormal coordinates is:

$$T_{\sigma\sigma} = \mu$$

$$T_{\varrho\varrho} = T_{\Theta\Theta} = T_{\Phi\Phi} = P$$

When α and β are different: $T_{\alpha\beta} = 0$

Here, μ is the body's mass density, and P is its pressure; both are functions of the radial coordinate r.

To integrate over the volume of a spherically symmetric ball of radius R, we perform a triple integral with the radial coordinate going from 0 to R, the polar angle θ going from 0 to π, and the azimuthal angle ϕ going from 0 to 2π. This integral sums the product of each tiny *volume element* multiplied by the value of the integrand in that volume.

In rectilinear coordinates, the volume element is (dx dy dz), the product of the incremental change in the three rectilinear coordinates. In spherical coordinates, the volume element is ($r^2 \sin\theta\, d\phi\, d\theta\, dr$). This is because a point in space moves a distance dr when the radial coordinate increases by dr (same as dx, dy, or dz), but the point moves a distance r $d\theta$ when θ increases by $d\theta$, and by r $\sin\theta\, d\phi$ when ϕ increases by $d\phi$.

To find a body's total mass M, we integrate its mass density μ as follows:

$$M = \int_0^R \int_0^\pi \int_0^{2\pi} \mu\ r^2 \sin\theta\, d\phi\, d\theta\, dr$$

$$M = 4\pi \int_0^R \mu\ r^2\, dr$$

$$M = (4\pi/3)\ R^3 <\mu>$$

Here, $<\mu>$ is the average mass density.

Let's next equate $G_{\sigma\sigma}$ with $8\pi\,T_{\sigma\sigma}$ and integrate over the volume of a ball of radius r that is concentric with the gravitating body of mass M, radius R, and uniform mass density μ. Assume initially that $r < R$.

$$4\pi \int_0^r G_{\sigma\sigma}\, r^2\, dr = 4\pi \int_0^r 8\pi\, T_{\sigma\sigma}\, r^2\, dr$$

$$\int dr\ \partial[r(1 - 1/\Sigma^2)]/\partial r = \int 8\pi\,\mu\, r^2\, dr$$

$$r\,(1 - 1/\Sigma^2) = 2\,(4\pi\, r^3/3)\ \mu$$

If $r < R$, the integration ball is entirely within mass M, and the right hand side of this equation equals twice the volume contained within r multiplied by mass density μ, all of which equals twice the mass contained within a distance r from the body's center.

If we integrate instead to $r = R$, the surface of mass M, we obtain:

$$r\,(1 - 1/\Sigma^2) = 2\,M$$

If we then integrate beyond the surface of mass M, to $r > R$, the quantity on the right does not change, because all we are adding is empty space. For $r > R$, we obtain:

$$r(1 - 1/\Sigma^2) = 2M$$

$$(1 - 1/\Sigma^2) = 2M/r$$

$$1/\Sigma^2 = 1 - 2M/r$$

Now, let's solve for Λ. In empty space outside the massive body, all components of the mass-energy tensor T are zero, and therefore so all components of the Einstein tensor G. Hence:

$$0 = G^{\varrho}{}_{\varrho} = G^{\sigma}{}_{\sigma}$$

$$0 = G^{\varrho}{}_{\varrho} - G^{\sigma}{}_{\sigma} = 2\,\Lambda{,}r / r\,\Sigma^2\Lambda + 2\,\Sigma{,}r / r\,\Sigma^3$$

Multiply by $r\,\Sigma^2/2$: $$0 = \Lambda{,}r/\Lambda + \Sigma{,}r/\Sigma$$

$$\Lambda{,}r/\Lambda = -\Sigma{,}r/\Sigma$$

$$\ln\Lambda = -\ln\Sigma + C$$

Here, C is an arbitrary integration constant. But recall that both Λ and Σ must approach 1 at large r. Hence at large r, $\ln\Lambda = \ln\Sigma = 0$, which means C must be zero. Finally, we obtain:

$$\Lambda = 1/\Sigma$$

$$\Lambda^2 = 1/\Sigma^2 = 1 - 2M/r$$

We have now derived the Schwarzschild metric:

$$ds^2 = -(1 - 2M/r)\,dt^2 + dr^2/(1 - 2M/r)$$
$$+ r^2\,d\theta^2 + r^2\sin^2\theta\,d\phi^2$$

Restoring Newton's constant G and the speed of light c, the key factors become:

$$\Lambda^2 = 1/\Sigma^2 = 1 - 2GM/rc^2$$

§A7.3 Curvature Tensors in Orthonormal Basis

For your convenience, I provide here the key tensors outside a massive body in the Schwarzschild metric in an orthonormal basis, with the actual values of Λ and Σ. Recall the metric and the basis vector definitions:

$$ds^2 = -d\sigma^2 + d\varrho^2 + d\Theta^2 + d\Phi^2$$

$$d\sigma = \Lambda\,dt$$

$$d\varrho = \Sigma\,dr$$

$$d\Theta = r\,d\theta$$

$$d\Phi = r\sin\theta\,d\phi$$

$$\Lambda^2 = 1 - 2M/r$$

$$\Sigma^2 = 1/(1 - 2M/r)$$

The derivatives of Λ and Σ are:

$$\Lambda{,}r = (1 / 2\Lambda)(-2M)(-1 / r^2) = +M / r^2 \Lambda$$

$$\Lambda{,}rr = M(-2 / r^3) / \Lambda - (M / r^2)(1 / \Lambda^2)(M / r^2 \Lambda)$$
$$= \{(-2M r \Lambda^2) - (M^2)\} / r^4 \Lambda^3$$
$$= \{-2M r + 4M^2 - M^2\} / r^4 \Lambda^3$$
$$= (-2M r + 3M^2) / r^4 \Lambda^3$$

$$\Sigma{,}r = (1 / \Lambda){,}r = -\Lambda{,}r / \Lambda^2 = -M / r^2 \Lambda^3$$

The nonzero connection coefficients are:

$$\Gamma^{\sigma}{}_{\varrho\sigma} = \Gamma^{\varrho}{}_{\sigma\sigma} = +M / r^2 \Lambda$$
$$\Gamma^{\varrho}{}_{\Theta\Theta} = -\Gamma^{\Theta}{}_{\varrho\Theta} = -\Lambda / r$$
$$\Gamma^{\varrho}{}_{\Phi\Phi} = -\Gamma^{\Phi}{}_{\varrho\Phi} = -\Lambda / r$$
$$\Gamma^{\Theta}{}_{\Phi\Phi} = -\Gamma^{\Phi}{}_{\Theta\Phi} = -\cot\theta / r$$

The Riemann components are:

$$R^{\sigma}{}_{\varrho\sigma\varrho} = -\Lambda{,}rr\ \Lambda + \Lambda{,}r\ \Sigma{,}r\ \Lambda^2$$
$$R^{\sigma}{}_{\varrho\sigma\varrho} = \{-(-2M r + 3M^2) / r^4 \Lambda^3\}\Lambda$$
$$+ (M / r^2 \Lambda)(-M / r^2 \Lambda^3)\Lambda^2$$
$$R^{\sigma}{}_{\varrho\sigma\varrho} = \{+2M r - 3M^2 - M^2\} / r^4 \Lambda^2$$
$$R^{\sigma}{}_{\varrho\sigma\varrho} = \{+2M r(1 - 2M / r)\} / r^4 \Lambda^2$$
$$R^{\sigma}{}_{\varrho\sigma\varrho} = +2M / r^3$$

$$R^{\sigma}{}_{\Theta\sigma\Theta} = -\Lambda{,}r / r \Lambda$$
$$R^{\sigma}{}_{\Theta\sigma\Theta} = -(M / r^2 \Lambda) / r \Lambda$$
$$R^{\sigma}{}_{\Theta\sigma\Theta} = -M / r^3 (1 - 2M / r)$$

$$R^{\sigma}{}_{\Phi\sigma\Phi} = R^{\sigma}{}_{\Theta\sigma\Theta} = -M / r^3 (1 - 2M / r)$$

$$R^{\varrho}{}_{\Theta\varrho\Theta} = +\Sigma{,}r\ \Lambda^3 / r$$
$$R^{\varrho}{}_{\Theta\varrho\Theta} = (-M / r^2 \Lambda^3)\ \Lambda^3 / r = -M / r^3$$

$$R^{\varrho}{}_{\Phi\varrho\Phi} = R^{\varrho}{}_{\Theta\varrho\Theta} = -M / r^3$$

$$R^{\Theta}{}_{\Phi\Theta\Phi} = 1 / r^2 - \Lambda^2 / r^2 = +2M / r^3$$

$$R^{\varrho}{}_{\Phi\Theta\Phi} = +\cot\theta / \Sigma r^2 = +\cot\theta\ (1 - 2M / r)^{1/2} / r^2$$

The Ricci components are:

$$R^{\sigma}{}_{\sigma} = R^{\sigma\varrho}{}_{\sigma\varrho} + R^{\sigma\Theta}{}_{\sigma\Theta} + R^{\sigma\Phi}{}_{\sigma\Phi}$$

$$R^{\sigma}{}_{\sigma} = (+2M / r^3) + (-M / r^3) + (-M / r^3) = 0$$

$$R^{\varrho}{}_{\varrho} = R^{\sigma\varrho}{}_{\sigma\varrho} + R^{\varrho\Theta}{}_{\varrho\Theta} + R^{\varrho\Phi}{}_{\varrho\Phi}$$

$$R^{\varrho}{}_{\varrho} = (+2M / r^3) + (-M / r^3) + (-M / r^3) = 0$$

$$R^{\Theta}{}_{\Theta} = R^{\sigma\Theta}{}_{\sigma\Theta} + R^{\varrho\Theta}{}_{\varrho\Theta} + R^{\Theta\Phi}{}_{\Theta\Phi}$$

$$R^{\Theta}{}_{\Theta} = (-M / r^3) + (-M / r^3) + (+2M / r^3) = 0$$

$$R^{\Phi}{}_{\Phi} = R^{\sigma\Phi}{}_{\sigma\Phi} + R^{\varrho\Phi}{}_{\varrho\Phi} + R^{\Theta\Phi}{}_{\Theta\Phi}$$

$$R^{\Phi}{}_{\Phi} = (-M / r^3) + (-M / r^3) + (+2M / r^3) = 0$$

$$R = R^{\sigma}{}_{\sigma} + R^{\varrho}{}_{\varrho} + R^{\Theta}{}_{\Theta} + R^{\Phi}{}_{\Phi} = 0$$

All components of G and T are zero in the empty space outside the massive body.

§A7.4 Curvature Tensors in Coordinate Basis

Many problems in general relativity employ the Schwarzschild metric. Below, in a coordinate basis, are the connection coefficients needed to compute geodesic paths.

$$ds^2 = -\Lambda^2 dt^2 + dr^2 / \Lambda^2 + r^2 d\theta^2 + r^2 \sin^2\theta \, d\phi^2$$

$$\Lambda^2 = 1 - 2M / r$$

$$\Lambda{,}r = (1 / 2\Lambda)(-2M)(-1 / r^2) = (+M / \Lambda r^2)$$

The nonzero derivatives of the metric components are:

$$g_{tt,r} = (-2\Lambda \ \Lambda{,}r) = (-2\Lambda)(+M / \Lambda r^2)$$

$$g_{tt,r} = (-2M / r^2)$$

$$g_{rr,r} = (-2\Lambda{,}r / \Lambda^3) = (-2 / \Lambda^3)(+M / \Lambda r^2)$$

$$g_{rr,r} = (-2M / \Lambda^4 r^2)$$

$$g_{\theta\theta,r} = (+2r)$$

$$g_{\phi\phi,r} = (+2r \sin^2\theta)$$

$$g_{\phi\phi,\theta} = (+2r^2 \sin\theta \cos\theta)$$

The nonzero, independent, connection coefficients for a spherically symmetric metric are derived in Section §A6.5. Raising the first index and deleting those proportional to time derivatives, we have:

$$\Gamma^{t}{}_{tr} = +g^{tt} g_{tt,r} / 2 = (+M / r^2 \Lambda^2)$$

$$\Gamma^r{}_{tt} = - g^{rr} g_{tt,r} / 2 = (+ M \Lambda^2 / r^2)$$
$$\Gamma^r{}_{rr} = + g^{rr} g_{rr,r} / 2 = (- M / r^2 \Lambda^2)$$
$$\Gamma^r{}_{\theta\theta} = - g^{rr} g_{\theta\theta,r} / 2 = (- r \Lambda^2)$$
$$\Gamma^r{}_{\phi\phi} = - g^{rr} g_{\phi\phi,r} / 2 = (- r \sin^2\theta \Lambda^2)$$
$$\Gamma^\theta{}_{r\theta} = + g^{\theta\theta} g_{\theta\theta,r} / 2 = (+ 1 / r)$$
$$\Gamma^\theta{}_{\phi\phi} = - g^{\theta\theta} g_{\phi\phi,\theta} / 2 = (- \sin\theta \cos\theta)$$
$$\Gamma^\phi{}_{r\phi} = + g^{\phi\phi} g_{\phi\phi,r} / 2 = (+ 1 / r)$$
$$\Gamma^\phi{}_{\theta\phi} = + g^{\phi\phi} g_{\phi\phi,\theta} / 2 = (+ \cot\theta)$$

The nonzero, independent, Riemann components for the Schwarzschild metric are derived in Section §A6.5. These are:

$$\Sigma,r = (1 / \Lambda),r = (- 1 / \Lambda^2) \Lambda,r = (- 1 / \Lambda^2) (+ M / \Lambda r^2)$$
$$\Sigma,r = (- M / \Lambda^3 r^2)$$

$$R^t{}_{rtr} = - \Lambda,rr / \Lambda + \Lambda,r (- M / \Lambda^3 r^2)$$
$$R^t{}_{rtr} = - (+ M / \Lambda r^2),r / \Lambda + (+ M / \Lambda r^2) (- M / \Lambda^3 r^2)$$
$$R^t{}_{rtr} = - (- M / \Lambda^2 r^2) (+ M / \Lambda r^2) / \Lambda$$
$$+ (+ 2 M / \Lambda r^3) / \Lambda - (M^2 / \Lambda^4 r^4)$$
$$R^t{}_{rtr} = + (M^2 / \Lambda^4 r^4) + (2 M / \Lambda^2 r^3) - (M^2 / \Lambda^4 r^4)$$
$$R^t{}_{rtr} = (2 M / \Lambda^2 r^3)$$

$$R^t{}_{\theta t\theta} = - r \Lambda,r \Lambda$$
$$R^t{}_{\theta t\theta} = - r (+ M / \Lambda r^2) \Lambda = (- M / r)$$

$$R^t{}_{\phi t\phi} = - r \sin^2\theta \Lambda,r \Lambda$$
$$R^t{}_{\phi t\phi} = - r \sin^2\theta (+ M / \Lambda r^2) \Lambda = (- M \sin^2\theta / r)$$

$$R^r{}_{\theta r\theta} = + r (- M / \Lambda^3 r^2) \Lambda^3 = (- M / r)$$

$$R^r{}_{\phi r\phi} = + r \sin^2\theta (- M / \Lambda^3 r^2) \Lambda^3 = (- M \sin^2\theta / r)$$

$$R^r{}_{\phi\theta\phi} = + r \sin\theta \cos\theta \Lambda^2$$

$$R^\theta{}_{\phi\theta\phi} = + \sin^2\theta (1 - \Lambda^2) = (+ 2 M \sin^2\theta / r)$$

§A7.5 Curvature Tensor Practice

This section provides answers, but not step-by-step derivations, for all curvature tensors for the Schwarzschild metric. This is a good opportunity to gain experience doing the calculations yourself — try spreading your wings and leap up into the air, with a safety net.

I will also introduce some new mathematical approaches that are often used. Hopefully, this will stretch your mind just a bit, and increase your mental flexibility without causing too much pain.

In a spherical coordinate basis, centered on a mass M, the metric can be diagonalized, making the most general form of the spherically symmetric interval:

$$ds^2 = -e^{2P}\, dt^2 + e^{2L}\, dr^2 + r^2\, d\theta^2 + r^2 \sin^2\theta\, d\phi^2$$

Here, P and L are two unknown functions that, due to the static, spherical symmetry, are only functions of radial coordinate r. At very large r, far from the gravitating mass, spacetime curvature approaches zero, and therefore both P and L must approach 0, matching the metric of flat spacetime.

Choosing to express the time and radial components of the metric as exponentials simplifies the calculations that follow, and provides a new experience.

The mathematics of curvature tensors is often somewhat simpler in an orthonormal non-coordinate basis, defined by:

$$d\sigma = e^{P}\ dt$$

$$d\varrho = e^{L}\ dr$$

$$d\Theta = r\ d\theta$$

$$d\Phi = r\ \sin\theta\ d\phi$$

This reduces the interval to:

$$ds^2 = -d\sigma^2 + d\varrho^2 + d\Theta^2 + d\Phi^2$$

Ready?

The first step is calculating coordinate commutators $c_{\alpha\beta}{}^{\mu}$:

General Rule: $$[\partial_\alpha, \partial_\beta] = \partial_\alpha\, \partial_\beta - \partial_\beta\, \partial_\alpha = \partial_\mu\ c_{\alpha\beta}{}^{\mu}$$

The nonzero c's are listed below. Since $g_{\sigma\sigma} = -1$ and $g_{\varrho\varrho} = g_{\Theta\Theta} = g_{\Phi\Phi} = +1$, the c's change sign when a σ index is raised or lowered, but do not change when other indices are raised or lowered. All c's change sign when the first two indices are swapped. In this case, all nonzero c's have two equal indices.

$$c_{\sigma\varrho}{}^{\sigma} = c_{\varrho\sigma\sigma} = c^{\sigma}{}_{\varrho\sigma} = +(P{,}r\ e^{-L})$$

$$c_{\varrho\sigma}{}^{\sigma} = c_{\sigma\varrho\sigma} = -(P{,}r\ e^{-L})$$

$$c_{\varrho\Theta\Theta} = c_{\varrho\Phi\Phi} = (-e^{-L} / r)$$

$$c_{\Phi\varrho\Phi} = c_{\Theta\varrho\Theta} = (+e^{-L} / r)$$

$$c_{\Theta\Phi\Phi} = -c_{\Phi\Theta\Phi} = (-\cot\theta / r)$$

Next are the connection coefficients, which are easier in an orthonormal basis, where all derivatives of the metric components are zero.

General orthonormal rule: $$\Gamma_{\alpha\beta\gamma} = (c_{\alpha\beta\gamma} + c_{\alpha\gamma\beta} - c_{\beta\gamma\alpha}) / 2$$

Since $c_{\alpha\alpha\beta} = 0$: $\quad \Gamma_{\alpha\alpha\beta} = (c_{\alpha\alpha\beta} + c_{\alpha\beta\alpha} - c_{\alpha\beta\alpha}) / 2 = 0$

Since $c_{\alpha\beta\alpha} = -c_{\beta\alpha\alpha}$: $\quad \Gamma_{\alpha\beta\alpha} = (c_{\alpha\beta\alpha} + c_{\alpha\alpha\beta} - c_{\beta\alpha\alpha}) / 2 = c_{\alpha\beta\alpha}$

$$\Gamma_{\beta\alpha\alpha} = (c_{\beta\alpha\alpha} + c_{\beta\alpha\alpha} - c_{\alpha\alpha\beta}) / 2 = c_{\beta\alpha\alpha} = -\Gamma_{\alpha\beta\alpha}$$

From above, we see that $\Gamma_{\alpha\alpha\beta} = 0$ whenever the first pair of indices are the same. Also, $\Gamma_{\alpha\beta\gamma} = 0$ if all indices are different, because all $c_{\alpha\beta\gamma} = 0$ if all indices are different.

The nonzero connection coefficients are:

$$\Gamma^{\sigma}{}_{\varrho\sigma} = +\Gamma^{\varrho}{}_{\sigma\sigma} = (+\,P{,}r\; e^{-L})$$
$$\Gamma^{\varrho}{}_{\Theta\Theta} = -\Gamma^{\Theta}{}_{\varrho\Theta} = (-\,e^{-L} / r)$$
$$\Gamma^{\varrho}{}_{\Phi\Phi} = -\Gamma^{\Phi}{}_{\varrho\Phi} = (-\,e^{-L} / r)$$
$$\Gamma^{\Theta}{}_{\Phi\Phi} = -\Gamma^{\Phi}{}_{\Theta\Phi} = (-\cot\theta / r)$$

Here are two tricky derivatives that are needed shortly.

$$\Gamma^{\sigma}{}_{\varrho\sigma,\varrho} = e^{-2L}\,(P{,}rr - P{,}r\; L{,}r)$$
$$\Gamma^{\varrho}{}_{\Theta\varrho,\Theta} = +\,e^{-2L}\,(L{,}r / r + 1 / r^2)$$

Next are the Riemann tensor components:

General orthonormal rule: $\quad R^{\alpha}{}_{\beta\gamma e} = \Gamma^{\alpha}{}_{\beta e,\gamma} - \Gamma^{\alpha}{}_{\beta\gamma,e} + \Gamma^{\alpha}{}_{\mu\gamma}\,\Gamma^{\mu}{}_{\beta e} - \Gamma^{\alpha}{}_{\mu e}\,\Gamma^{\mu}{}_{\beta\gamma} - \Gamma^{\alpha}{}_{\beta\mu}\,c_{\gamma e}{}^{\mu}$

The nonzero Riemann coefficients are:

$$R^{\sigma}{}_{\varrho\sigma\varrho} = e^{-2L}\,(-\,P{,}rr + P{,}r\; L{,}r - P{,}r\; P{,}r)$$
$$R^{\sigma}{}_{\Theta\sigma\Theta} = -\,e^{-2L}\; P{,}r / r$$
$$R^{\sigma}{}_{\Phi\sigma\Phi} = -\,e^{-2L}\; P{,}r / r$$
$$R^{\varrho}{}_{\Theta\varrho\Theta} = +\,e^{-2L}\,(L{,}r / r)$$
$$R^{\varrho}{}_{\Phi\varrho\Phi} = +\,e^{-2L}\,(L{,}r / r)$$
$$R^{\Theta}{}_{\Phi\Theta\Phi} = (1 - e^{-2L}) / r^2$$

Now the nonzero Ricci components:

$$R^{\sigma}{}_{\sigma} = e^{-2L}\,(-\,P{,}rr + P{,}r\; L{,}r - P{,}r\; P{,}r - 2\,P{,}r / r)$$
$$R^{\varrho}{}_{\varrho} = e^{-2L}\,(-\,P{,}rr + P{,}r\; L{,}r - P{,}r\; P{,}r + 2\,L{,}r / r)$$
$$R^{\Theta}{}_{\Theta} = +\,e^{-2L}\,(-\,P{,}r / r + L{,}r / r) + (1 - e^{-2L}) / r^2$$
$$R^{\Phi}{}_{\Phi} = +\,e^{-2L}\,(-\,P{,}r / r + L{,}r / r) + (1 - e^{-2L}) / r^2$$

Now the curvature scalar:

$$R = R^{\sigma}{}_{\sigma} + R^{\varrho}{}_{\varrho} + R^{\Theta}{}_{\Theta} + R^{\Phi}{}_{\Phi}$$

$$R = 2\,e^{-2L}(-\,P_{,rr} + P_{,r}\,L_{,r} - P_{,r}\,P_{,r} - 2\,P_{,r}/r + 2\,L_{,r}/r) + 2\,(1 - e^{-2L})/r^2$$

Finally, the Einstein tensor:

$$G^{\beta}{}_{\beta} = R^{\beta}{}_{\beta} - R/2$$

$$G^{\sigma}{}_{\sigma} = -\,e^{-2L}(2\,L_{,r}/r) - (1 - e^{-2L})/r^2$$

$$G^{\varrho}{}_{\varrho} = +\,e^{-2L}(2\,P_{,r}/r) - (1 - e^{-2L})/r^2$$

$$G^{\Theta}{}_{\Theta} = +\,e^{-2L}(+\,P_{,r}/r - L_{,r}/r + P_{,rr} - P_{,r}\,L_{,r} + P_{,r}\,P_{,r})$$

$$G^{\Phi}{}_{\Phi} = G^{\Theta}{}_{\Theta}$$

Now, we employ Einstein's field equations. In empty space, outside the massive body, all components of the mass-energy tensor T, and therefore the Einstein tensor G, are zero. Hence:

$$0 = G^{\varrho}{}_{\varrho} - G^{\sigma}{}_{\sigma} = 2\Lambda_{,r}/r\,\Sigma^2\Lambda + 2\Sigma_{,r}/r\,\Sigma^3$$

$$0 = +\,e^{-2L}(2\,P_{,r}/r) + e^{-2L}(2\,L_{,r}/r)$$

$$e^{-2L}(2\,P_{,r}/r) = -\,e^{-2L}(2\,L_{,r}/r)$$

$$P_{,r} = -\,L_{,r}$$

Integrating over r, we obtain:

$$P = -L + C$$

Here, C is an arbitrary integration constant. But recall that both P and L must approach 0 at large r. Hence, C must be zero, and:

$$P = -L$$

Next, we rearrange the equation for $G^{\sigma}{}_{\sigma}$. We start by lowering the first index, which simply means multiplying everything by –1.

$$G_{\sigma\sigma} = g_{\sigma\mu}\,G^{\mu}{}_{\sigma} = -\,G^{\sigma}{}_{\sigma}$$

$$G_{\sigma\sigma} = +\,e^{-2L}(2\,L_{,r}/r) + (1 - e^{-2L})/r^2$$

$$G_{\sigma\sigma} = (1/r^2)\,(2\,L_{,r}\,r\,e^{-2L} - e^{-2L} + 1)$$

$$G_{\sigma\sigma} = (1/r^2)\,\partial[\,r - r\,e^{-2L}\,]/\partial r$$

Proceeding in the same manner as Section §A7.2, we employ Einstein's field equations:

$$G_{\sigma\sigma} = 8\pi\,T_{\sigma\sigma} = 8\pi\,\mu$$

Here, μ is the gravitating body's mass density. In general, μ is a function of the radial coordinate r.

Let's next integrate over r, ensuring that we include all of mass M.

$$4\pi \int_0^r G_{\sigma\sigma}\ r^2\ dr = 4\pi \int_0^r 8\pi\, T_{\sigma\sigma}\ r^2\ dr$$

$$\int dr\ \partial\,[\,r - r\ e^{-2L}\,]\,/\partial r = \int 8\pi\ \mu\ r^2\ dr$$

$$[\,r - r\ e^{-2L}\,] = 2\,(4\pi\, r^3 / 3)\ \mu$$

$$[\,r - r\ e^{-2L}\,] = 2\,M$$

$$[\,1 - e^{-2L}\,] = 2\,M / r$$

$$e^{-2L} = 1 - 2\,M / r$$

$$e^{+2P} = 1 - 2\,M / r$$

$$e^{+2L} = 1 \,/\, (\,1 - 2\,M / r\,)$$

$$ds^2 = -\ e^{2P}\, dt^2 + e^{2L}\, dr^2 + r^2\, d\theta^2 + r^2 \sin^2\theta\ d\phi^2$$

$$ds^2 = -\ (\,1 - 2\,M / r\,)\, dt^2 + dr^2 \,/\, (\,1 - 2\,M / r\,) + r^2\, d\theta^2 + r^2 \sin^2\theta\ d\phi^2$$

Appendix 8
Derivation of FLRW Equation

The Friedman-Lemaître-Robertson-Walker (FLRW) equation describes the evolution of our universe. It presumes the 3-D space of the universe has uniform curvature — the same at every point, but not necessarily at every moment in time. In topology, this 3-D space is considered a hypersurface within our 4-D spacetime. This hypersurface is "spherically" symmetric, like the 2-D surface of a 3-D ball, but with one more dimension. It is characterized by a "radius" A(t) and three angular axes χ, θ, and ϕ.

This Appendix derives the metric for a hypersurface of uniform curvature.

There are three categories of curvature:

- Positive (>0) — like a sphere
- Zero — like a plane
- Negative — like a hyperboloid

In the context of spacetime curvature, positive means greater than zero.

The most general form of the metric for these three types is:

$$ds^2 = -dt^2 + A^2(t) \{d\chi^2 + C^2 d\theta^2 + C^2 \sin^2\theta \ d\phi^2 \}$$

Where we define:

Hypersurface radius:	A(t) is an unknown function of time
For curvature > 0:	$C = \sin\chi$
For zero curvature:	$C = \chi$
For curvature < 0:	$C = \sinh\chi$

The parameter C is a function of χ only, not time. In whatever curvature category the universe begins, it will always remain in that same category, although the magnitude of its curvature may change. The curvature magnitude of any "spherical" surface decreases as its "radius" increases, and vice versa, in any number of dimensions.

Let's employ an orthonormal basis (see Section §A6.6):

$$ds^2 = -dt^2 + db^2 + de^2 + df^2$$

Where:

$$db = A(t) \ d\chi$$

$$de = A(t) \ C \ d\theta$$

$$df = A(t) \ C \ \sin\theta \ d\phi$$

$$\partial/\partial b = \partial\chi/\partial b \ \partial/\partial\chi = A^{-1} \ \partial/\partial\chi$$

$$\partial/\partial e = \partial\theta/\partial e \ \partial/\partial\theta = A^{-1} \ C^{-1} \ \partial/\partial\theta$$

The basis commutators are:

$$[\partial_t,\partial_b] = \partial/\partial t\ \partial/A\partial\chi - \partial/A\partial\chi\ \partial/dt$$
$$= (-A{,}t / A)\ \ \partial/A\partial\chi$$

$$[\partial_t,\partial_e] = \partial/\partial t\ \partial/(A\ C\ \partial\theta) - \partial/(A\ C\ \partial\theta)\ \partial/dt$$
$$= (-A{,}t / A)\ \ \partial/(A\ C\ \partial\theta)$$

$$[\partial_t,\partial_f] = \partial/\partial t\ \partial/(A\ C\ \sin\theta\ \partial\phi) - \partial/(A\ C\ \sin\theta\ \partial\phi)\ \partial/dt$$
$$= (-A{,}t / A)\ \ \partial/(A\ C\ \sin\theta\ \partial\phi)$$

$$[\partial_b,\partial_e] = \partial/(\ A\ \partial\chi)\ \partial/(A\ C\ \partial\theta) - \partial/(A\ C\ \partial\theta)\ \partial/(A\ \partial\chi)$$
$$= (-C{,}\chi / A\ C)\ \ \partial/(A\ C\ \partial\theta)$$

$$[\partial_b,\partial_f] = \partial/(A\ \partial\chi)\ \partial/(A\ C\ \sin\theta\ \partial\phi)$$
$$- \partial/(A\ C\ \sin\theta\ \partial\phi)\ \partial/(A\ \partial\chi)$$
$$= (-C{,}\chi / A\ C)\ \ \partial/(A\ C\ \sin\theta\ \partial\phi)$$

$$[\partial_e,\partial_f] = \partial/(A\ C\ \partial\theta)\ \partial/(A\ C\ \sin\theta\ \partial\phi)$$
$$- \partial/(A\ C\ \sin\theta\ \partial\phi)\ \partial/(A\ C\ \partial\theta)$$
$$= (-\cot\theta / A\ C)\ \ \partial/(A\ C\ \sin\theta\ \partial\phi)$$

The commutators determine the commutation coefficients:

$$e_\alpha\ c_{\beta\mu}{}^{\alpha} = [\partial_\beta\, ,\, \partial_\mu\,]$$

$$c_{tbb} = -c_{btb} = -A{,}t / A$$
$$c_{tee} = -c_{ete} = -A{,}t / A$$
$$c_{tff} = -c_{ftf} = -A{,}t / A$$
$$c_{bee} = -c_{ebe} = -C{,}\chi / A\ C$$
$$c_{bff} = -c_{fbf} = -C{,}\chi / A\ C$$
$$c_{eff} = -c_{fef} = -\cot\theta / A\ C$$

Note the simplifying symmetry properties:

$$c_{a\beta\gamma'f} = 0 \text{ for all } a\beta\gamma$$
$$c_{a\beta\gamma'e} = 0 \text{ unless } a\beta\gamma = eff$$
$$c_{aee} = c_{aff} \text{ for } a = t \text{ or } b$$
$$c_{t\beta\beta'b} = 0$$

any c with one f index $= 0$

any c with one e index $= 0$, except c_{eff}

any c with more than one t indices $= 0$

any c with three different indices $= 0$

Next, we calculate the connection coefficients (see Section §A6.6) in terms of the commutation coefficients:

$$\Gamma_{tbt} = c_{tbt} = 0$$
$$\Gamma_{tet} = c_{tet} = 0$$
$$\Gamma_{tft} = c_{tft} = 0$$
$$\Gamma_{bet} = \{ c_{bet} + c_{bte} - c_{etb} \} / 2 = 0 + 0 - 0$$
$$\Gamma_{bft} = \{ c_{bft} + c_{btf} - c_{ftb} \} / 2 = 0 + 0 - 0$$
$$\Gamma_{eft} = \{ c_{eft} + c_{etf} - c_{fte} \} / 2 = 0 + 0 - 0$$

$$\Gamma_{tbb} = c_{tbb}$$
$$\Gamma_{teb} = \{ c_{teb} + c_{tbe} - c_{ebt} \} / 2 = 0 + 0 - 0$$
$$\Gamma_{tfb} = \{ c_{tfb} + c_{tbf} - c_{fbt} \} / 2 = 0 + 0 - 0$$
$$\Gamma_{beb} = c_{beb} = 0$$
$$\Gamma_{bfb} = c_{bfb} = 0$$
$$\Gamma_{efb} = \{ c_{efb} + c_{ebf} - c_{fbe} \} / 2 = 0 + 0 - 0$$

$$\Gamma_{tbe} = \{ c_{tbe} + c_{teb} - c_{bet} \} / 2 = 0 + 0 - 0$$
$$\Gamma_{tee} = c_{tee}$$
$$\Gamma_{tfe} = \{ c_{tfe} + c_{tef} - c_{fet} \} / 2 = 0 + 0 - 0$$
$$\Gamma_{bee} = c_{bee}$$
$$\Gamma_{bfe} = \{ c_{bfe} + c_{bef} - c_{feb} \} / 2 = 0 + 0 - 0$$
$$\Gamma_{efe} = c_{efe} = 0$$

$$\Gamma_{tbf} = \{ c_{tbf} + c_{tfb} - c_{ft} \} / 2 = 0 + 0 - 0$$
$$\Gamma_{tef} = \{ c_{tef} + c_{tfe} - c_{eft} \} / 2 = 0 + 0 - 0$$
$$\Gamma_{tff} = c_{tff}$$
$$\Gamma_{bef} = \{ c_{bef} + c_{bfe} - c_{efb} \} / 2 = 0 + 0 - 0$$
$$\Gamma_{bff} = c_{bff}$$
$$\Gamma_{eff} = c_{eff}$$

The 6 nonzero connection coefficients are:

$$\Gamma_{tbb} = c_{tbb} = - A{,}t / A$$
$$\Gamma_{tee} = c_{tee} = - A{,}t / A$$
$$\Gamma_{bee} = c_{bee} = - C{,}\chi / A\,C$$
$$\Gamma_{tff} = c_{tff} = - A{,}t / A$$
$$\Gamma_{bff} = c_{bff} = - C{,}\chi / A\,C$$
$$\Gamma_{eff} = c_{eff} = - \cot\theta / A\,C$$

From the connection and commutation coefficients, the Riemann components are:

General orthonormal rule: $$R^{\alpha}{}_{\beta\gamma e} = \Gamma^{\alpha}{}_{\beta e,\gamma} - \Gamma^{\alpha}{}_{\beta\gamma,e} + \Gamma^{\alpha}{}_{\mu\gamma}\,\Gamma^{\mu}{}_{\beta e} - \Gamma^{\alpha}{}_{\mu e}\,\Gamma^{\mu}{}_{\beta\gamma} - \Gamma^{\alpha}{}_{\beta\mu}\,c_{\gamma e}{}^{\mu}$$

$$R^{t}{}_{btb} = \Gamma^{t}{}_{bb,t} - \Gamma^{t}{}_{bt,b} + \Gamma^{t}{}_{\mu t}\,\Gamma^{\mu}{}_{bb} - \Gamma^{t}{}_{\mu b}\,\Gamma^{\mu}{}_{bt} - \Gamma^{t}{}_{b\mu}\,c_{tb}{}^{\mu}$$
$$R^{t}{}_{btb} = (-1)\,(- A{,}t / A){,}t - 0 + (0)\,\Gamma^{\mu}{}_{bb} - \Gamma^{t}{}_{bb}\,(0) - \Gamma^{t}{}_{bb}\,c_{tb}{}^{b}$$
$$R^{t}{}_{btb} = (A{,}tt / A) + (- A{,}t\,A{,}t / A^2) + (A{,}t / A)\,(A{,}t / A)$$
$$R^{t}{}_{btb} = (A{,}tt / A)$$

$$R^{t}{}_{bte} = \Gamma^{t}{}_{be,t} - \Gamma^{t}{}_{bt,e} + \Gamma^{t}{}_{\mu t}\,\Gamma^{\mu}{}_{be} - \Gamma^{t}{}_{\mu e}\,\Gamma^{\mu}{}_{bt} - \Gamma^{t}{}_{b\mu}\,c_{te}{}^{\mu}$$
$$R^{t}{}_{bte} = 0 - 0 + (0)\,\Gamma^{\mu}{}_{be} - \Gamma^{t}{}_{ee}\,(0) - \Gamma^{t}{}_{bb}\,(0) = 0$$

$$R^{t}{}_{btf} = \Gamma^{t}{}_{bf,t} - \Gamma^{t}{}_{bt,f} + \Gamma^{t}{}_{\mu t}\,\Gamma^{\mu}{}_{bf} - \Gamma^{t}{}_{\mu f}\,\Gamma^{\mu}{}_{bt} - \Gamma^{t}{}_{b\mu}\,c_{tf}{}^{\mu}$$
$$R^{t}{}_{btf} = 0 - 0 + (0)\,\Gamma^{\mu}{}_{bf} - \Gamma^{t}{}_{ff}\,(0) - \Gamma^{t}{}_{bb}\,(0) = 0$$

$$R^{t}{}_{bbe} = \Gamma^{t}{}_{be,b} - \Gamma^{t}{}_{bb,e} + \Gamma^{t}{}_{\mu b}\,\Gamma^{\mu}{}_{be} - \Gamma^{t}{}_{\mu e}\,\Gamma^{\mu}{}_{bb} - \Gamma^{t}{}_{b\mu}\,c_{be}{}^{\mu}$$
$$R^{t}{}_{bbe} = 0 - 0 + \Gamma^{t}{}_{bb}\,(0) - \Gamma^{t}{}_{ee}\,(0) - (0)\,c_{be}{}^{e} = 0$$

$$R^{t}{}_{bbf} = \Gamma^{t}{}_{bf,b} - \Gamma^{t}{}_{bb,f} + \Gamma^{t}{}_{\mu b}\,\Gamma^{\mu}{}_{bf} - \Gamma^{t}{}_{\mu f}\,\Gamma^{\mu}{}_{bb} - \Gamma^{t}{}_{b\mu}\,c_{bf}{}^{\mu}$$
$$R^{t}{}_{bbf} = 0 - 0 + \Gamma^{t}{}_{bb}\,(0) - \Gamma^{t}{}_{ff}\,(0) - \Gamma^{t}{}_{bb}\,(0) = 0$$

$$R^{t}{}_{bef} = \Gamma^{t}{}_{bf,e} - \Gamma^{t}{}_{be,f} + \Gamma^{t}{}_{\mu e}\,\Gamma^{\mu}{}_{bf} - \Gamma^{t}{}_{\mu f}\,\Gamma^{\mu}{}_{be} - \Gamma^{t}{}_{b\mu}\,c_{ef}{}^{\mu}$$
$$R^{t}{}_{bef} = 0 - 0 + \Gamma^{t}{}_{ee}\,(0) - \Gamma^{t}{}_{ff}\,(0) - \Gamma^{t}{}_{bb}\,(0) = 0$$

$$R^{t}{}_{ete} = \Gamma^{t}{}_{ee,t} - \Gamma^{t}{}_{et,e} + \Gamma^{t}{}_{\mu t}\,\Gamma^{\mu}{}_{ee} - \Gamma^{t}{}_{\mu e}\,\Gamma^{\mu}{}_{et} - \Gamma^{t}{}_{e\mu}\,c_{te}{}^{\mu}$$
$$R^{t}{}_{ete} = (A{,}t\,/\,A){,}t - 0 + (0)\,\Gamma^{t}{}_{ee} - \Gamma^{t}{}_{ee}\,(0) - \Gamma^{t}{}_{ee}\,c_{te}{}^{e}$$
$$R^{t}{}_{ete} = (A{,}tt\,/\,A) + (-\,A{,}t\,A{,}t\,/\,A^{2}) - (+\,A{,}t\,/\,A)\,(-\,A{,}t\,/\,A)$$
$$R^{t}{}_{ete} = (A{,}tt\,/\,A)$$

$$R^{t}{}_{etf} = \Gamma^{t}{}_{ef,t} - \Gamma^{t}{}_{et,f} + \Gamma^{t}{}_{\mu t}\,\Gamma^{\mu}{}_{ef} - \Gamma^{t}{}_{\mu f}\,\Gamma^{\mu}{}_{et} - \Gamma^{t}{}_{e\mu}\,c_{tf}{}^{\mu}$$
$$R^{t}{}_{etf} = 0 - 0 + (0)\,\Gamma^{\mu}{}_{ef} - \Gamma^{t}{}_{ff}\,(0) - \Gamma^{t}{}_{ee}\,(0) = 0$$

$$R^{t}{}_{ebe} = \Gamma^{t}{}_{ee,b} - \Gamma^{t}{}_{eb,e} + \Gamma^{t}{}_{\mu b}\,\Gamma^{\mu}{}_{ee} - \Gamma^{t}{}_{\mu e}\,\Gamma^{\mu}{}_{eb} - \Gamma^{t}{}_{e\mu}\,c_{be}{}^{\mu}$$
$$R^{t}{}_{ebe} = 0 - 0 + \Gamma^{t}{}_{bb}\,\Gamma^{b}{}_{ee} - \Gamma^{t}{}_{\mu e}\,(0) - \Gamma^{t}{}_{ee}\,c_{be}{}^{e}$$
$$R^{t}{}_{ebe} = (A{,}t\,/\,A)\,(-\,C{,}\chi\,/\,A\,C) - (A{,}t\,/\,A)\,(-\,C{,}\chi\,/\,A\,C) = 0$$

$$R^{t}{}_{ebf} = \Gamma^{t}{}_{ef,b} - \Gamma^{t}{}_{eb,f} + \Gamma^{t}{}_{\mu b}\,\Gamma^{\mu}{}_{ef} - \Gamma^{t}{}_{\mu f}\,\Gamma^{\mu}{}_{eb} - \Gamma^{t}{}_{e\mu}\,c_{bf}{}^{\mu}$$
$$R^{t}{}_{ebf} = 0 - 0 + \Gamma^{t}{}_{bb}\,(0) - \Gamma^{t}{}_{ff}\,(0) - \Gamma^{t}{}_{ee}\,(0) = 0$$

$$R^{t}{}_{eef} = \Gamma^{t}{}_{ef,e} - \Gamma^{t}{}_{ee,f} + \Gamma^{t}{}_{\mu e}\,\Gamma^{\mu}{}_{ef} - \Gamma^{t}{}_{\mu f}\,\Gamma^{\mu}{}_{ee} - \Gamma^{t}{}_{e\mu}\,c_{ef}{}^{\mu}$$
$$R^{t}{}_{eef} = 0 - 0 + \Gamma^{t}{}_{ee}\,(0) - \Gamma^{t}{}_{ff}\,(0) - \Gamma^{t}{}_{ee}\,(0) = 0$$

$$R^{t}{}_{ftf} = \Gamma^{t}{}_{ff,t} - \Gamma^{t}{}_{ft,f} + \Gamma^{t}{}_{\mu t}\,\Gamma^{\mu}{}_{ff} - \Gamma^{t}{}_{\mu f}\,\Gamma^{\mu}{}_{ft} - \Gamma^{t}{}_{f\mu}\,c_{tf}{}^{\mu}$$
$$R^{t}{}_{ftf} = (A{,}t\,/\,A){,}t - 0 + (0)\,\Gamma^{\mu}{}_{ff} - \Gamma^{t}{}_{ff}\,(0) - \Gamma^{t}{}_{ff}\,c_{tf}{}^{f}$$
$$R^{t}{}_{ftf} = (A{,}tt\,/\,A) + (-\,A{,}t\,A{,}t\,/\,A^{2}) - (A{,}t\,/\,A)\,(-\,A{,}t\,/\,A)$$
$$R^{t}{}_{ftf} = (A{,}tt\,/\,A)$$

$$R^{t}{}_{fbe} = \Gamma^{t}{}_{fe,b} - \Gamma^{t}{}_{fb,e} + \Gamma^{t}{}_{\mu b}\,\Gamma^{\mu}{}_{fe} - \Gamma^{t}{}_{\mu e}\,\Gamma^{\mu}{}_{fb} - \Gamma^{t}{}_{f\mu}\,c_{be}{}^{\mu}$$
$$R^{t}{}_{fbe} = 0 - 0 + \Gamma^{t}{}_{bb}\,(0) - \Gamma^{t}{}_{ee}\,(0) - \Gamma^{t}{}_{ff}\,(0) = 0$$

$$R^{t}{}_{fbf} = \Gamma^{t}{}_{ff,b} - \Gamma^{t}{}_{fb,f} + \Gamma^{t}{}_{\mu b}\,\Gamma^{\mu}{}_{ff} - \Gamma^{t}{}_{\mu f}\,\Gamma^{\mu}{}_{fb} - \Gamma^{t}{}_{f\mu}\,c_{bf}{}^{\mu}$$
$$R^{t}{}_{fbf} = 0 - 0 + \Gamma^{t}{}_{bb}\,\Gamma^{b}{}_{ff} - \Gamma^{t}{}_{ff}\,(0) - \Gamma^{t}{}_{ff}\,c_{bf}{}^{f}$$
$$R^{t}{}_{fbf} = (A{,}t\,/\,A)\,(-\,C{,}\chi\,/\,A\,C) - (A{,}t\,/\,A)\,(-\,C{,}\chi\,/\,A\,C) = 0$$

$$R^{t}{}_{fef} = \Gamma^{t}{}_{ff,e} - \Gamma^{t}{}_{fe,f} + \Gamma^{t}{}_{\mu e}\,\Gamma^{\mu}{}_{ff} - \Gamma^{t}{}_{\mu f}\,\Gamma^{\mu}{}_{fe} - \Gamma^{t}{}_{f\mu}\,c_{ef}{}^{\mu}$$
$$R^{t}{}_{fef} = 0 - 0 + \Gamma^{t}{}_{ee}\,\Gamma^{e}{}_{ff} - \Gamma^{t}{}_{ff}\,(0) - \Gamma^{t}{}_{ff}\,c_{ef}{}^{f}$$
$$R^{t}{}_{fef} = (A{,}t\,/\,A)\,(-\cot\theta\,/\,A\,C) - (A{,}t\,/\,A)\,(-\cot\theta\,/\,A\,C) = 0$$

$$R^{b}{}_{ebe} = \Gamma^{b}{}_{ee,b} - \Gamma^{b}{}_{eb,e} + \Gamma^{b}{}_{\mu b}\,\Gamma^{\mu}{}_{ee} - \Gamma^{b}{}_{\mu e}\,\Gamma^{\mu}{}_{eb} - \Gamma^{b}{}_{e\mu}\,c_{be}{}^{\mu}$$
$$R^{b}{}_{ebe} = 0 - 0 + (0)\,\Gamma^{\mu}{}_{ee} - (0)\,\Gamma^{e}{}_{eb} - \Gamma^{b}{}_{ee}\,c_{be}{}^{e}$$
$$R^{b}{}_{ebe} = -\,(-\,C{,}\chi\,/\,A\,C)\,(-\,C{,}\chi\,/\,A\,C)$$

$$R^b{}_{ebf} = \Gamma^b{}_{ef,b} - \Gamma^b{}_{eb,f} + \Gamma^b{}_{\mu b}\,\Gamma^\mu{}_{ef} - \Gamma^b{}_{\mu f}\,\Gamma^\mu{}_{eb} - \Gamma^b{}_{e\mu}\,c_{bf}{}^\mu$$
$$R^b{}_{ebf} = 0 - 0 + (0)\,\Gamma^\mu{}_{ef} - \Gamma^b{}_{ff}\,(0) - \Gamma^b{}_{ee}\,(0) = 0$$

$$R^b{}_{eef} = \Gamma^b{}_{ef,e} - \Gamma^b{}_{ee,f} + \Gamma^b{}_{\mu e}\,\Gamma^\mu{}_{ef} - \Gamma^b{}_{\mu f}\,\Gamma^\mu{}_{ee} - \Gamma^b{}_{e\mu}\,c_{ef}{}^\mu$$
$$R^b{}_{eef} = 0 - 0 + \Gamma^b{}_{ee}\,(0) - \Gamma^b{}_{ff}\,(0) - \Gamma^b{}_{ee}\,(0) = 0$$

$$R^b{}_{fbf} = \Gamma^b{}_{ff,b} - \Gamma^b{}_{fb,f} + \Gamma^b{}_{\mu b}\,\Gamma^\mu{}_{ff} - \Gamma^b{}_{\mu f}\,\Gamma^\mu{}_{fb} - \Gamma^b{}_{f\mu}\,c_{bf}{}^\mu$$
$$R^b{}_{fbf} = (-\,C,\chi\,/\,A\,C),b - 0 + (0)\,\Gamma^\mu{}_{ff} - \Gamma^b{}_{ff}\,(0) - \Gamma^b{}_{ff}\,c_{bf}{}^f$$
$$R^b{}_{fbf} = (-\,C,\chi\,/\,A\,C),\chi\,(1\,/\,A) - (-\,C,\chi\,/\,A\,C)\,(-\,C,\chi\,/\,A\,C)$$
$$R^b{}_{fbf} = (-\,C,\chi\chi\,/\,A^2\,C) + (C,\chi\,C,\chi/\,A^2\,C^2) - (C,\chi\,/\,A\,C)^2$$
$$R^b{}_{fbf} = (-\,C,\chi\chi\,/\,A^2\,C)$$

$$R^b{}_{fef} = \Gamma^b{}_{ff,e} - \Gamma^b{}_{fe,f} + \Gamma^b{}_{\mu e}\,\Gamma^\mu{}_{ff} - \Gamma^b{}_{\mu f}\,\Gamma^\mu{}_{fe} - \Gamma^b{}_{f\mu}\,c_{ef}{}^\mu$$
$$R^b{}_{fef} = 0 - 0 + \Gamma^b{}_{ee}\,\Gamma^e{}_{ff} - \Gamma^b{}_{ff}\,(0) - \Gamma^b{}_{ff}\,c_{ef}{}^f$$
$$R^b{}_{fef} = (-\,C,\chi\,/\,A\,C)\,(-\,\cot\theta\,/\,A\,C)$$
$$- (-\,C,\chi\,/\,A\,C)\,(-\,\cot\theta\,/\,A\,C) = 0$$

$$R^e{}_{fef} = \Gamma^e{}_{ff,e} - \Gamma^e{}_{fe,f} + \Gamma^e{}_{\mu e}\,\Gamma^\mu{}_{ff} - \Gamma^e{}_{\mu f}\,\Gamma^\mu{}_{fe} - \Gamma^e{}_{f\mu}\,c_{ef}{}^\mu$$
$$R^e{}_{fef} = (-\,\cot\theta\,/\,A\,C),\theta\,(1/\,A\,C) - 0 + \Gamma^e{}_{te}\,\Gamma^t{}_{ff} + \Gamma^e{}_{be}\,\Gamma^b{}_{ff}$$
$$- \Gamma^e{}_{ff}\,(0) - \Gamma^e{}_{ff}\,c_{ef}{}^f$$
$$R^e{}_{fef} = (1 + \cot^2\theta\,)\,/\,(\,A^2\,C^2\,) + (A,t\,/\,A)\,(+\,A,t\,/\,A)$$
$$+ (C,\chi\,/\,A\,C)\,(-\,C,\chi\,/\,A\,C) - (-\,\cot\theta\,/\,A\,C)^2$$
$$R^e{}_{fef} = 1\,/\,(\,A^2\,C^2\,) - (C,\chi\,/\,A\,C)^2 + (A,t\,/\,A)^2$$

The nonzero Riemann components are:

$$R^t{}_{btb} = R^t{}_{ete} = R^t{}_{ftf} = +\,A,tt\,/\,A$$

$$R^b{}_{ebe} = R^b{}_{fbf} = (-\,C,\chi\chi\,/\,A^2\,C) + (A,t\,/\,A)^2$$

$$R^e{}_{fef} = (1 - C,\chi\,C,\chi\,)\,/\,(A\,C)^2 + (A,t\,/\,A)^2$$

Let's now evaluate the expressions involving C for each curvature category.

For **positive curvature**: define $k = +1$

$$C = \sin\chi$$

$$C{,}\chi = \cos\chi$$

$$-\,C{,}\chi\chi \,/\, C = \sin\chi \,/\, \sin\chi = +1 = k$$

$$(1 - C{,}\chi\, C{,}\chi) \,/\, C^2 = \sin^2\chi \,/\, \sin^2\chi = +1 = k$$

$$R^b{}_{ebe} = R^b{}_{fbf} = R^e{}_{fef} = k \,/\, A^2 + (A{,}t \,/\, A)^2$$

For **zero curvature**: define $k = 0$

$$C = \chi$$

$$C{,}\chi = 1$$

$$-\,C{,}\chi\chi \,/\, C = 0 = k$$

$$(1 - C{,}\chi\, C{,}\chi) \,/\, C^2 = 0 = k$$

$$R^b{}_{ebe} = R^b{}_{fbf} = R^e{}_{fef} = k \,/\, A^2 + (A{,}t \,/\, A)^2$$

For **negative curvature**: define $k = -1$

$$C = \sinh\chi$$

$$C{,}\chi = \cosh\chi$$

$$-\,C{,}\chi\chi \,/\, C = -\sinh\chi \,/\, \sinh\chi = -1 = k$$

$$(1 - C{,}\chi\, C{,}\chi) \,/\, C^2 = -\sinh^2\chi \,/\, \sinh^2\chi = -1 = k$$

$$R^b{}_{ebe} = R^b{}_{fbf} = R^e{}_{fef} = k \,/\, A^2 + (A{,}t \,/\, A)^2$$

With these three definitions of k, we have the same Riemann components for all three curvature categories. The nonzero Riemann components are:

$$R^t{}_{btb} = R^t{}_{ete} = R^t{}_{ftf} = +\,A{,}tt \,/\, A$$

$$R^b{}_{ebe} = R^b{}_{fbf} = R^e{}_{fef} = k \,/\, A^2 + (A{,}t \,/\, A)^2$$

Next come the Ricci components and the curvature scalar:

$$R^t{}_t = R^{tb}{}_{tb} + R^{te}{}_{te} + R^{tf}{}_{tf} = +\,3\,A{,}tt \,/\, A$$

$$R^t{}_b = R^t{}_e = R^t{}_f = 0$$

$$R^b{}_b = R^{tb}{}_{tb} + R^{be}{}_{be} + R^{bf}{}_{bf}$$
$$= +\,A{,}tt \,/\, A + 2\,k \,/\, A^2 + 2\,(A{,}t \,/\, A)^2$$

$$R^b{}_e = R^b{}_f = 0$$

$$R^e{}_e = R^{te}{}_{te} + R^{be}{}_{be} + R^{ef}{}_{ef}$$
$$= +A{,}tt / A + 2k / A^2 + 2(A{,}t / A)^2$$

$$R^e{}_f = R^{te}{}_{tf} + R^{be}{}_{bf} = 0$$

$$R^f{}_f = R^{tf}{}_{tf} + R^{bf}{}_{bf} + R^{ef}{}_{ef}$$
$$= +A{,}tt / A + 2k / A^2 + 2(A{,}t / A)^2$$

$$R = R^t{}_t + R^b{}_b + R^e{}_e + R^f{}_f$$
$$= +6A{,}tt / A + 6k / A^2 + 6(A{,}t / A)^2$$

$$-R/2 = -3A{,}tt / A - 3k / A^2 - 3(A{,}t / A)^2$$

Finally, we have the Einstein tensor components:

$$G^a{}_\beta = R^a{}_\beta - g^a{}_\beta R/2$$

The diagonal components are: $G^a{}_a = R^a{}_a - R/2$

$$G^t{}_t = +3A{,}tt / A - 3A{,}tt / A - 3k / A^2 - 3(A{,}t / A)^2$$

$$G^b{}_b = +A{,}tt / A + 2k / A^2 + 2(A{,}t / A)^2$$
$$- 3A{,}tt / A - 3k / A^2 - 3(A{,}t / A)^2$$

From above, all off diagonal Ricci components are zero, and the three angular Ricci diagonal components are equal, hence so are the corresponding Einstein components.

The nonzero Einstein tensor components for a hypersurface of constant curvature are:

$$G^t{}_t = -3kA^{-2} - 3A^{-2}(A{,}t)^2$$
$$G^b{}_b = G^e{}_e = G^f{}_f = -2A{,}tt / A - kA^{-2} - A^{-2}(A{,}t)^2$$

Where:

$k = +1$ for positive (>0) curvature

$k = 0$ for zero curvature

$k = -1$ for negative curvature

Index

We employ the following notation in this Index:

§12 refers to all of Chapter 12.
§12.0 refers to the start of Chapter 12.
§12.3 refers to Chapter 12, Section 3.
§12.3 f refers to Chapter 12, Section 3 plus one or more following sections.

List of Physical Constants

π	3.141 592 653 59...	pi
e	2.718 284 590 45...	base of natural logarithm
c	299 792 458 m / s	speed of light
eV	$1.602\ 176\ 5 \times 10^{-19}$ joules	electron-volt
E λ	1.239 84193 eV-μm	Product of any photon's energy E and wavelength λ
g	9.806 65 m / s^2	acceleration of gravity at Earth's surface
G	6.6741×10^{-11} m^3 / kg-s^2	Newton's gravitational constant
G M Earth	$1.986\ 004\ 42 \times 10^{14}$ m^3 / s^2	G times Earth's mass
G M Sun	$1.327\ 124\ 40 \times 10^{20}$ m^3 / s^2	G times Sun's mass
h	$6.626\ 070\ 0 \times 10^{-34}$ joule-sec	Planck's constant
ħ	$1.054\ 571\ 8 \times 10^{-34}$ joule-sec	Reduced Planck's constant = h / 2 π
ħ c	0.197 326 97 eV-μm	Product of ħ and c
H	74 300 km / hr per 10^6 lt.yrs	Hubble's constant of universal expansion
k	$1.386\ 49 \times 10^{-23}$ joules / K	Boltzmann's constant
light-year	$9.460\ 730 \times 10^{12}$ m	Distance light travels in 1 year
m_e	$9.109\ 383\ 5 \times 10^{-31}$ kg	Mass of electron = 0.510 998 946 MeV / c^2
m_n	$1.672\ 621\ 9 \times 10^{-27}$ kg	Mass of neutron = 939.565 41 MeV / c^2
m_p	$1.672\ 621\ 9 \times 10^{-27}$ kg	Mass of proton = 938.272 08 MeV / c^2
M_E	5.972×10^{24} kg	Mass of Earth
M_E	4.435 mm	Mass of Earth in gravitational units (G = c = 1)
M_M	7.342×10^{22} kg	Mass of Earth's Moon
M_S	$1.988\ 6 \times 10^{30}$ kg	Mass of Sun
M_S	1.4767 km	Mass of Sun in gravitational units
P_L	$1.616\ 2 \times 10^{-35}$ m	Planck length = $\sqrt{(G ħ / c^3)}$
P_M	$2.176\ 5 \times 10^{-8}$ kg	Planck mass = $\sqrt{(ħ c / G)}$
P_T	$5.391\ 1 \times 10^{-44}$ s	Planck time = $\sqrt{(G ħ / c^5)}$
q_p	$1.602\ 176\ 565 \times 10^{-19}$	Charge of proton in coulombs
r_p	0.88×10^{-15} m	Radius of proton
R_E	6 371 km	Radius of Earth
R_{EO}	1.496×10^8 km	Radius of Earth's orbit around Sun
R_M	1 737 km	Radius of Moon
R_{MO}	384 399 km	Radius of Moon's orbit around Earth (semi-major axis)
R_S	696 392 km	Radius of Sun
year	$3.155\ 76 \times 10^7$ seconds	Seconds in 1 Year

Other Great Books by Robert Piccioni

The Feynman Simplified Series — ebooks

- ***4A***: ***Math For Physicists,*** all the math needed to master the Feynman Lectures, and more.
- ***4B***: ***The Best of Feynman,*** Feynman's most important principles and equations
- ***1A, 1B, 1C and 1D*** cover Feynman's Volume 1
- ***2A, 2B, 2C and 2D*** cover Feynman's Volume 2
- ***3A, 3B, and 3C*** cover Feynman's Volume 3

The Feynman Simplified Series — paperbacks

- ***Part 1*** covers Feynman Volume 1
- ***Part 2*** covers Feynman Volume 2
- ***Part 3*** covers Feynman Volume 3
- ***Part 4***: ***Math For Physicists,*** and ***The Best of Feynman***

Printed Books, each top-rated by Amazon readers:

- *Everyone's Guide to Atoms, Einstein, and the Universe*
- *Can Life Be Merely An Accident?*
- *A World Without Einstein*
- *Quantum Mechanics, Strange Theory of the Micro-World*
- *Our Universe: The Known & The Unknown*
- *Science & Faith, Road to Harmony*

The Everyone's Guide Series of short ebooks

- *Einstein: His Struggles, and Ultimate Success*, plus
    - *Special Relativity*; 3 Volumes
    - *General Relativity*: 5 Volumes, introduction to Gravitational Waves
- *Quantum Mechanics*: 5 Volumes, introduction to Entanglement
- *Higgs, Bosons, & Fermions…*, introduction to Particle Physics
- Cosmology
    - *Our Universe*: 5 Volumes, everything beyond the Sun
    - *Our Place in the Universe*, a gentle overview
    - *Black Holes, Supernovae & More*
    - *We are Stardust*
- *Timeless Atoms*
- *Science & Faith*

Meet the Author

Congratulations for embarking on a most exciting adventure—the quest to understand Nature.

I'd like to tell you something about myself and share some thoughts.

First, the obligatory bio: I have a B.S. in physics from Caltech, a Ph.D. in high-energy physics from Stanford University, and was on the faculty of Harvard University. After "retiring", I taught at the Osher Institutes at UCLA and CSUCI, where students honored me as "Teacher of the Year." In between, I ran eight high-tech companies, and hold several device patents in medical, semiconductor, and smart-energy fields.

My goal is to help more people appreciate and enjoy science.

One doesn't have to be a world-class musician to appreciate great music, and I believe the same is true for science — everyone can enjoy the exciting discoveries and intriguing mysteries of our universe. I've given over 400 presentations on science to general audiences of all ages and interests, and have written 10 printed books and 40 ebooks. My books have won national and international competitions, are among the highest rated books in this field on Amazon.com, and three have been the 1st, 2nd, and 3rd best selling books in their categories.

I was privileged to learn physics from the some of the greatest scientists of our age. I had personal relationships with four Nobel Laureates, and have enjoyed enlightening discussions with a dozen more. These great scientists were eager to share the joy of physics, and I absorbed everything I could. I would like to honor them by sharing their profound knowledge and insights with you.

What makes me a good teacher is that I am smart enough to understand what the greatest physicists have discovered, but not so smart that I learned all this easily. Indeed, I worked intensively for many years to master general relativity, quantum mechanics, tensor calculus, and cosmology. I have, therefore, first-hand experience, sympathy, and deep respect for mere mortals facing the challenge of understanding all this amazing science. Having seen the vistas from these peaks of intellectual achievement, I know the best paths up to the mountain tops, and I am eager to share.

I hope to bring the wonders of science to a wider audience.

Please help me make ***Einstein's Theories of Relativity*** even better. Contact me with any suggestions, comments, or questions at my website:

www.guidetothecosmos.com

While you are there, check out my other books, and my free newsletters and audio/video recordings.

Printed in Dunstable, United Kingdom

64101930R00181